After the
Dinosaurs

LIFE OF THE PAST
James O. Farlow, editor

After the Dinosaurs

The Age of Mammals

Donald R. Prothero

Indiana University Press

Bloomington and Indianapolis

This book is a publication of

Indiana University Press
601 North Morton Street
Bloomington, IN 47404-3797 USA

http://iupress.indiana.edu

Telephone orders   800-842-6796
Fax orders   812-855-7931
Orders by e-mail   iuporder@indiana.edu

The paper used in this publication meets
the minimum requirements of American
National Standard for Information
Sciences—Permanence of Paper for
Printed Library Materials, ANSI
Z39.48-1984.

Manufactured in the United States of
America

Library of Congress Cataloguing-in-Publication Data

Prothero, Donald R.
    After the dinosaurs : the age of mammals / Donald R. Prothero.
        p. cm.—(Life of the past)
    Includes bibliographical references and index.
    ISBN 0-253-34733-5 (cloth : alk. paper)   1. Mammals, Fossil.   2.
Paleontology—Cenozoic.   I. Title. II. Series.
    QE881.P76 2006
    569—dc22

                                                     2005033287

1 2 3 4 5 11 10 09 08 07 06

This book is dedicated to my sons
Erik, Zachary, and Gabriel Prothero
May their Cenozoic future be bright

The opening of the next great period in the life of the earth, the Cainozoic period, was a period of upheaval and extreme volcanic activity. Now it was that the vast masses of the Alps and Himalayas and the mountain backbone of the Rockies and Andes were thrust up, and that the rude outlines of our present oceans and continents appeared. The map of the world begins to display a first dim resemblance to the map of to-day.

At the outset of the Cainozoic period the climate of the world was austere. It grew generally warmer until a fresh phase of great abundance was reached, after which conditions grew hard again and the earth passed into a series of extremely cold cycles, the Glacial Ages, from which apparently it is now slowly emerging.

H. G. Wells, *A Short History of the World*, 1922

CONTENTS

The Mesozoic Era, or Age of Dinosaurs, is enormously popular in the public eye, with numerous books and television shows documenting the fascinating lives and times of these tremendous creatures. But for the last 65 million years, the dinosaurs (except for their bird descendants) have been extinct. In their place evolved an enormous variety of land creatures, especially the mammals, which are equally bizarre and fascinating to the public and paleontologist alike. The Age of Mammals, or the Cenozoic Era, has not received nearly same amount of attention as the Mesozoic. Yet there is an amazing story of the rapid evolution of thousands of species of mammals, including gigantic hornless rhinos, sabertoothed cats, mastodonts and mammoths, and many other fascinating creatures (including our own ancestors). This story is part of a larger story of global climate change: from the greenhouse conditions of the Mesozoic, the world warmed up dramatically about 55 million years ago and then began to cool down, so that glacial ice returned by 33 million years ago. The vegetation of the world went through an equally fascinating transformation, from tropical jungles in Montana and forests at the poles to grasslands and savannas across the entire world by 7 million years ago. And life in the sea, although less familiar to us, also underwent dramatic changes reflecting global climate change, including the evolution of such creatures as giant great white sharks, seals, sea lions, and dolphins and whales.

Yet there are remarkably few accounts of the Cenozoic for the nonspecialist. The first synthesis was Henry Fairfield Osborn's *The Age of Mammals in Europe, Asia, and North America* (1910), which represented what was known almost a century ago. Since then, a few books have focused on the evolution of Cenozoic mammals exclusively, such as Björn Kurtén's *The Age of Mammals* (1971) and Jordi Agusti and Mauricio Anton's *Mammoths, Sabertooths, and Hominids: 65 Million Years of Mammalian Evolution in Europe* (2002). Only Charles Pomerol's *The Cenozoic Era: Tertiary and Quaternary* (1982) described not only mammals but also the climatic story and the evolution of marine life, but it is twenty-four years out of date and also out of print. Since that time, we have learned a tremendous amount about the dating and correlation of Cenozoic rocks, the changes in Cenozoic climate, and especially the evolution of Cenozoic life, from the plants and plankton to the mammals. This book attempts to bring all these new discoveries into a common context for the intelligent lay reader and the scientist who is not a specialist in Cenozoic geology and paleontology. I have attempted to write at a fairly general level, although some geological and biological concepts are assumed (or introduced in the first chapter). However, I have also tried to show the evidence for our conclusions about Cenozoic events and to provide full scientific references for those who wish to go on to the primary literature. I hope that geologists, paleontologists, and general readers will find this book as useful for the twenty-first century as Osborn's book was for the twentieth century.

Donald R. Prothero
La Crescenta, California
June 2005

Acknowledgments

A project such as this would not have been possible without the aid of many people. I thank Jim Farlow for supporting the project, and Bill Berggren and Jim Farlow for reading the manuscript carefully and correcting many errors. I thank Bob Sloan, Miki Bird, Jane Quinet, and Kevin Marsh at Indiana University Press for their editorial help, and Carlotta Shearson for careful copy editing. I thank Mark Hallett for the striking cover art and for providing other images. Numerous colleagues sent me photographs and illustrations, and they are acknowledged in the appropriate places. I thank the national Science Foundation and the Petroleum Research Fund of the American Chemical Society for their financial support, which makes this research possible. Most important of all, I thank my amazing wife, Teresa, and my sons, Erik, Zachary, and Gabriel, for their love and support. They have made all the blood, sweat, and tears of producing this book worthwhile.

After the Dinosaurs

Figure 1.1. Reconstruction of late Eocene brontotheres. Painting by Z. Burian.

1

Introduction

Fossil hunting is by far the most fascinating of all sports. It has some danger, enough to give it zest and probably about as much as in the average modern engineered big-game hunt, and the danger is wholly to the hunter. It has uncertainty and excitement and all the thrills of gambling with none of its vicious features. The hunter never knows what his bag may be, perhaps nothing, perhaps a creature never before seen by human eyes. It requires knowledge, skill, and some degree of hardihood. And its results are so much more important, more worthwhile, and more enduring that those of any other sport! The fossil hunter does not kill, he resurrects. And the result of this sport is to add to the sum of human pleasure and to the treasure of human knowledge.

    George G. Simpson, *Attending Marvels*, 1934

Finding Fossils

The sun blazed down on the two men as they slowly walked up and down the ravines of the badlands. They walked stooped over with their eyes glued to the ground. The temperature was over 104°F (40°C), and there was no shade anywhere in the desolate landscape. They had been working like this all day and yet had only a few fossil jaws and teeth to show for their time and effort. Wide, floppy hats and loose, light-colored clothing kept off the sun, but they dared not wear dark glasses, despite the glare from the ground. To find the fossils they were seeking, they needed to detect subtle differences in the color and surface texture of the rocks on the ground, and dark glasses made this difficult. Many of the things they picked up were shiny black pebbles or concretions that resembled fossils. Frequently, they found chunks of fossil bone, which were clearly identifiable by their spongy texture in cross-section. Most of these pieces of bone were too broken to be identified. Others were scraps of fossil turtle shell, which had little scientific value. Occasionally they got lucky and found an isolated mammal tooth or two. These were worth saving, since the pattern of the tooth crowns of most mammals is distinctive. Fossil teeth are sometimes easy to spot, for instance, when the tooth enamel is black and shiny and stands out on the baked tan muds.

The men were hoping to find remains of the largest animals of the Eocene, the elephant-sized brontotheres, which were distantly related to horses and rhinos but had two blunt battering-ram horns on their noses (fig. 1.1). If the men were really lucky, they might find two or more brontothere teeth together, or a partial jaw with three or more teeth in it. Even a complete jaw and skull of a common animal, however, is not as valuable as a single tooth of a rare animal, which may be known only from a few scraps. Every isolated tooth of a rare fossil gets immediate attention when it is brought back to a museum. Sometimes it is described and published before anything else in the collection.

The two scientists were in luck today. One stooped down and noticed a small pile of bone fragments (fig. 1.2). In the midst of the pile, the skull and lower jaw of a fossil mammal protruded from the ground, lying on its side. Although the skull and jaw were nearly complete, they did not cause a lot of excitement. They belonged to a common fossil mammal, an oreodont (discussed in chapter 5), which must have roamed this area in herds of thousands over 30 million years ago (fig. 1.3). Oreodonts have no living descendants; they are distantly related to camels, yet they looked nothing like today's ships of the desert. Although there were already hundreds of unstudied oreodont specimens back in the museum, this oreodont skull was worth collecting because it was so complete.

The collectors carefully dug a trench around the specimen until it rested on a pedestal of rock. Since the specimen was fragile, they made a cast of plaster bandages around the skull. Once the cast had dried, they carefully pried it up and turned it over. The skull had come out in one piece without breaking! After a few more strips of plaster bandage had been wrapped around the exposed surface, it was ready to carry back to the truck.

A complete oreodont skull was a good day's work but nothing to write home about. As the men were working their way back to the truck, however, one of them spotted another ridge of fossil bone protruding from the ground. Although only a few inches were exposed, the thickness and curvature suggested that it was the back of a large jaw (fig. 1.4). A few minutes of careful excavation of the exposed part revealed that the specimen was indeed a very large one and that it continued into the hillside. The two men returned to the truck and carefully drove it up to the ravine as close as four-wheel drive could reach. First, they used the heavy-duty truck jack to lift a huge slab of sandstone from over the specimen and slide the slab off the cliff. Then they used picks and brooms to carefully dig a trench around the specimen, exposing it on all sides. When they

Figure 1.2. (A) Digging an oreodont skull from the ground and (B) covering it with a plaster bandage. Photos by the author.

A

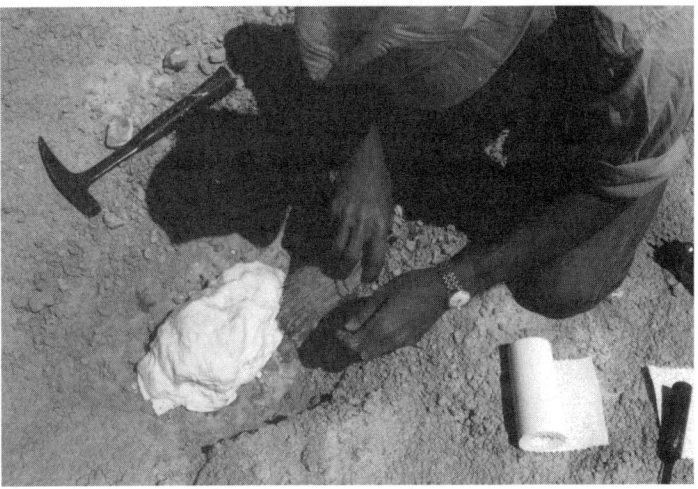

B

Figure 1.3. Reconstruction of the oreodont Merycoidodon. *Although about the size of a sheep, it was more closely related to camels and was the most abundant animal in the Eocene and Oligocene beds of the Big Badlands. Painting by B. Horsfall, in Scott 1913.*

Introduction • 3

Figure 1.4. Digging out and jacketing a brontothere jaw from the Eocene badlands near Lusk, Wyoming. (A) The jaw is fully exposed and trenched by many hours of hard labor with a pick. The skull is just visible in the quarry face behind the jaw. (B) The jaw is now covered by a thick jacket of plaster to protect it from damage during transport. (C) Once the jacket dries, the jaw is pried from the ground with a pick and turned over, so the underside is exposed and ready for the final plaster jacket. Back in the laboratory, the preparator will cut open the jacket with a saw (the same kind used by doctors to cut casts from broken bones) and then carefully scrape away all the sediment while preserving the bone. Photos by the author.

were done, they could see that they had a complete set of the lower jaws of a fossil brontothere (fig. 1.1). The jaws were almost two feet long and in excellent condition, but still fragile. With all the surrounding rock, the specimen weighed several hundred pounds, so it could not be moved easily. To protect it for transport, the two scientists mixed up a small tub of plaster of Paris and tore burlap bags into strips. After dipping the strips into the plaster, they smoothed them over the fossil, overlapping each strip so that a solid bandage was formed. After about half an hour, the plaster jacket was finished and drying quickly in the hot sun. Next came the hard part. The jacket surrounded the specimen on nearly all sides, but it was still attached to the ground. More digging isolated the plaster jacket on a higher pedestal of rock. Carefully, the two scientists dug the pedestal from underneath the specimen. At last, they wedged the pickaxe underneath the cast, prying it from the ground and flipping it over. The underside of the jaw was revealed in almost perfect condition, with very few broken pieces. After carefully trimming the ragged edge of the jacket, they covered the exposed side with more plaster and burlap. This brontothere was ready to be transported to the museum for study.

Not all fossils are so large or glamorous. In some areas, the fossils are so small that they cannot be seen from more than a foot away. The only way to collect them is to crawl on your hands and knees, with your eyes six inches from the ground. If the ground is rich in small teeth and bones, it is more efficient to use a large crew of students or volunteers. The greater the number of trained eyes covering the ground, the better. In such deposits, a few teeth are considered an excellent find since the fossils are badly crushed and seldom yield a complete skull. However, these tiny, isolated teeth are important because, for most mammals, teeth are our only record of their early evolution.

If fossil hunting sounds like grueling, backbreaking work, it is. Most fossil hunting bears little resemblance to the glamorous misconceptions we see in the movies. Scientist who study fossils, paleontologists, must put up not only with difficult conditions but also with days and weeks of looking without finding anything. To persevere in the face of such disappointment and discouragement, paleontologists must really love their work. However, one excellent find in a field season is often enough to make thousands of hours of toiling in the sun worthwhile.

Many a youngster has dug large holes in the backyard, unsuccessfully looking for the dinosaurs from the children's books. How do paleontologists know where to dig? First of all, they must know where to look. Fossils are nearly always found in sedimentary rocks, which are formed from sand or mud or fossil shells. Only a small fraction of the earth's sedimentary rocks carry fossils, so it helps to look in rocks that are known to be fossil bearing. Rock strata that were laid down in the ocean rarely produce fossils of land animals. Only sandstones and mudstones that were originally sands and muds on a river floodplain or in a lake will yield fossils of land mammals or dinosaurs. The rocks also have to be of the correct age. If they are more than 65 million years old, they will not produce many mammals, but they might produce dinosaurs. If the rocks are younger than 65 million years, however, no dinosaurs will be found, since they all became extinct at that time. Paleontologists must take all these factors is into account when they study the geology of an area, or learn of a fossil locality from some other collector.

Once you're in the right place, you have to know *how* to look. Slowly scanning the ground a few inches at a time is a suitable pace, even if it takes tremendous patience. Finally, you have to know what to look for. Paleontologists develop a mental filter, known as a "search image," that screens out all the nonfossils and fossil-like objects they see. Only the genuine glint of enamel or spongy texture of bone catches the eye among all the objects on the desert floor. Once paleontologists have spotted bone or enamel, they must also have the training to recognize and identify what they've found. If it's really worthwhile, it deserves special treatment. To develop this kind of skill generally requires years of education and many more years of practice in the field, collecting and identifying hundreds of specimens. Since most finds are fragmentary, paleontologists must know the skeleton of each animal so well that any piece is instantly recognized. Only a few of the handful of paleontologists employed today have all these skills so well developed that they are master collectors. Good fossil collectors are a rare breed these days, but what they have found is extremely impressive considering their small numbers. From their years of collecting, we have fossils in museums that tell us the story of the evolution of dinosaurs, elephants, horses, rhinos, and many other important fossil animal groups.

These methods are standard for collecting fossil vertebrates (animals with backbones, like fish, reptiles, amphibians, birds, and mammals), which are generally rare and difficult to collect. By contrast, invertebrates (animals without backbones) are generally much more common—at least those with hard shells or skeletons, like clams, snails, sea urchins, and corals. Obviously, soft-bodied animals without skeletons, like worms and jellyfish, seldom fossilize. In many places, fossils occur as dense shell beds with thousands to millions of shells packed in close together. Here, collecting is much easier, and the collector need worry only about damaging the fragile shells as they are collected, and about keeping good records of everything that is collected. More often, however, marine shales and sandstones have relatively few fossils, so collecting in these locations is the same kind of backbreaking work I have just described, hiking over miles of landscape, looking for the rare shell.

Yet another set of conditions applies to microfossils, the skeletons of tiny organisms usually less than a millimeter in size (fig. 1.5). Most microfossils are the shells of single-celled organisms, such as the amoeba-like foraminifera and radiolaria that float in the plankton and settle on the sea bottom. Other microfossils come from plant-like single-celled organisms (such as diatoms). Still others are from multicellular animals that happen to be microscopic in size, such as the tiny snail-like pteropods that float in the plankton, or the minute crustaceans known as ostracodes, which litter the sea bottom with trillions of their tiny kidney-bean-shaped shells that hinge over their backs. In any case, microfossils are usually not rare. Some oceanic sediments are composed of nothing but microfossils, so even a sample of a few grams yields thousands of shells. In most marine sediments around the world, microfossils are abundant, so the experienced micropaleontologist need scoop only a few grams of sample into a bag, take it back to the lab, and look through the microscope. Better still, microfossils are so abundant that they can even be recovered from samples drilled from deep underground in the search for oil. For years, oil companies hired micropa-

Figure 1.5. Microfossils. The large shells made of bubble-shaped chambers are planktonic foraminifera, and the smaller fossils with the mesh-like skeletons are radiolaria. Photo courtesy of Scripps Institution of Oceanography.

leontologists because they could use the tiny microfossils found throughout the deep drill holes to determine how old the sediment was, or how deep the water once was at the site of deposition. In addition, many microfossils are sensitive to the oceanic conditions in which they lived. They often track changes not only in water depth but also in oceanic temperature and chemistry. As we shall see later in this book, the study of microfossils and the chemicals trapped in their skeletons is the key to understanding how ancient oceans and climates have evolved over time.

Dating Rocks

Paleontologists work in a world with a time frame completely different from ordinary everyday history. From various methods, we now know that the earth is about 4.6 billion years old, a staggering number in human terms. It is such an immense amount of time that some sort of analogy is necessary to make it comprehensible. Suppose we were to compress all 4.6 billion years of earth history into a single calendar year. On this scale, each of the 365 "calendar days" equals 12 million years, and each minute of the "calendar" is 8561 years long! The earth forms on New Year's Day in this calendar. The first recognizable life—consisting of tiny, single-celled bacteria and blue-green cyanobacteria—does not appear until February 21. Complex, multicellular life, such as jellyfish, trilobites, and corals, does not appear until November 12. The first amphibians crawl out on land on November 28. The first tiny mammals and the first bird, *Archaeopteryx,* appear during the peak of the Age of Dinosaurs, the Jurassic Period, on December 17. The final extinction of the dinosaurs and the beginning of the Age of Mammals occur on the day after Christmas. The first ape-like primates that are members of our own family, the hominids, do not appear until eight hours before New Year's Eve. Neanderthal Man, the classic Stone Age "caveman," appears ten minutes before New Year's Eve, as the countdown begins at parties everywhere. Recorded history begins less than one minute before New Year's Eve, as the conductor raises his baton to start *Auld Lang Syne.* Within a second before midnight, Charles Darwin's *On the Origin of Species* is published, and the American Civil War is fought. Virtually all of human history, especially the last few millennia, is drowned out by the drunks who blow their noisemakers a fraction of a second too early!

On the scale of geologic time, human affairs appear pretty insignificant. Geologists are accustomed to dealing with such large amounts of time and routinely deal with thousands and millions of years. For most geologic problems, events of less than thousands of years in duration cannot even be distinguished in the layers of sedimentary rocks. When dealing with events that occurred hundreds of millions or billions of years ago, even a million years here or there is negligible. A sense of "deep time" (as John McPhee labeled it) is important to all of us, not just to the geologists. Most geologists, however, find it practical to deal with time not in millions of years but in relative time terms. Just as historians use "Elizabethan" or "Edwardian" to refer to periods in English history, so geologists use "Cambrian" and "Cretaceous" to refer to distinct episodes in earth history.

For the purposes of this book, most of these time terms will not be necessary. The last 65 million years, known as the Age of Mammals in popular parlance, is formally known as the Cenozoic Era. The Cenozoic is divided into a number of epochs (fig. 1.6), beginning with the Paleocene approximately 65 million years ago and running to the present. The Paleocene, which lasted from 65 to 55 million years ago, is followed by the Eocene (55–34 million years ago), the Oligocene (34–23 million years ago), the Miocene (23–5 million years ago), the Pliocene (5–1.8 million years ago), and the Pleistocene, or ice ages (1.8 million years to 10,000

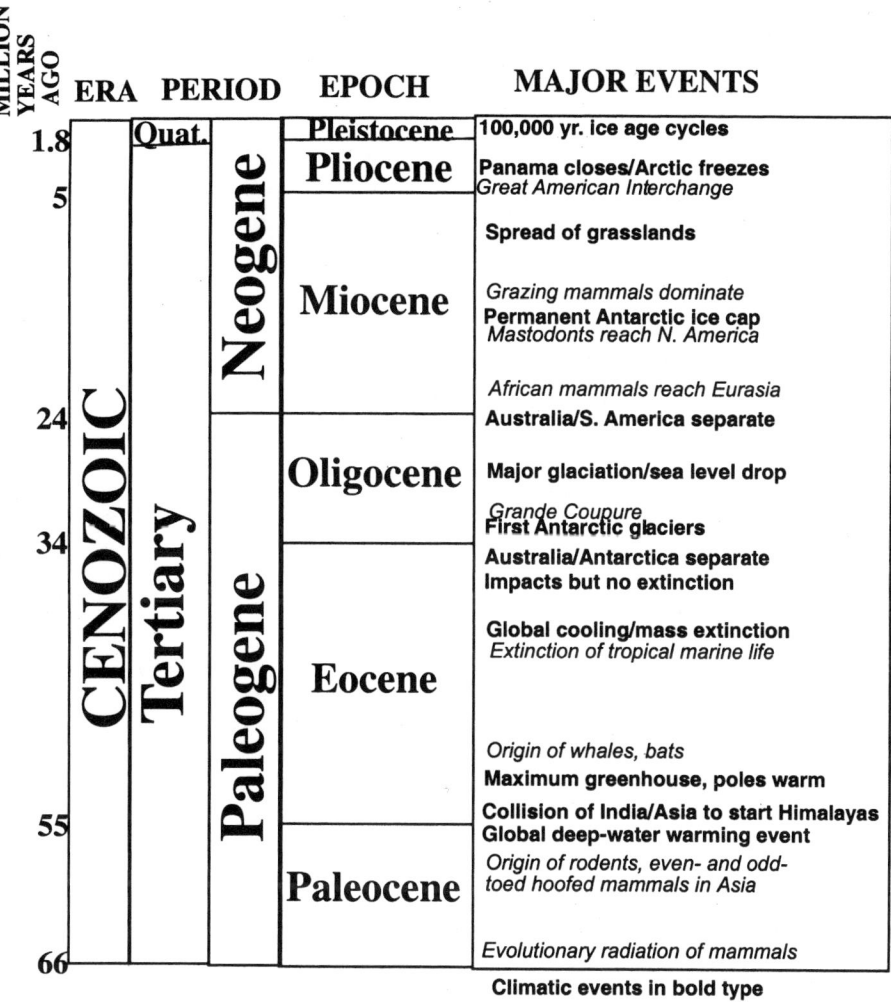

MILLION YEARS AGO	ERA	PERIOD		EPOCH	MAJOR EVENTS
1.8		Quat.		Pleistocene	**100,000 yr. ice age cycles**
5		Neogene		Pliocene	**Panama closes/Arctic freezes** *Great American Interchange*
				Miocene	**Spread of grasslands** *Grazing mammals dominate* **Permanent Antarctic ice cap** *Mastodonts reach N. America* *African mammals reach Eurasia*
24	CENOZOIC	Tertiary	Paleogene	Oligocene	**Australia/S. America separate** **Major glaciation/sea level drop** *Grande Coupure* **First Antarctic glaciers**
34				Eocene	**Australia/Antarctica separate** **Impacts but no extinction** **Global cooling/mass extinction** *Extinction of tropical marine life* *Origin of whales, bats* **Maximum greenhouse, poles warm**
55				Paleocene	**Collision of India/Asia to start Himalayas** **Global deep-water warming event** *Origin of rodents, even- and odd- toed hoofed mammals in Asia* *Evolutionary radiation of mammals*
66					

Climatic events in bold type
Biotic events in italics

Figure 1.6. Cenozoic timescale. Abbreviation: Quat. = Quaternary.

years ago). The period since the last retreat of the glaciers, which includes the present interglacial warming, is called the Holocene, or Recent (10,000 years ago to present). Although these terms may seem intimidating at first, using them is much easier than trying to talk about the age of an event in terms of millions of years.

How did we establish these divisions, and where did these terms come from? Since the late 1600s, geologists have been able to establish the relative ages of fossils and rocks (i.e., this fossil is younger than or older than that fossil) by the *principle of superposition*. First proposed by the Danish physician Nicolaus Steno in 1669, this principle states that in any layered sequence of rocks (layered sediments or lava flows), the oldest rocks are at the bottom of the stack, and the rocks get progressively younger as you move up the pile. Clearly, the rocks at the top of the stack could not have accumulated unless there were already rocks on the bottom of the stack to build upon. A good analogy is a stack of papers on a messy desk. Those at the top were put there recently, whereas those at the bottom of the stack may have been laid there months ago and have been gradually buried by more recent activity.

The next breakthrough came in the late 1700s, when geologists began to try to decipher the superposed stacks of sandstones, shales, and limestones in England and Europe and to reconstruct the history of the earth. Some thought that the entire stack was produced during the biblical creation week and then modified by Noah's flood. In 1760, Italian geologist Giovanni Arduino referred to the ancient granitic rocks and metamorphic rocks found at the bottom of the stack in most places in the world, and in the cores of uplifted mountain ranges, as Primitive or Primary, since they were supposedly produced in the original creation of the earth. Above the Primary rocks were layered sedimentary rocks, usually tilted and deformed and found in mountain ranges, which were called Secondary and were supposedly produced as Noah's flood retreated from the mountains. Above these were Tertiary rocks, which were still horizontal and often poorly consolidated, supposedly produced from the final stages of the retreat of Noah's flood. Of these terms, only Tertiary survives in modern timescales as a term for the first 63 million years of the Cenozoic, although some authors have tried to replace it with a less archaic term. For example, many geologists prefer to use Paleogene for the first 42 million years of the Cenozoic (the Paleocene, Eocene, and Oligocene epochs) and Neogene for the last 23 million years (Miocene, Pliocene, and Pleistocene epochs). However, the terms Tertiary and Quaternary persist, even though their flood-geology connotations are no longer considered valid.

Flood geology began to break down when geologists looked closer at the rocks. Soon they began to find supposedly Primitive granites that had once been molten and had cut across and intruded through Secondary sedimentary rocks, showing that the granites had to be younger than the sedimentary rocks. In addition, sandstones, shales, and limestones all over Europe looked similar, so matching them up from one place to another was difficult. The next breakthrough occurred in the 1790s, when William Smith, an untutored engineer for a canal company in southern England, began collect fossils from the fresh canal excavations. He noticed that each rock formation had its own distinctive suite of fossils and that no two formations had identical fossil contents. He soon became so good at recognizing this pattern of *faunal succession* that he could amaze the wealthy gentlemen-collectors by telling them exactly where their fossil collections came from. More importantly, faunal succession helped him map the rock formations and determine their precise sequence, because each sandstone or shale or limestone had a different fossil assemblage from the formations above it and below it. By the 1820s, geologists had mapped most of the formations of England and Wales on the basis of their fossil content and had begun to coin the terms, such as Carboniferous and Cretaceous, that make up the geologic timescale we used today.

But faunal succession is not enough. The sequence of strata in southern England is fairly thick and complete, but there are still gaps in the record, known as *unconformities*. Even the mile-thick pile of sedimentary rocks in the Grand Canyon represents only about 25% of the time between 250 and 550 Ma (mega-annum, or million years before present), and none of the time before or after. Nowhere on earth is there a complete record that spans all of geologic time. Thus, geologists had to use faunal succession to correlate rock sequences from one place to another. This practice of using fossils to correlate strata is known as *biostratigraphy*. Each distinctive fossil assemblage is unique to a given period of geologic time and can be used to correlate one local stratigraphic section with another. For example, the Cenozoic sequence in the Rocky Mountain region is the most complete terrestrial sequence anywhere in the world, but nowhere is it complete. Over a century ago, paleontologists had to patch together local sections from different areas to give a complete timescale of mammal evolution in North America (fig. 1.7). For example, the upper part of the section in the San Juan Basin of New Mexico overlaps in age and fossil content with the Wasatch Formation in Wyoming; together these two sections give us a composite section spanning most of the Paleocene and early Eocene. The upper Wasatch Formation, in turn, overlaps with the lower part of the Huerfano section in Colorado, and the upper part of the Huerfano section overlaps in fossil content and age with the base of the Bridger Basin section in Wyoming. The top of the Bridger Basin section, in turn, overlaps the base of the Uinta Basin section. These sections, knitted together over a wide region using successions of land mammal fossils, represent an almost complete record of most of the Paleocene and Eocene in North America. By correlating these and several other sections across the region, we can get a detailed picture of the geological, climatic, and faunal events for this span of time.

How was the modern geological timescale developed? Unfortunately, it was a rather haphazard, unplanned process. The scale was not set up by a single person in a systematic fashion, so that everything would be organized and represent a complete sequence. The time terms were proposed for distinctive rock units at different times and places by different geologists, so the timescale just grew and evolved. For example, the now familiar term Jurassic was named by the explorer and naturalist Alexander von Humboldt in 1795 for the distinctive sequence of rocks in the Jura Mountains of the French Alps. The Cretaceous was named by William Conybeare and William Phillips in 1822 and is based on the Latin word *creta,* or "chalk," since the Cretaceous beds include the famous chalk deposits of the White Cliffs of Dover. As we saw, the term Tertiary ("third" in Latin) was left over from Arduino's flood geology of 1760, but

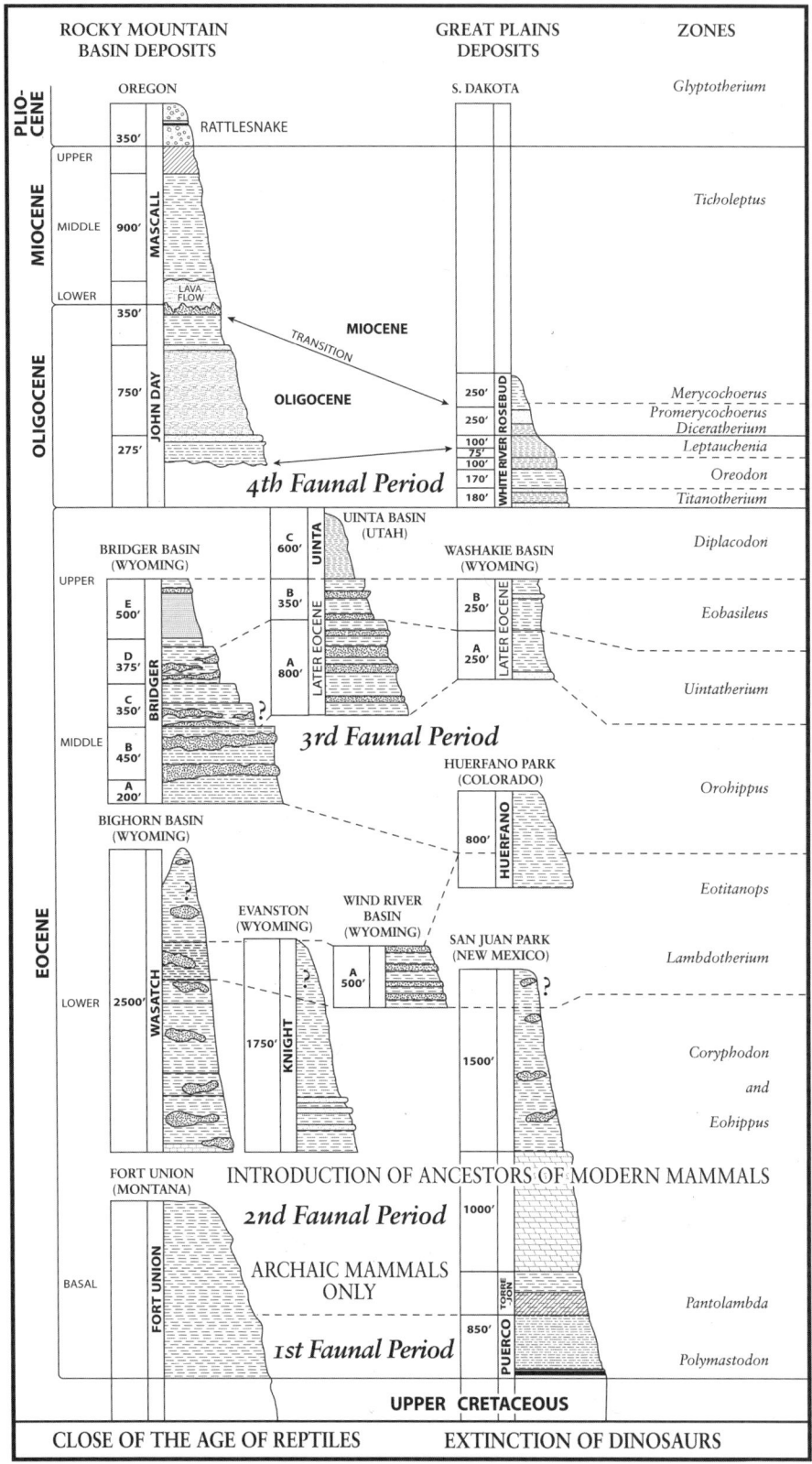

in 1829, Paul Desnoyers proposed the term Quaternary ("fourth" in Latin) for the poorly consolidated post-Tertiary deposits of the Seine Basin in France. At the time, they were thought to be deposits formed after Noah's flood; but by 1837, Louis Agassiz was attributing them to a great ice age, and since then the terms Quaternary and ice age have been closely linked.

By the 1830s, geologists noticed that there were dramatic differences between the oldest strata (then known as Transition and Carboniferous), with their peculiar fossils of brachiopods and corals and sea lilies, and the younger strata (already divided into the Triassic, Jurassic, and Cretaceous), which were full of ammonites. In 1838, Adam Sedgwick applied the term Paleozoic ("ancient life" in Greek) to these oldest fossiliferous rocks, which would soon be divided by him and by geologist Roderick Murchison into the Cambrian, Silurian, Devonian, Permian, and so on. In 1840, geologist John Phillips wrote an article for the *Penny Cyclopaedia* in which he used the term Mesozoic ("middle life" in Greek) for the ammonite-bearing beds of the Triassic, Jurassic, and Cretaceous, and the term Cenozoic ("recent life" in Greek) for the younger beds without ammonites that had been called Tertiary and Quaternary. This three-fold division of the fossil record into Paleozoic, Mesozoic, and Cenozoic eras was no accident, because the great Permian extinction at the end of the Paleozoic wiped out 95% of species on earth. This extinction radically changed the life on the seafloor that arose in the Triassic, producing a very different looking Mesozoic fauna. Likewise, the Mesozoic and Cenozoic are bounded by the second largest extinction known, the Cretaceous-Tertiary extinction. This event is abbreviated K/T in geological shorthand, because on geological maps Cretaceous is abbreviated with a "K" (from the German *Kreide* for "chalk"; the "C" was already preempted by the Carboniferous). The "T" is for Tertiary. (Recently, a number of geologists have advocated replacing K/T with K/P for Cretaceous-Paleogene, because the Paleogene has now been formally defined and Tertiary is an obsolete usage. However, in this book I will continue to use the more familiar abbreviation K/T.) The K/T event wiped out not only the ammonites but also the great marine reptiles, and the dinosaurs on the land (discussed further in chapter 2).

In contrast to this chaotic growth of most of the timescale, pioneering geologist Charles Lyell attempted to subdivide the Cenozoic in a planned, logical fashion. In the third volume of his revolutionary work *Principles of Geology* (1833), Lyell tried to replace Tertiary and Quaternary with a finer-scale subdivision of the Cenozoic based on the percentage of recent molluscan fossils in the fauna. The French conchologist Gérard-Paul Deshayes had studied more than 8,000 species (40,000 specimens) and noticed that the mollusks look more and more modern in younger strata. Lyell used this work to propose four "periods" (now called epochs) for the Tertiary. The Eocene ("dawn of the recent" in Greek) had only 3.5% of living mollusks; the Miocene ("less recent" in Greek) contained 17% modern species; the older Pliocene ("more recent" in Greek) had 33–50% modern species; and the newer Pliocene had 90% living mollusks in its fossils.

Rudwick (1978) has shown that Lyell was thinking of the change in molluscan fossils as a continuously ticking "clock" (fig. 1.8), so that by identifying the percentage of modern species in a fossil collection, one could subdivide the Tertiary into many fine numerical increments. But in practical terms, the system was flawed. First of all, molluscan turnover was not a continuous clock-like process but rather an episodic one, with periods of stability and mass extinction (as we shall see in later chapters). Second, Lyell and Deshayes's "species" are difficult to use today because some have been combined, and other "species" have been split into many species by later scientists, or raised to higher (generic) rank. Even with up-to-date species lists, Lyell's molluscan "clock" would be hard to use. Stanley et al. (1980) calculated that only 50% of the molluscan species (by modern definitions) were in existence at the beginning of the Pliocene, only 5% existed at the beginning of the Miocene, and almost none were present in the Eocene.

Lyell's noble attempt at a logical, clock-like subdivision of the Cenozoic had a bigger problem: it was not compatible with the system of subdividing geologic time that was already in existence. The rest of the timescale was built by biostratigraphic analysis of local sections, so the "Lutetian" Stage of the Eocene is based on a set of strata in the Paris Basin with a distinctive assemblage of mollusks. Lyell's clock model does not mesh well with this system. Rather than placing discrete boundaries on real rock units in the field, Lyell's clock model had no real boundaries, only arbitrary subdivisions of a continuum of molluscan fossils. In Lyell's mind, the Miocene was not a division of time between 5 Ma and 23 Ma (as we now define it) but a discrete moment when approximately 17% of the molluscan species were modern forms. Thus, there were no precise boundaries for his units. His system baffled geologists who tried to apply traditional stratigraphic methods to an essentially chronological concept. Lyell indicated that several areas and their fossils were typical of each of his "periods," which led to much confusion as later stratigraphers quarreled over what typified the Eocene or Miocene. Although Lyell's concepts originated in the Italian Tertiary section, Deshayes's collections were from the Paris Basin. Much of the Paris Basin fauna is restricted to that area, so the "type" fauna is difficult to recognize outside France.

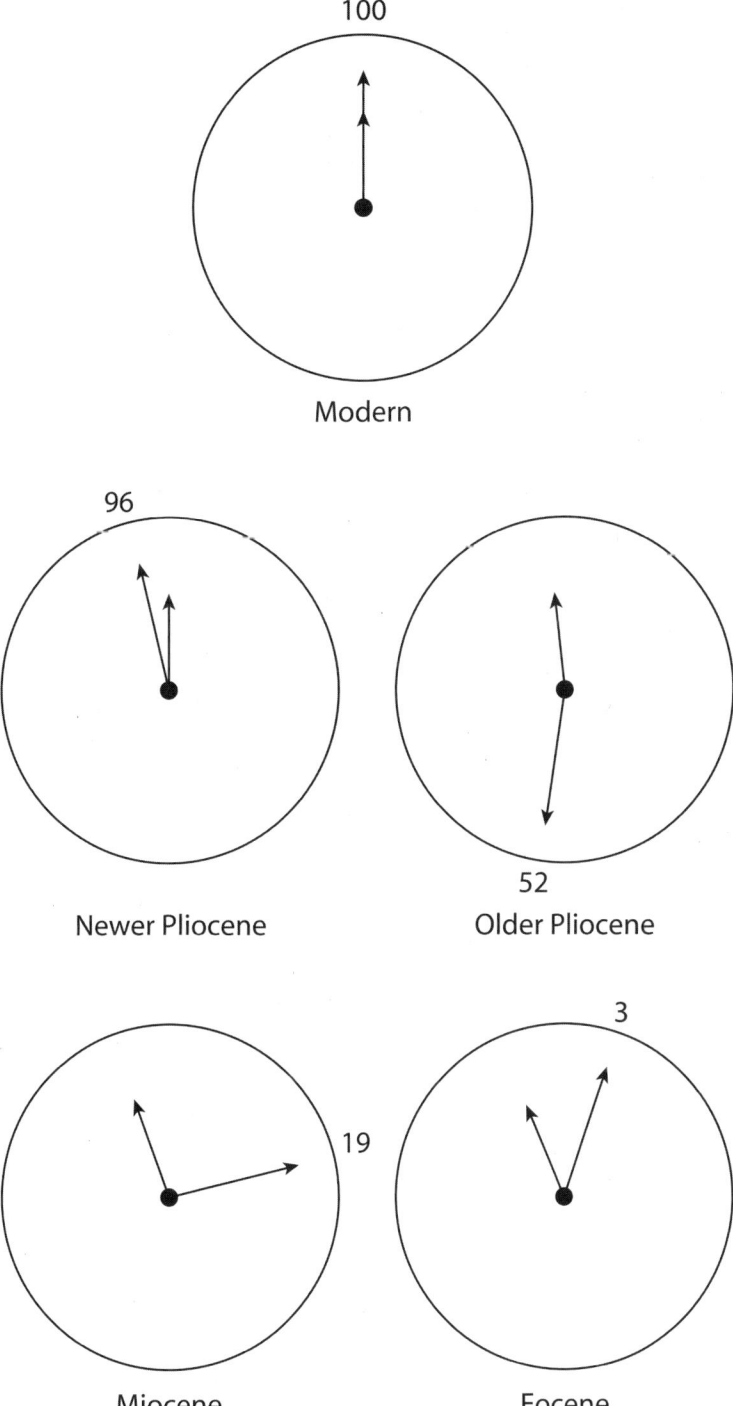

100

Modern

96

Newer Pliocene

Older Pliocene

52

Miocene

19

Eocene

3

Figure 1.8. Lyell's conceptions of the Cenozoic epochs as "moments" on the "clock" of molluscan turnover. The numbers indicate the percentages of modern species in fossil collections from each epoch. After Rudwick 1978.

For this reason, paleontologists argued for more than a century about how to define and subdivide the Eocene, Miocene, and Pliocene. In the Paris Basin, the upper Eocene was poorly fossiliferous and had been labeled the lower Miocene by some geologists. In 1854, Heinrich Ernst von Beyrich coined the term Oligocene ("few recent" in Greek, because there were fewer recent fossils than in the Miocene) for a sequence of rocks in northern Germany and Belgium that were more fossiliferous and apparently younger than the upper Eocene rocks of the Paris Basin. The fossils of von Beyrich's Oligocene were more advanced than those of the French Eocene, but not as modern as those of the Miocene, so this epoch was placed between Lyell's Eocene and Miocene. Unfortunately, the type Oligocene is in a different basin and does not overlie the type Eocene, so it is difficult to decide where one ends and the other begins.

The origin of the Paleocene was similarly confusing. In 1874, paleobotanist W. P. Schimper recognized a series of fossil plants in the Paris Basin that he decided were distinct from those of Lyell's Eocene. He called these Paleocene ("ancient recent" in Greek). Unfortunately, fossil plants are relatively rare and difficult to correlate around the world, so the term Paleocene did not catch on until the mammals and marine mollusks had also been studied and compared. Most works published in the early twentieth century still used lower Eocene for what we now call the Paleocene (e.g., fig. 1.7), so the reader must be careful when interpreting these early figures and texts. The United States Geological Survey did not formally recognize the Paleocene until 1939. However, since that time, the standard epochs of the Cenozoic—Paleocene, Eocene, Oligocene, Miocene, Pliocene, Pleistocene—have become internationally accepted and are the most useful way to subdivide the last 65 million years.

These are relative time terms, established based on local stratigraphic sections in Europe and correlated with their molluscan fossils. How do we correlate them outside Europe? How do we establish their numerical age? Each of these problems was a major field of study unto itself, and the answers have been slow in coming; there were many false starts before geologists arrived at methods that can date almost any Cenozoic rock around the world.

The first problem is correlating the classic Eocene, Oligocene, and Miocene rocks of western Europe with the rocks rest of the world. How can we decide if rocks in Utah, or on the deep seafloor, are Eocene or Oligocene or Miocene? The classic type sections in Europe turned out to be rather poor choices for the foundation of a timescale. Most of the subdivisions, or stages, of the Eocene and Oligocene in western Europe were based on relatively thin, incomplete shallow marine strata, with large gaps, or unconformities, between them (fig. 1.9). For years, geologists argued about whether a particular stage was sequential with the next, or partially overlapped it, or whether there was a gap between the two. In many cases, two successive stages were not in the same basin, or did not lie superposed upon each other, so determining whether they were successive or overlapping in age was impossible. In other cases,

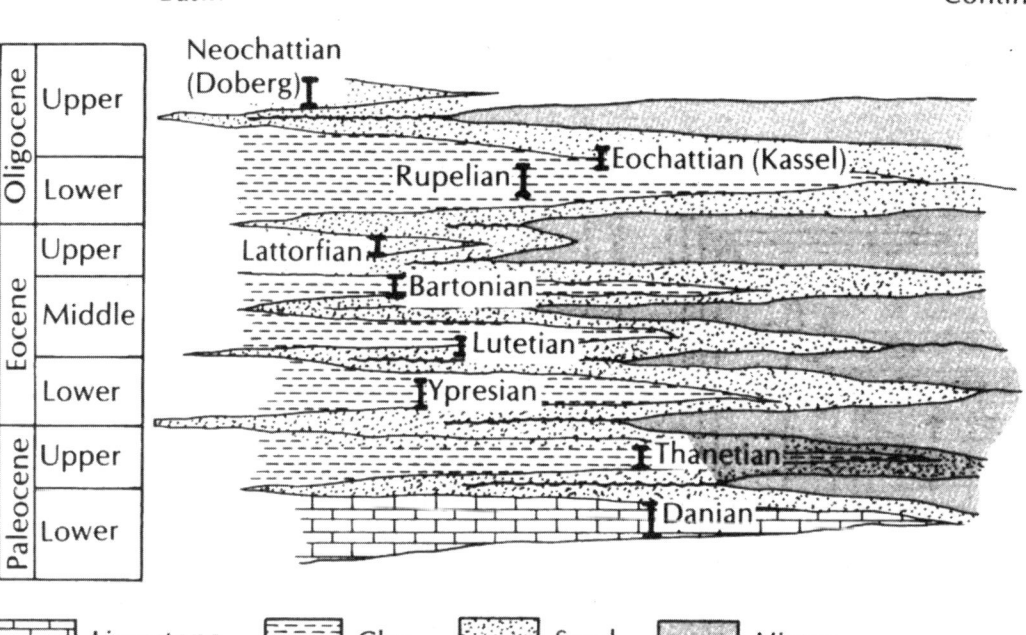

Figure 1.9. Depositional history of the type areas of the Paleogene stages and ages in northwestern Europe. Note that the incomplete section is composed of an irregular pattern of transgressive and regressive shallow-marine sequences with major gaps, or hiatuses (gray shaded pattern). Each of the stratotype sections is shown by a short vertical bar and represents only a small portion of Cenozoic time, with major gaps between the sections. The endemism of molluscan faunas further complicated intercontinental correlation of these stratotypes. Eventually the more-complete sequences of deep-sea cores became the standard, and the biostratigraphy of the Cenozoic is now largely based on these cores and their microfossils. After Hardenbol and Berggren 1978. AAPG © 1978. Reprinted by permission of the AAPG whose permission is required for further use.

there was a clear gap between the types of two stages. Should the lower stage be extended upward to cover the gap, or should the upper stage be extended downward? Compounding the problem was that the biostratigraphic index fossils for these stages were shallow-marine mollusks, which are mostly species restricted to Europe and rarely could be compared to mollusks (or any other fossils) elsewhere. As a result, even up until the 1970s, there was much confusion about what was Eocene, Oligocene, Miocene, or Pliocene outside western Europe. We have since learned that a lot of the estimates made in the early and middle part of twentieth century were far off.

The solution came in the 1960s and 1970s, when micropaleontologists began to study the fossils of deep-marine sediments. These form a much more complete and continuous record of geologic time than any other sedimentary environment because they form where little erosion can take place and because there is a steady rain of tiny planktonic microfossils accumulating on the seafloor. By the 1960s and 1970s, the Deep Sea Drilling Project was recovering and studying hundreds of cores drilled from the sea bottom, which were full of tiny microfossils that could be analyzed in every centimeter of the core. Not only were these microfossils abundant, but they could be correlated worldwide because they lived over most of the world's ocean wherever the temperature of the water was right. By the early 1970s, the sequence of microfossil zones was the global standard for telling biostratigraphic age in the Cenozoic, not only for deep-sea cores but also for marine rocks that had been uplifted into mountains. Once the deep-sea zonation had been established, micropaleontologists (e.g., Berggren 1971) examined samples from the classic type sections of shallow-marine sediments in western Europe and could then determine where each European stage fit on the global micropaleontological timescale (fig. 1.9). This method allows us to use relative age terms like middle Eocene or late Oligocene on a global basis wherever the appropriate microfossils are available.

But how do we attach numerical ages in millions of years ago to Cenozoic events? Ideally, we would like to employ the standard geologic method, radiometric dating, to get numerical ages. This method uses the decay of an unstable radioactive parent atom, such as ^{40}K or ^{87}Rb, as a clock. The rate of decay of this atom to its stable daughter atom (^{40}K decays to ^{40}Ar, ^{87}Rb decays to ^{87}Sr, and so on) is known, so by measuring how much of the daughter atom has accumulated, we can tell how long the system has been decaying and, therefore, how old it is. However, the system works only in crystals that have been formed by cooling from a molten state, which happens mainly in igneous rocks. When we measure the amounts of parent and daughter atoms, we assume that the atoms were locked into a crystal when it formed by cooling and that none of them have leaked out of the crystal. But the crystals in the mineral grains in a sedimentary rock, such as sandstone, didn't form by cooling from molten magma. Instead, they formed from rocks that cooled long ago and were then uplifted and exposed to weathering and erosion, forming sand grains; these grains were re-deposited much later as sediments, which eventually solidified to form a sedimentary rock. If you try to date the individual minerals in a sandstone, you are dating the ages of the parent rocks that supplied each grain, but not the sandstone itself, which is much younger. With few exceptions, you cannot numerically date any sedimentary rock directly.

So how do we get numerical ages on sedimentary rocks, which contain most of the fossils needed to get relative ages? The ideal scenario involves a volcanic ash layer formed from a molten magma (so it can be dated by the radioactive decay of its ^{40}K) but laid down between fossiliferous sedimentary layers, so relative age can be established. Since the advent of the potassium-argon dating method in the late 1950s, geologists have been trying to date volcanic ashes at a furious pace and have dated many of the ashes found in key fossil-bearing sections that establish the ages of Cenozoic stages and zones. These have been incorporated into the standard timescale (the most recent revision was published by Berggren and others in 1995), so when you see the timescales in the chapters of this book, they give the numerical ages of the standard zones of microfossils, marine megafossils (mostly mollusks), and land mammals. These then precisely determine the ages of almost any Cenozoic rock, on land or in the ocean.

One other important tool is valuable in dating Cenozoic sedimentary rocks—magnetic stratigraphy. Almost all sediment has a tiny fraction of a percent of magnetic grains (usually of the iron oxides magnetite and hematite) within it. As the soupy sediment begins to compact and consolidate and turn into stone, the magnetic fields of these grains become aligned with the earth's magnetic field at the time the sediment was deposited. Since the 1960s, we have known that the earth's magnetic field has changed direction many times. For example, 780,000 years ago, the earth's field was reversed from its present direction, so that if you held up a compass, the needle would have pointed south, not north. The field flips back and forth between normal and reversed polarity every few million years, sometimes multiple times in a million years. It takes only 4,000 to 5,000 years for the flip to take place, and then the field is stable for hundreds of thousands to millions of years. These flip-flops of the earth's magnetic field form a random and irregular pattern through time, like a bar code. If we examine something that records a long period of geologic time (such as a thick sequence of sediments) and measure the direction of the field every few centimeters or meters, we might be able to match the pattern of the field changes in our local section to the global magnetic polarity timescale. This method, known as magnetic stratigraphy, requires that hundreds of small oriented samples of rock be taken and measured many times (after many cleaning steps) in a machine known as a magnetometer. However, if the results are good, we have a record of the earth's field that allows us to match each local section to the global timescale precisely. The pattern is not unique in and of itself, since the earth's field has flipped back and forth many thousands of times. What we need is a radiometric date or, if that's not available, a distinctive fossil assemblage to tie the pattern of magnetic flip-flops to the global magnetic timescale. The correlation can be very precise, so each polarity change can be dated to the nearest 100,000 years (good precision for rocks as old as 65 Ma). Even better, the method works on almost any sedimentary rock in any environment, from the deep sea to the land. By contrast, we cannot use marine microfossils on terrestrial rocks, or land mammals in the deep sea, and volcanic ashes are only sporadically available. But if we can combine magnetic stratigraphy with any other method of dating (fossils or volcanic ashes), we can obtain precise dates on nearly any section where such dates were impossible only a few years ago. This is how most of the rocks and fossils discussed in this book were dated.

What's in a Name?

Paleontologists and biologists must also use different names for the animals, as well as for the ages of their fossilized remains. Most familiar living animals today have common names that are widely understood, so that we know a white rhino from a black rhino from an Indian rhino. Yet in many parts of the English-speaking world, the same common name can have different meanings. In most of the United States, for example, a gopher is a digging rodent, but in the southeastern states, a gopher can be a tortoise. Many animals have different common names in different parts of the country. In countries where English is not the native language, the animals have names in the local language. To get around this problem, biologists long ago adopted scientific names that are universal, regardless of region or native language. In 1758, when the naming system was first widely adopted, Latin was the universal language of scholars, so all scientific names are Latin in form or are derived from Latinized words from Greek or some other language. A scientist will always understand *Geomys* to mean the rodent gopher, and *Gopherus* to mean the gopher tortoise. By convention, the name of any species is a compound of two words, always found together. These names are always italicized in print. The first word is the genus name (plural: genera), which is always capitalized. The second word is the trivial (species) name, which is never capitalized. Four example, the correct scientific name of our species is *Homo* (genus) *sapiens* (species), which means "thinking man." Another related species in our genus is *Homo erectus* ("erect man"), our probable ancestor. Similarly, the Indian and Javan rhinoceros are in the same genus (*Rhinoceros*) but are different species. The Indian rhino is *Rhinoceros unicornis,* and the Javan rhino is *Rhinoceros sondaicus.* The black rhino is in a different genus, *Diceros,* which has only one living species, *Diceros bicornis.*

Most fossils discussed in this book have no popular names. The fossils are known only by their scientific names and are always italicized in this book. At first, these long scientific names may seem hard to pronounce and remember. If you break them down syllable by syllable, however, they are not so intimidating.

Generic and trivial names are not the only names used to identify and classify an organism. Every genus belongs to a larger subdivision of life called a family. For example, humans belong in the family Hominidae, and true rhinos in the family Rhinocerotidae. All zoological family names can be recognized by the "-idae" ending. All the families, in turn, can be included in orders. Thus, the Hominidae can be grouped with the other families of apes, monkeys, lemurs, and tarsiers in the order Primates. Rhinos belong with the tapirs, horses, and various extinct groups in the order

Perissodactyla, or the odd-toed hoofed mammals. Orders are subdivisions of a larger group, the class. Both perissodactyls and primates are mammals, or members of class Mammalia. Classes are grouped into an even larger group, the phylum (plural, phyla). For example, mammals, birds, amphibians, reptiles, and fishes are all members of the phylum Chordata, which includes all animals with a spinal cord. Finally, the major phyla are grouped into the great kingdoms of life: the kingdom Animalia, the kingdom Plantae, the kingdom Fungi, and so on. This hierarchical classification not only serves as a useful tool but also indicates closeness of evolutionary relationship. Animals in the same genus are more closely related to each other than they are to animals in any other genus, and so on. The division of kingdoms into phyla, and phyla into classes, and so on, is actually a reflection of the branching tree of life.

For Further Reading

For those without much background in geology, the fundamentals can be found in the following books:

Marshak, S., and D. R. Prothero. 2001. *Earth: Portrait of a Planet*. New York: W. W. Norton (basic physical geology text).

Prothero, D. R., and R. H. Dott Jr. 2003. *Evolution of the Earth*. 7th ed. New York: McGraw-Hill (basic earth history text).

More details about the principles of stratigraphy, as well as details about how the Cenozoic timescale was developed, can be found in the following books:

Prothero, D. R. 1990. *Interpreting the Stratigraphic Record*. New York: W. H. Freeman (basic principles of stratigraphy).

Prothero, D. R. 1994. *The Eocene-Oligocene Transition: Paradise Lost*. New York: Columbia University Press. (Chapter 2 covers the recent developments in the dating of the geological time scale.)

Prothero, D. R., and F. Schwab. 2003. *Sedimentary Geology*. 2nd ed. New York: W. H. Freeman. (The final chapter discusses the principles of chronostratigraphy and the development of the timescale.)

Figure 2.1. Mammals cavorting on Triceratops *skull. Painting by Mark Hallett.*

2

The End of the Dinosaurs?

> Mass extinction is box office, a darling of the popular press,
> the subject of cover stories and television documentaries, many
> books, even a rock song. . . . At the end of 1989, the Associ-
> ated Press designated mass extinction as one of the "Top 10
> Scientific Advances of the Decade." Everybody has weighed in,
> from the *Economist* to *National Geographic*.
>
> David Raup, *Extinction: Bad Genes or Bad Luck?* 1991

Out with a Bang . . .

Dinosaurs are big business these days. First discovered and named less
than 170 years ago, today they are a major part of popular culture, with
hugely successful movies (especially the three *Jurassic Park* movies, which
are among the most profitable of all time) and enormous amounts of mer-
chandise available for sale. Dozens of television programs with computer-
generated animated dinosaurs can be found on cable channels like the
Discovery Channel or the Learning Channel, and even on network television, playing many times a year. Nearly all children under the age of ten in
the United States and many other countries are fascinated with dinosaurs
and can rattle off dozens of polysyllabic dinosaur names that baffle their
parents.

Naturally, such a huge amount of interest has fueled even greater speculation as to why dinosaurs are no longer with us. That's a question that
has baffled scientists, too, ever since it became clear that dinosaurs were
extinct. As a result, almost as many explanations have been proposed as
there have been people who have proposed them. These explanations
range from the plausible (it got too hot or too cold, or the atmosphere
changed so that there was too much oxygen or too little carbon dioxide)
to the more exotic (volcanoes erupted and changed the atmosphere, or sea
level dropped and changed the climate) to the implausible (medical prob-

lems ranging from slipped discs to diseases, or evolutionary senescence) to the outright bizarre (they all got depressed and died, or aliens from outer space wiped them out). Some explanations seem plausible until the facts are considered. For example, some have blamed dinosaur extinction on the development of flowering plants, which were supposedly more difficult to digest and could have caused constipation or indigestion—except that flowering plants first evolved in the Early Cretaceous, about 60 million years before the dinosaurs died out. In fact, several scientists have suggested that the duckbill dinosaurs and horned dinosaurs, with their complex battery of grinding teeth, evolved to exploit this new resource of rapidly growing flowering plants. Others have blamed extinction on competition from the mammals (fig. 2.1), which allegedly ate all the dinosaur eggs—except that mammals and dinosaurs appeared at the same time in the Late Triassic, about 190 million years ago, and there is no reason to believe that mammals suddenly acquired a taste for dinosaur eggs after 120 million years of coexistence. Some explanations (such as the dinosaurs all died of diseases) fail because there is no way to scientifically test them, and they cannot move beyond the realm of speculation and guesswork.

This focus on explaining dinosaur extinction misses an important point: the extinction at the end of the Cretaceous Period was a global event that killed off organisms up and down the food chain. It wiped out many kinds of plankton in the ocean and many marine organisms that lived on the plankton at the base of the food chain. These included a variety of bizarre clams and snails, and especially the ammonites, a group of shelled squid-like creatures that dominated the Mesozoic seas and had survived many previous mass extinctions. The K/T event marked the end of the marine reptiles, such as the mosasaurs and the plesiosaurs, which were the largest creatures that had ever lived in the seas and which ruled the seas long before whales evolved. On land, there was also a crisis among the land plants, in addition to the disappearance of dinosaurs. So any event that can explain the destruction of the base of the food chain (plankton in the ocean, plants on land) can better explain what happened to organisms at the top of the food chain, such as the dinosaurs. By contrast, any explanation that focuses strictly on the dinosaurs (like most of those above) completely misses the point. The Cretaceous extinctions were a global phenomenon, and dinosaurs were just a part of a bigger picture.

According to one popular story, the Age of Dinosaurs ended suddenly 65 million years ago, when a giant rock from space plummeted to the earth (fig. 2.2). Estimated to be 10 to 15 kilometers in diameter, this bolide (either a comet or an asteroid) was traveling at cosmic speeds of 20–70 kilometers per second, or 45,000–156,000 mph. Such a huge mass traveling at such tremendous speeds carries an enormous amount of energy, estimated to be the equivalent of 100 million megatons of TNT. That's far more energy than contained in all the world's nuclear weapons

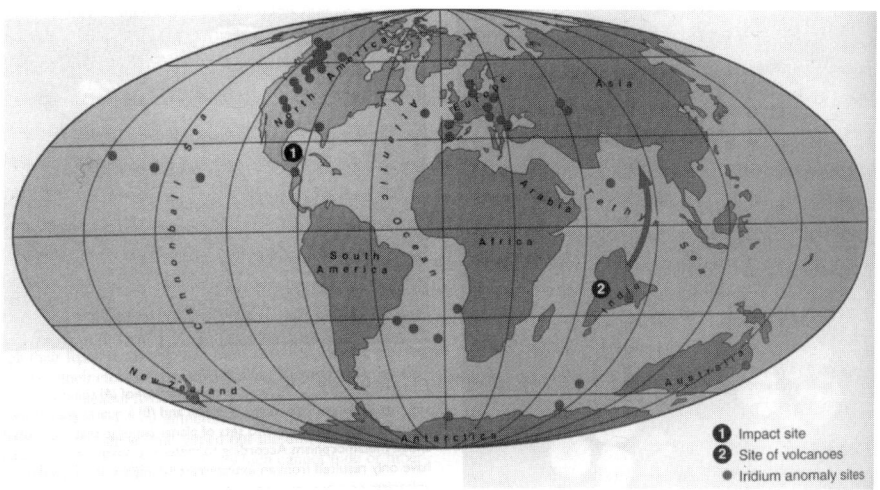

Figure 2.2. Locations of some of the many K/T iridium anomalies (dots), as well as the site of the Chicxulub impact in the Yucatan (no. 1) and the site of the Deccan eruptions in India (no. 2). After Prothero and Dott 2003.

at the height of the Cold War. When the bolide struck, it generated a huge shock wave that leveled everything for thousands of kilometers around the impact and caused most of the landscape to burst into flames. The bolide struck an area of the Yucatan Peninsula of Mexico known as Chicxulub (pronounced "CHICK-zu-loob"), excavating a crater 15–20 kilometers deep and at least 170 kilometers in diameter. The impact displaced huge volumes of seawater and produced a tsunami (incorrectly known as a "tidal wave" since there are no tides involved) that rose to a level of 100 meters and spread across the Caribbean, even rolling as far as 20 kilometers onto the land. Meanwhile, the bolide itself excavated 100 cubic kilometers of rock and debris from the site, which rose to an altitude of 100 kilometers. Most of it fell back immediately, but some of it remained as dust in the atmosphere for months. This material, along with the smoke from the fires, shrouded the earth, creating a form of nuclear winter. According to computerized climate models, global temperatures fell to near the freezing point, photosynthesis halted, and most plants on land and in the sea died. According to some scenarios, the impact happened to hit the sulfur-rich gypsum deposits in the bedrock of the Yucatan. This supposedly caused an acid rain bath that lasted for decades, killing off more plants and most animals on land.

The impact hypothesis was simply a wild speculation until 1978, when Berkeley geologist Walter Alvarez, working on a thick sequence of deep-marine limestones near Gubbio, in the central Apennines of Italy, took a sample of a clay layer that marks the end of the Cretaceous. When nuclear chemists Frank Asaro and Helen Michel analyzed it, they found that it contained an unusually high concentration of the rare platinum-group element iridium. Alvarez's father, Nobel-prize-winning physicist Luis Alvarez, suggested the explanation that the iridium was evidence of extraterrestrial influx, and he proposed the asteroid-impact scenario, which was published in 1980 (Alvarez et al. 1980). Walter Alvarez's find was the first concrete evidence that brought the ideas about the Cretaceous extinctions out of the realm of speculation and into that of testable science.

. . . Or a Whimper?

Naturally, when such this startling idea was first presented, scientists were skeptical and sought to test it further. At first, there was doubt about the iridium in the clay layer itself. Could it have been concentrated by ordinary geochemical processes? But within a year or two, iridium had been found in a number of sites around the world, both marine and nonmarine, which ruled out some sort of marine chemical concentration process. By 1983, however, a number of scientists pointed out that there was another event occurring at the end of the Cretaceous: the eruption of the Deccan lavas in western India and southern Pakistan (fig. 2.2). These eruptions produced over 10,000 cubic kilometers of lava, with individual flows as thick as 150 meters (although most are 10–50 meters thick); in western India alone, the total thickness exceeds 2,400 meters. Such huge eruptions could have thrown enormous amounts of volcanic ash into the atmosphere, producing a nuclear winter effect almost like that predicted by the impact hypothesis. They could also have pumped a lot of carbon dioxide from the mantle into the atmosphere, changing the chemistry of the atmosphere and oceans. In addition, deep mantle-derived lavas (such as lava from Kilauea on Hawaii) even produce iridium, which is enriched in mantle rocks as it is in extraterrestrial rocks (just not in crustal rocks). Finally, the dating of the Deccan lavas has shown conclusively that they were in peak eruption in the last half million years of the Cretaceous and that the eruptions continued into the Paleocene.

The eruptions of the Deccan lavas are a well-known fact, and their age and effects are not in dispute. The impact hypothesis, however, took longer to evaluate. As the 1980s progressed, scientists found more and more iridium layers, and also many sedimentary beds that contained evidence of droplets of material ejected from the crater, and quartz grains that had undergone high-pressure shock. In many sites around the Caribbean (especially in Cuba, in Haiti, and on the Mexican coast of the Gulf right through the Brazos River, Texas), geologists found evidence of a huge tsunami, with giant blocks ripped up and tumbled about and deposited catastrophically. In 1990, the Chicxulub crater was discovered, although its existence had been documented years earlier but had not been connected to the Cretaceous impact. The crater itself was completely filled, buried by younger deposits and then overgrown by jungle, so it is visible only on gravity surveys. Once drill cores were brought up, they produced shattered rocks characteristic of the fallback into craters, and the dates on the crater rocks were in the 65–66 Ma range, exactly at the end of the Cretaceous.

So now we have two events at the end of the Cretaceous that could have caused the great extinction: the Chicxulub impact and the eruptions of the Deccan lavas. However, there is a third global event that has to be considered: sea level dropped dramatically during the latest Cretaceous, causing shorelines to retreat and exposing large areas of continental shelf that had once been prime habitat for marine organisms. So now we are looking at three possible causes. Which is the primary cause? Two of the events were relatively gradual (with effects that would have been spread out over hundreds of thousands of years); the third event (the impact) was geologically instantaneous, with most effects taking place over hours to weeks to months at the most.

The best way to evaluate the relative importance of all three events is too look at the direct evidence from the fossil record itself. Which organisms were most severely affected, and which ones were relatively unaffected? Was the extinction gradual (suggesting volcanism or sea level retreat as a major cause) or instantaneous (supporting the impact hypothesis)?

How Do We Evaluate the Evidence?

It should be a relatively straightforward task to look at the distribution of fossils and declare either that they all disappear at once or that they disappear gradually from the rocks right up to the end of the Cretaceous. But, as scientists have learned the hard way, it is not that simple. The fossil record is not a perfect reflection of every organism that has ever lived on this planet. The chances that any one species of animal or plant will be fossilized is less than a fraction of a percent by most estimates, so that some 99% of the species that have lived on the earth have never been fossilized. A much higher percentage of the insects and other animals without hard skeletons that fossilize easily are missing from the fossil record. Once an organism dies, its body must be buried quickly before decomposers break it down completely, or scavengers chew it up, or river or ocean currents roll it around and break it up. After it is buried, it can be transformed by chemicals in the groundwater, it can dissolve away completely, or it can be destroyed by the high heat and pressures that rocks experience with deep burial in the earth's crust. If it survives all these ordeals, then the fossil needs the unusual good fortune of having been exposed again at the earth's surface during the last few centuries, so that paleontologists can find it and collect it and preserve it in a museum. Consider the huge area of eroding rocks exposed on the earth today, with their fossils weathering out and being continuously destroyed; there are at most a few thousand paleontologists in the entire world to search these places and find the fossils before they are lost forever, after their millions of years of burial. You can see how extraordinarily fortunate it is that we have any fossils at all, let alone that our fossil record is as good as it is.

Consequently, when we plot the distribution and abundance of fossils through a sequence of rocks to test gradual versus catastrophic hypotheses, we must keep many possibilities in mind. The first is that not every time interval when an organism lived will be represented by its fossils. If a group of animals is rare to begin with (such as dinosaurs), their fossil record will be patchy and incomplete, and their chances of preservation are poor. This preservation question was the major dispute during the 1980s, when the impact-extinction hypothesis first appeared. A number of paleontologists pointed out that dinosaur fossils were unknown in the last 3 meters of strata representing the very end of the Cretaceous in eastern Montana (the only place we have a good record of the latest Cretaceous on land). Did this mean that dinosaurs were extinct before the impact? Or were dinosaur fossils so rare that they were just not preserved in this final interval? In 1982, Phil Signor and Jere Lipps pointed out that in a record like that of the dinosaurs, determining whether an extinction was truly gradual or instantaneous would be impossible (fig. 2.3). Because the fossils are rare, and might disappear gradually over time owing to poor preservation, even a record of instantaneous extinction would end up looking like a gradual extinction in the fossil record. Ironically, there are processes that can work in reverse of this bias. Just as the Signor-Lipps effect can cause an instantaneous extinction to look gradual, a long interval when no rocks are deposited will cause most of the fossil lineages to terminate at this horizon (known as an unconformity, or a gap in the rock record). Geologists are frequently confronted by horizons in the rock record where fossils all appear to go extinct at the same instant in time, but these situations are usually due to an unconformity artificially truncating the end of the organisms' ranges in time, so that the extinctions appear to be instantaneous.

There are other pitfalls as well, such as the "Lazarus effect," in which organisms temporarily disappear from the fossil record and then reappear somewhere in the fossil record of the Paleocene, as if risen from the dead, like Lazarus in the Bible. In many cases, organisms may be rare and simply not preserved in the terminal interval of the Cretaceous at a particular location, but they lived on elsewhere on the earth without leaving fossils. The Lazarus effect is a major problem when we focus on the extinction record of a local area. Unless we look at the global distribution of an organism for millions of years after its supposed extinction, we cannot be sure whether it was truly extinct or just absent from the local fossil record. Likewise, the fossils themselves can fool us. If they are durable objects (like shark teeth or dinosaur teeth), they can be eroded out of Cretaceous rocks and then redeposited in sediments that were formed long after the Cretaceous. This phenomenon misled a number of paleontologists in early studies in the Late Cretaceous of Montana. A few paleontologists

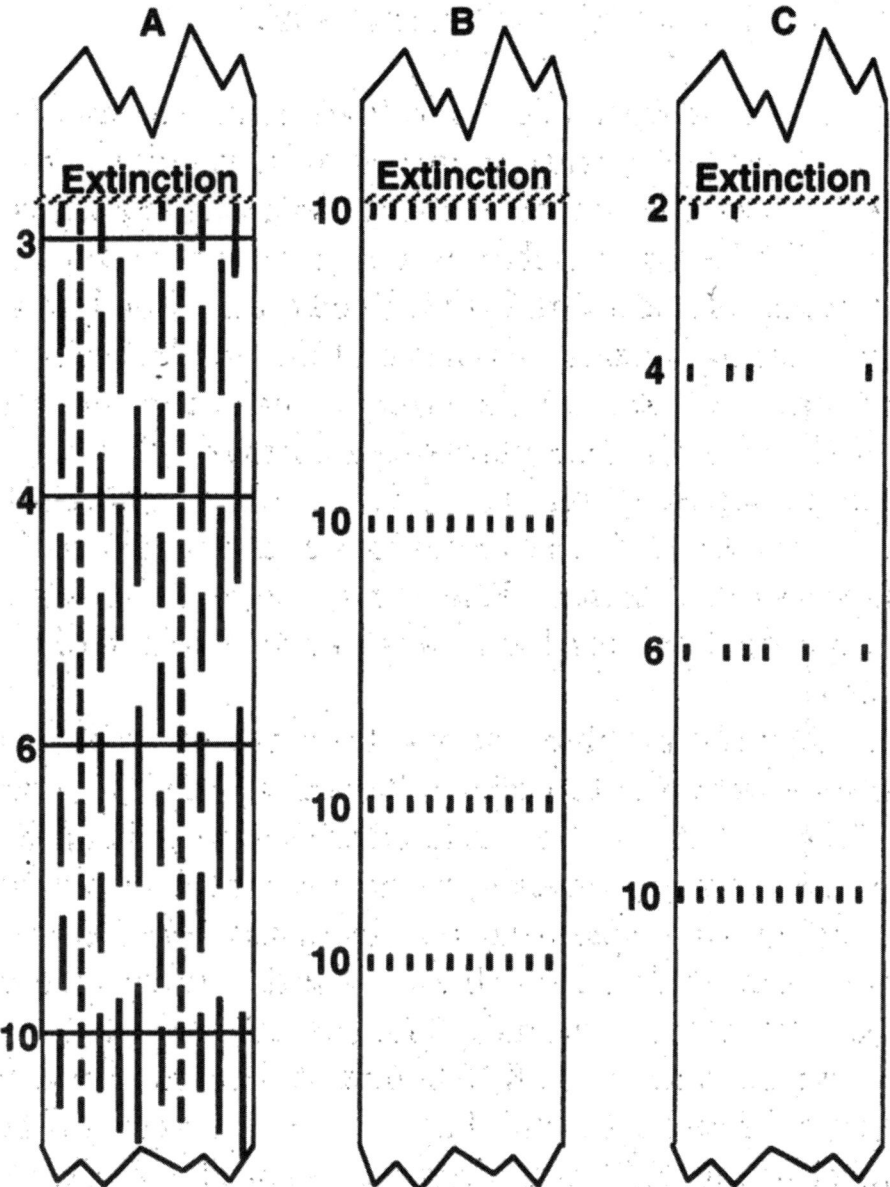

Figure 2.3. The Signor-Lipps effect. Signor and Lipps (1982) argued that an abrupt extinction in the fossil record could look like a gradual extinction owing to errors in sampling. (A) In this geologic section, there are ten species, each with a distinct and irregular pattern of preservation. If we take samples at just four levels (horizontal lines), we appear to have 10, 6, 4, and 3 species at each successive level, and abrupt extinction appears to be gradual. (B) In this section, the sampling shows no decline in species diversity through four levels, so we can be confident that this is a truly abrupt extinction. (C) In this section, there appears to be a gradual decline in diversity, which could be real or just a result of the Signor-Lipps effect. From Archibald 1996.

(Rigby et al. 1987) found pristine *Tyrannosaurus* teeth in Paleocene river-channel sediments and claimed that dinosaurs lived on into the Paleocene. However, Eaton et al. (1989) showed that these same river-channel sands yielded teeth of Cretaceous marine sharks (which certainly did not swim in these freshwater rivers, let alone survive into the Paleocene), and that dinosaur teeth were much more durable than people had supposed. Eaton's work effectively demolished the idea that there were Paleocene tyrannosaurs, because reworking of fossils into younger sediments can be a problem. Dave Archibald (1996) calls this the "Zombie effect," since these objects keep moving around after they're dead, like zombies in a cheap horror film.

With all these caveats in mind, let us look at the end of the Cretaceous and see what pattern we can determine.

The End of the Cretaceous: The Marine Realm

Before we examine glamorous and controversial taxa like the dinosaurs, let us look at organisms that have a good fossil record. We shall start with the base of the food chain in the marine realm: the plankton. Because we are talking about thousands of tiny shells in every cubic centimeter of deep-sea sediment, the problem is not poor preservation or missing specimens but only how we interpret the pattern of specimens in numerous deep-sea cores around the world. The food chain is based on the planktonic photosynthetic protists (fig. 2.4), such as diatoms, dinoflagellates, and the coccolithophorids, which secrete skeletons (silica in diatoms, organic material in dinoflagellates, and calcite in coccolithophorids) around their protoplasm but are photosynthetic plants nonetheless. Coccolithophorids secrete dozens of tiny button-shaped plates to surround their spherical cells; these tiny plates (only a few microns in diameter) separate when the cells die, and accumulate on the seafloor as coccoliths. During the Cretaceous, the shallow-marine seas around the world supported trillions of coccolithophorids, and the calcareous ooze made of coccoliths that accumulated at the bottom of these shallow seas became the rock known as chalk. As mentioned in chapter 1, the Cretaceous gets its name from the chalks of the White Cliffs of Dover and other areas in northern Europe (*creta* is the Latin word for "chalk").

The fossil records of the diatoms and the dinoflagellates do not show much of an extinction event (MacLeod et al. 1997); these organisms passed right through the K/T boundary with little change. This would seem to argue for a gradual extinction event, but the impact advocates dismiss this evidence by arguing that diatoms and dinoflagellates could have survived the long darkness as resting spores, and re-emerged when the dust clouds receded. Coccolithophorids, in contrast, died out dramatically at the K/T boundary, which suggests the extinction event had a significant effect on the surface plankton (Pospichal 1996; MacLeod et al. 1997).

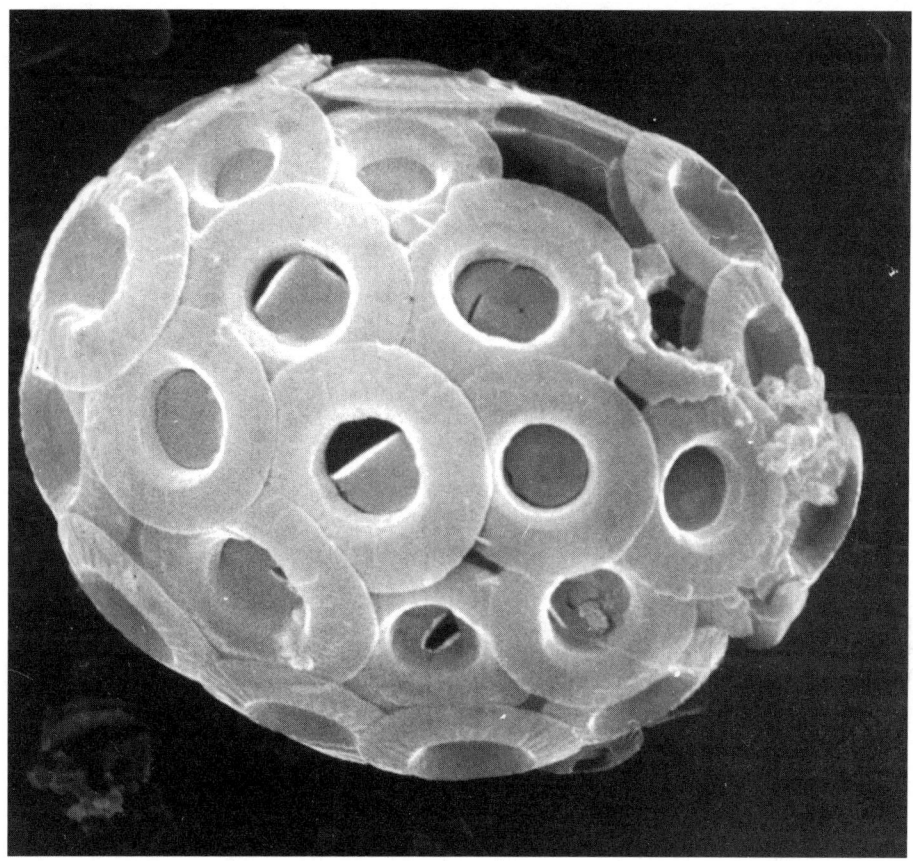

Figure 2.4. Coccoliths are button-shaped plates that cover spherical planktonic algae known as coccolithophorids. These spheres break apart and cover the ocean floor by the trillions, making the chalk deposits of the Cretaceous. Each coccolith is only tens of microns in diameter, about a tenth of the size of the radiolaria and foraminifera shown in fig. 1.5. Photo courtesy of W. Siesser.

Next up the food chain were the amoeba-like protistans that eat diatoms and coccoliths. Known as the foraminiferans and radiolarians (fig. 1.5), they secrete skeletons of calcite and silica, respectively. Like the diatoms and dinoflagellates, the radiolarians show little sign of a K/T extinction (MacLeod et al. 1997), and the impact advocates have had trouble explaining this fly in their ointment. Because, unlike plants, radiolaria do not have resting spores, how so many of them survived in the deep sea when the world had supposedly become hellish is not explained by the impact hypothesis. In fact, the richness of radiolarians actually increased across the K/T boundary, which suggests that marine productivity actually increased in the waters around the Antarctic, and completely contradicts the idea that the oceans became stagnant and stratified as a result of the K/T impact (MacLeod et al. 1997).

The End of the Dinosaurs? • 31

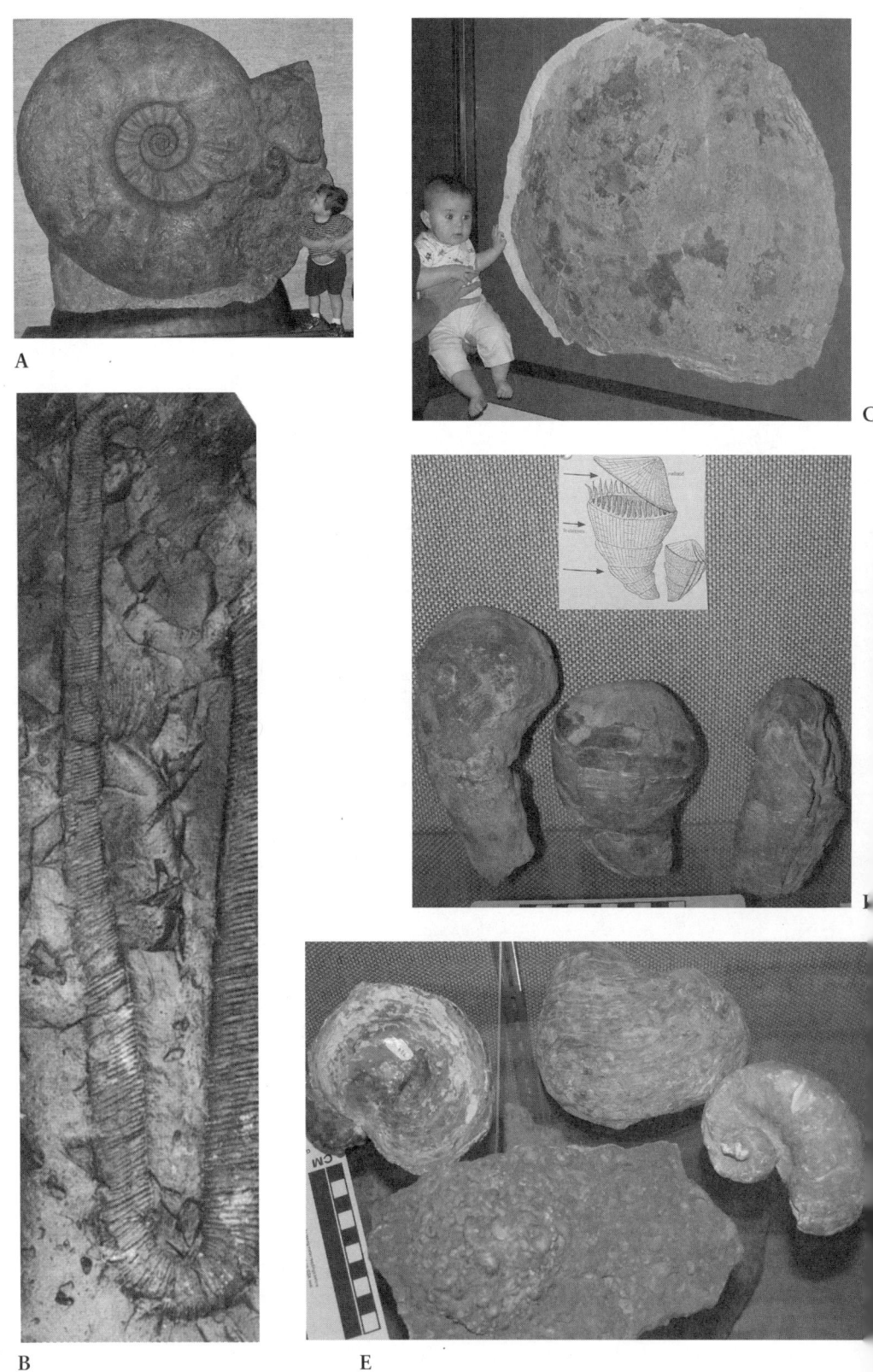

A

B

C

E

Figure 2.5. Representative Cretaceous mollusks. (A) The giant ammonite Parapuzosia *had a squid-like head protruding from opening in the shell next to the child. (B) Other ammonites uncoiled from the normal flat plane and had odd shapes, like this hairpin-shaped* Hamites. *(C) The bivalves include the giant flat clam* Inoceramus, *which reached over 2 meters in diameter. (D) The strangest of all bivalves were the reef-forming rudistids, which had one shell shaped like a cone embedded in the sea bottom and an upper shell that functioned as a lid. (E) The weirdly asymmetrical spiral oyster* Exogyra *was also characteristic of the Cretaceous. Photo B courtesy of P. Ward; other photos by the author.*

The most studied group of microfossils is the foraminifera. The benthic foraminifera that live on or in the sediment of the seafloor again show almost no extinction (MacLeod et al. 1997), probably because events on the ocean's surface seldom affect them directly. They survive on detritus that sinks down to the deep ocean and would not be affected by anything on the surface that did not last long enough to affect their food supply or did not change the chemistry or temperature of the deep-ocean currents.

The free-floating planktonic foraminifera are a different story. For many years, Gerta Keller and Norman MacLeod and their colleagues have argued that the planktonic foraminifera show a gradual extinction pattern, with two-thirds of the species going extinct 300,000 years before the K/T boundary, almost a third of the foraminiferans going right through the boundary and surviving into the early Paleocene, and only relatively few dying out precisely at the iridium layer. However, other micropaleontologists, such as Jan Smit, Brian Huber, and Dick Olsson, have challenged this interpretation, arguing that the planktonic foraminifera do show a pattern consistent with the impact. At present, it is not clear who is right, but there is some consensus that a large number of planktonic foraminifera died out before the impact (consistent with the effects of the Deccan lavas), yet more species apparently died out at the K/T boundary than Keller and MacLeod are willing to admit.

Next up the food chain are the colonial organisms, such as the reef corals. In the past, it was believed that the number of these species dropped dramatically at the end of the Cretaceous and that reef corals took a long time to recover in the Paleocene (Coates and Jackson 1985). Impact advocates have argued that this is because reefs would be sensitive to the shallow-water effects of the impact, as well as to the loss of so much of the plankton on which they feed. But several authors (e.g., Rosen and Turnsek 1989; Rosen 2000) have shown that coral diversity was actually declining much earlier in the Late Cretaceous and that much of their apparent absence in the early Paleocene can be explained by Lazarus taxa that eventually reappear by the late Paleocene. Hence, it is no longer clear that corals were that strongly affected by the K/T impact, contrary to previous claims.

Further up the food chain are the marine mollusks (fig. 2.5). Because they have hard shells, they are well represented in the Late Cretaceous, so their fossil record has been well studied. Here, even the impactors have conceded defeat. Two of the most distinctive groups of Cretaceous clams are the large, flat "dinner plate" clams, or inoceramids (fig. 2.5C), some of which reached a meter in diameter, and the cone-shaped colonial rudistids (fig. 2.5D), which made up most of the tropical reefs in the Cretaceous. All the paleontologists who study these concede that they went extinct in the middle part of the latest Cretaceous, about 5 million years before the impact, with maybe one species of each surviving to the end of the Cretaceous to witness the bolide hit Chicxulub (Kauffman 1988; MacLeod 1994). Clearly, marine mollusks were affected by something that changed the oceans gradually and long before the impact, such as the climatic effects of the Deccan eruptions. In addition, most of the other marine snails and clams that have been studied show relatively little extinction, with only 55% of the clams (mostly the inoceramids and rudistids just mentioned) and only 35% of the marine snails dying out. Most of the studies have documented that this extinction appears to be gradual (Hansen et al. 1987, 1993; Bryan and Jones 1989; Zinsmeister et al. 1989) rather than concentrated at the K/T boundary, although naturally the impact advocates have tried to reinterpret these data to fit their biases. Sheehan and Hanson (1986) and Gallagher (1991) noted that organisms that depended heavily on the plankton for their food supply (such as certain species of clams and snails), or that had planktonic larvae, were the most severely affected (which is consistent with the idea that the surface plankton were most affected by the K/T events and that the benthos was relatively sheltered). By contrast, mollusks that either fed primarily on bottom detritus or had swimming or benthic larvae seemed to survive the K/T event disproportionately, presumably because the surface events did not extinguish their larvae and because the benthic food supply was also unaffected.

Besides clams and snails, the third large group of mollusks is the cephalopods, a group that today includes the squids, the octopuses, and the chambered nautilus. Mesozoic seas were dominated by the ammonites (figs. 2.5A, B), which were shelled cephalopods much like the living chambered nautilus. They are known to have been in decline throughout the latest Cretaceous, with only 8–16 species in the latest Cretaceous. These gradually disappear through the end of the latest Cretaceous in the rock sequence on the Bay of Biscay on the northern Spanish coast (Ward et al. 1991), with the last species disappearing 20 centimeters below the K/T boundary. The same pattern has been reported from the Antarctic (Zinsmeister and Feldmann 1993). The pattern appears to support a gradual extinction, with no ammonites around to see the rock from space, although naturally the impact advocates want to dismiss this pattern as due to the Signor-Lipps effect. However, the nautiloids went right through the K/T boundary with almost no documented extinction, and they were still thriving in small numbers through the Cenozoic. According to Kennedy (1993), the difference might have been that ammonites might have produced thousands of planktonic larvae that would have been sensitive to changes in the surface ocean, whereas the few larvae of nautiloids were benthic swimmers.

Yet another group of cephalopods were the squid-like belemnites, which left conical solid shells from inside their bodies that resemble large-caliber bullets. This group was also in decline through the entire Late Cretaceous, with only one species surviving into the latest Cretaceous and dying out before the K/T impact (MacLeod et al. 1997).

Another group of shelled invertebrates are the brachiopods, or "lamp shells." These are generally rare in the Mesozoic, but in a few places such as Denmark (fig. 3.1), they show an abrupt extinction at the end of the Cretaceous (Surlyk and Johansen 1984). However, their close relatives, the bryozoans, or "moss animals," show an interesting pattern. The conservative group of bryozoans, the cyclostomes, survived the K/T events with only minor extinction, whereas the more advanced cheilostomes suffered a major extinction (Hakansson and Thomsen 1979).

The last major group of shelled invertebrates is the echinoderms, including the sea stars, sea urchins, brittle stars, sea cucumbers, and crinoids, or "sea lilies." According to Birkelund and Hakansson (1982), there is virtually no change between the Cretaceous and Paleocene in the sea stars, brittle stars, or crinoids. In the early Paleocene, the crinoids actually flourished before the brachiopods, bryozoans, or clams could return. Smith and Jeffrey (1998, 2000) showed that the extinction in the echinoids (sea urchins, heart urchins, sea biscuits, and sand dollars) was also not as severe as previously claimed. Most of the taxa that lived on shallow-water carbonate shelves were hard hit, but the rest went through relatively unscathed. In fact, the greatest extinction did not happen until the middle Paleocene, when the deposition of thick chalky sediments required by certain echinoids ceased.

The final important group of marine life is the vertebrates. The record of fossil fish is fragmentary, but about 90% of the families survived, so there is no evidence of a mass extinction (MacLeod et al. 1997). The marine reptiles were once the top of the oceanic food chain in oceans dominated by the ammonites and fish. Some groups, such as the dolphin-like ichthyosaurs and long-necked paddling plesiosaurs, were in decline throughout the entire later Cretaceous and were probably extinct long before the impact. The huge seagoing monitor lizards known as mosasaurs, however, were a different story. They were flourishing in the Late Cretaceous, with as many as seventy species worldwide in the latest Cretaceous. However, once again the great sea level drop that eliminated most of the shallow seas at the end of the Cretaceous wiped out our record of mosasaurs at the end of the Cretaceous as well. They are rare in the few marine sections that do go up to the K/T boundary, so it is hard to tell if any were alive to witness the impact.

In summary, the marine record shows a mixture of patterns (fig. 2.6). Some extinctions, like those of the planktonic foraminifera and coccolithophorids, and possibly the brachiopods (which are shallow-water benthic filter feeders), are consistent with the effects of an impact. Others, like the benthic foraminifera, the dinoflagellates, diatoms, and radiolarians,

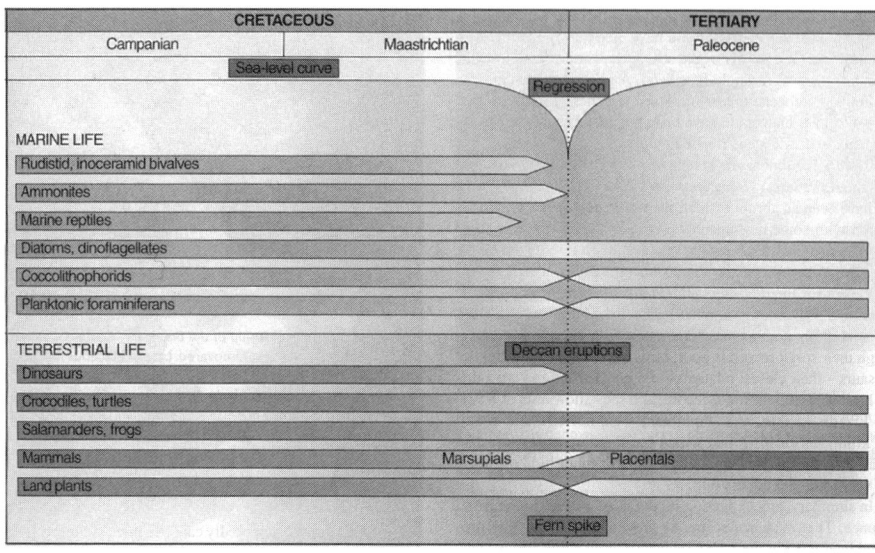

Figure 2.6. Pattern of diversity changes and extinction through the latest Cretaceous and early Cenozoic. Some groups gradually declined well before the K/T event, whereas others survived into the Paleocene with little or no effect. Only a few died out abruptly at the K/T boundary. From Prothero and Dott 2003.

and most of the clams, snails, nautiloids, echinoderms, and bryozoans, show relatively little change at the K/T boundary. Still others (such as the corals, inoceramids, rudistids, belemnites, ammonites, and marine reptiles) were in decline long before the latest Cretaceous, so it is not clear whether any survived to witness the impact—but they were clearly affected by causes that were protracted over the entire later Cretaceous, instead of being wiped out all at once by a rock from space. Such a mixed pattern shows that long-term effects (such as the climatic changes caused by the Deccan eruptions) were important and that only a small portion of the marine realm was healthy and thriving when the impact occurred. After these are accounted for, a large portion of the marine biosphere (the benthic foraminifera, the dinoflagellates, diatoms, and radiolarians, and most of the nautiloids, clams, snails, echinoderms, and bryozoans) still survived with little or no effect. To some extent, these survivors are explainable as bottom-dwellers or swimmers that did not require the plankton for their food supply or larvae. Even so, the evidence does not support any scenario that makes the oceans too hellish.

The End of the Cretaceous: The Terrestrial Realm

Naturally, most people are interested in the terrestrial realm, since the dinosaurs are the only familiar organisms that died out at the end of the Cretaceous. Here, we run into problems: only one place in the world, the Hell

Figure 2.7. The badlands of the Hell Creek Formation at Steve's Hill in Harding County, northwestern South Dakota. The dark bands are rich in coal from swampy conditions, and the lighter bands are sands and muds deposited on floodplains and rivers. The K/T boundary is near the top of this section. Photo courtesy of K. Johnson.

Creek Formation of eastern Montana and the western Dakotas (fig. 2.7), preserves a good record of the end of the Cretaceous on land. That section has been studied intensively by a number of vertebrate paleontologists who specialize in the fish, amphibians, reptiles (including dinosaurs), and mammals found in the section in abundance. Archibald and Bryant (1990) and Archibald (1996) summarized this work. The 111 non-marine species found in the latest Cretaceous show a peculiar pattern of extinction that cannot be simply explained by the effects of one impact (fig. 2.6). First of all, about 65% of the species survived the impact, so the effect was not as dramatic as previously argued. There was significant extinction in the sharks, in the pouched marsupial mammals, in the lizards, and of course in the dinosaurs. However, a number of scientists have argued that the dinosaurs were already on the decline through the entire later Cretaceous and that there may have been only one or two species of *Tyrannosaurus* and *Triceratops* alive at the end. The land plants show a change from a flora that produces *Aquilapollenites* pollen to a different Paleocene flora, and right at the K/T boundary is a layer rich in fern spores, nicknamed the "fern spike." Ferns are thought to have acted as "disaster taxa" that flourished in a world where most other organisms were exterminated. However, Sweet et al. (1990) showed that the floras changed in several steps, with several extinction events in the latest Cretaceous before the impact, so clearly the impact does not explain all floral change.

More revealing is what *did not* die out. Bony fish and amphibians marched through the K/T boundary almost unscathed, as did turtles, crocodilians, and the crocodile-like champsosaurs. Mammals also flourished, with the placental mammals gradually taking over as the pouched marsupials vanished. LaBandeira and Sepkoski (1993) showed that there was no extinction in the insects, a group that should have been the most sensitive to a global catastrophe predicted by the impact advocates. Kozisek (2003) pointed out that tropical honeybees cannot survive if the tropics become too cold or flowers disappear. Nor do the birds show much extinction, even though they too should have been vulnerable (Chiappe 1995). As Archibald (1996) points out, when the pattern is examined in detail, one can make predictions about whether the impact, the volcanic event, or sea level change best explains the total pattern of extinction and survival. The impact model fares the worst, with the fewest correct predictions of what goes extinct and what does not. Surprisingly, the effects of the global sea level drop explain things the best (especially the loss of the sharks, which is not explained under any impact scenario).

The details of what survives and what does not are particularly important. For example, some extreme impact scenarios postulate extensive acid rain bathing the earth for a long time after the impact. However, the survival of amphibians shows that this is simply a fantasy (Weil 1984). Amphibians breathe through their porous skins and are sensitive to slight changes in the acidity of their watery habitat. Even now, the slightly more acidic conditions of lakes and ponds due to human-induced acid rain are causing frogs and salamanders to die out rapidly. If the entire earth had been subjected to a huge acid bath, there simply would not be a frog or salamander alive on the earth today.

Other survivors have a story to tell as well. Some impact advocates argue that only animals with large body sizes were affected (i.e., dinosaurs), but many of the crocodiles and champsosaurs were almost as large as the big dinosaurs (and larger than the small dinosaurs), and none went extinct. Nor did all the smaller animals escape unscathed, since marsupials were among the smallest mammals, and they did not do so well. Others have argued that aquatic taxa survived by hiding in bodies of water during the initial hell-on-earth phase of the impact; but this argument fails to explain why the sharks died out completely while the bony fish did just fine (Robertson et al. 2004). Clearly, the popular but simplistic impact model is insufficient when the data are examined in detail. The impact did indeed occur, but much of the terrestrial extinction was apparently due to other causes before the impact. In addition, the hellish scenarios of the impact and its effects on the earth must be greatly exaggerated, because so many animals that *could not* survive such conditions (such as amphibians, crocodiles, insects, and freshwater fish) *did* survive.

But these problems do not daunt the impact advocates. In their minds, if the data do not agree, then there must be something wrong with the data, rather than with the impact model itself. Paleontologists had been saying for a long time that dinosaur diversity was declining through the Late Cretaceous, and only a few species were alive in the final million years before the impact. With a large crew of Earthwatch volunteers, Sheehan et al. (1991) intensively collected the upper part of the Hell Creek Formation, documenting every last dinosaur scrap right up to the last specimen 3 meters below the K/T boundary. They tried to argue that there was no statistically significant decline in dinosaur diversity before the K/T boundary. However, Hurlbert and Archibald (1996) showed that the statistical tests used by Sheehan et al. (1991) are inconclusive; dinosaurs may have indeed been declining long before the impact, but the data are insufficient to tell.

It has now been over twenty-five years since the impact model was first proposed, and the results are interesting. The geochemists and geophysicists, who are familiar with impact physics and iridium, but not with biology, were the first to be convinced, and they have not taken the biological data seriously. But the paleontologists and biologists, who know plants and animals best, have not jumped on the impact bandwagon so readily. At the 1985 meeting of the Society of Vertebrate Paleontology in Rapid City, South Dakota, reporter Malcolm Browne of the *New York Times* took an informal poll of the vertebrate paleontologists assembled at their annual convention. Even though it was already five years after the proposal of the original impact hypothesis, the vast majority found the impact model unconvincing as a complete explanation for the K/T extinctions. As time has gone on, the impact advocates have preemptively declared victory and claimed that there are no longer any doubts that the impact did the whole thing. But as books by Archibald (1996), Hallam and Wignall (1997), and Dingus and Rowe (1998) continue to argue, the story is not as simple as the impact advocates would like us to believe. In 1997, almost thirty years after the iridium anomaly was first discovered, the entire K/T fossil record was summarized by twenty-two distinguished British paleontologists (specialists in almost every group affected), and their overwhelming consensus was that the K/T impact had little effect on life, with the exceptions noted above (MacLeod et al. 1997). Even as this book was being written, Brysse (2004) did another survey of vertebrate paleontologists. Of those surveyed, 72% felt that the K/T extinctions were caused by gradual processes followed by an impact. Only 20% felt that the impact was the sole cause. (The remaining 8% had no opinion as to its cause, or questioned whether it was really a mass extinction at all.) The majority of the respondents also felt that the extinction was gradual, rather than instantaneous, and that the impact model was too simplistic to adequately explain the entire K/T mass extinction.

But Are the Dinosaurs Really Extinct?

In the public mind, the word "dinosaur" is almost synonymous with the idea of extinction. "As dead as the dinosaurs" goes the phrase, after all. Indeed, most of the huge terrestrial reptiles that are called "dinosaurs" by the public did vanish at the end of the Cretaceous (despite occasional claims that they survived into the Paleocene and then died out). But since the 1960s, another important idea has taken hold in paleontology. The huge creatures that we know as dinosaurs may be extinct, but they have living descendants—the birds. And if birds are dinosaurs, then dinosaurs are not extinct after all! They survived as smaller, flying animals with feathers, and today they are more successful than ever.

The idea that birds are dinosaurs is an old one. Almost as soon as the first fossils of the Late Jurassic bird *Archaeopteryx* were described in the 1860s, British biologist Thomas Henry Huxley and German embryologist Karl Gegenbaur noticed their great similarity to the small running dinosaur *Compsognathus* (the "compys" from *Jurassic Park*) from the same limestone quarries in Solnhofen, Germany, that produced *Archaeopteryx*. Huxley and Gegenbaur used this evidence, and that from embryology, to argue that birds were closely related to dinosaurs. However, by the 1880s, the birds-as-dinosaurs hypothesis fell out of favor as scientists focused on scenarios in which flight arose from gliding down from the trees, not from running on the ground. Some pointed out that the known dinosaur fossils lacked collarbones (which fuse into the Y-shaped "wishbone" of birds). If dinosaurs had lost their collarbones, then how could their supposed descendants regain them and use them as the wishbone spring that helps power their wings and flight muscles? And after all, the prevailing idea was that dinosaurs were huge, lumbering, slow, stupid creatures—how could they have evolved into quick, intelligent birds?

For the next seventy years, the issue remained unresolved. A grand total of eight specimens of *Archaeopteryx* were eventually found in the Jurassic limestones of Solnhofen, but few other Mesozoic birds were known. Then in the 1960s, Yale paleontologist John Ostrom made several important discoveries. He dug up and described the first good specimen of the running predatory dromaeosaur *Deinonychus* (the correct name for the "*Velociraptor*" of *Jurassic Park* fame). Contrary to the prevailing idea that dinosaurs were sluggish and stupid, this specimen showed that at least some dinosaurs were active, intelligent creatures with a high metabolism, and could balance on their hind legs. Next, Ostrom examined all the known specimens of *Archaeopteryx*. Some he found had been misidentified as pterodactyls, and one was originally identified as the dinosaur *Compsognathus* until the feather impressions showed up. If it was that easy to mistake *Archaeopteryx* for a dinosaur, Ostrom wondered, what does this say about the ancestry of birds? Ostrom compiled a list of evolutionary specializations found only in birds and dromaeosaur dinosaurs such as *Deinonychus*. For example, both birds and dromaeosaurs have a unique wrist structure known as the semilunate carpal. This half-moon-shaped wrist bone is formed by the fusion of most of the wrist elements. It allows birds to have a strong forward flexion of their wrist in the front part of the flight stroke, and dromaeosaurs presumably used the same motion when they extended their claws to grab prey. Likewise, the hind limb is full of shared specializations. For example, only birds, dinosaurs, and pterodactyls have a unique ankle known as the mesotarsal joint. Most vertebrates have a hinge between the shin bone and the first row of ankle bones, which allows them to turn their foot forward and backwards. But birds, pterosaurs, and dinosaurs have a joint between the first and second row of ankle bones, a condition found in no other animal. The first row of ankle bones often becomes simply a small cap of bone on the end of the shin bone, and in many dinosaurs, it fuses to the shin bone completely. You can see this even in modern birds. The next time you eat the drumstick (shin bone) of a chicken or turkey, notice the little cap of cartilage at the end of the drumstick. That is actually the first row of ankle bones, fused to the end of the shin bone—a unique dinosaurian feature of your Thanksgiving feast!

By the 1970s, the list of anatomical specializations shared only by birds and dinosaurs was impressive, and most paleontologists found them convincing. But as in any area of science, there are always skeptics. Ornithologists were strongly wedded to the idea that flight had to have originated from the trees down, and they could not imagine how a dinosaur running along the ground could evolve flight. They argued that the lack of a collarbone still stood in the way of bird origins from dinosaurs. And they argued that feathers are a unique structure evolved for flight, and do not make sense on a dinosaur.

But as the 1980s and 1990s progressed, more evidence poured in. Collarbones (which are delicate and seldom preserved) were found in new specimens from a number of predatory dinosaurs, so that argument is invalid. Hundreds of new specimens of Jurassic and Cretaceous birds have been found. The best are those from the famous Liaoning lake beds of China, which are so exquisitely preserved that they even show feathers, stomach contents, and other delicate structures. And sure enough, in addition to discoveries of numerous primitive birds, there have been many nonflying dinosaurs that had feathers. It seems clear now that many small dinosaurs had feathers, which evolved not for flight but for insulation, and were secondarily turned into flight structures much later. Finally, it is not true that flight can have evolved only from the trees down and that feathers had no use on a land animal that had not yet developed the ability to fly. For example, Ken Dial (2003) has shown that chukar partridges rarely lift off from the ground and fly. Instead, they use their feathers to help them run up steep slopes, flapping their wings to climb almost vertically. So there are plausible intermediate stages for feathers aiding flight after all. They were used for insulation only at first, then they helped in propulsion along the ground and up inclines, and eventually they would help in short glides before full powered flight evolved.

As of this writing, there is almost no doubt among vertebrate paleontologists that birds are descended from dinosaurs. If that is the case, then dinosaurs did not die out at the end of the Cretaceous after all. From now on, I shall refer to the traditional concept of the large reptiles commonly known as "dinosaurs" (but excluding birds) as the "non-avian dinosaurs."

Look out the window at the dinosaurs flying in the sky, or the dinosaur in your birdcage, or the dinosaur in your Thanksgiving feast or your next chicken dinner. Dinosaurs are all around you!

For Further Reading

For a balanced discussion of the K/T extinctions and bird origins, I recommend the following:

Archibald, J. D. 1996. *Dinosaur Extinctions and the End of an Era: What the Fossils Say.* New York: Columbia University Press.

Dingus, L., and T. Rowe. 1998. *The Mistaken Extinction: Dinosaur Evolution and the Origin of Birds.* New York: W.H. Freeman.

Hallam, A., and P. B. Wignall. 1997. *Mass Extinctions and Their Aftermath.* Oxford: Oxford University Press.

MacLeod, N., et al. 1997. The Cretaceous-Tertiary biotic transition. *Journal of the Geological Society, London* 154:265–292.

Shipman, P. 1999. *Taking Wing:* Archaeopteryx *and the Origin of Bird Flight.* New York: Simon and Schuster.

Figure 3.1. The sea cliffs at Stevns Klint, Denmark, showing the white Late Cretaceous chalks, the K/T boundary clay (the layer just below the overhang), and the grayish Danian chalks immediately above it. Photo courtesy of Jes Rust.

3

Brave New World:
The Paleocene

The Age of Reptiles ended because it had gone on long enough
and it was all a mistake in the first place.

Will Cuppy, *How to Become Extinct*, 1941

The Aftermath

Whatever the causes of the K/T extinctions, the world of the Paleocene
was very different from any that had preceded it for almost 160 million
years. Most of the major groups of animals that had lived in the latest
Cretaceous were still around, but others that had dominated the seas (ammonites, marine reptiles) and the land (non-avian dinosaurs) were not.
The early Paleocene is a classic example of a "recovery" interval, in which
the world rebounds from a major mass extinction event. A few places in
the world, such as the sea cliffs along the coast of Denmark (fig. 3.1), preserve an excellent record of this recovery period. After all the interest in
mass extinctions in the past twenty-five years, paleontologists are now beginning to focus more attention on recovery intervals because they tell us
much about ancient environments and evolution.

What did the earth look like at the beginning of the Cenozoic? Although much of that world would be familiar to us today, many changes
in the continents were still taking place (fig. 3.2). The North Atlantic,
which had begun to rift open at almost 220 Ma, was about two-thirds of
its modern size. The South Atlantic had opened in the Early Cretaceous
and was about half as wide as it is today. The southern continents that
once made up the Gondwana supercontinent had been breaking up since
the middle part of the Cretaceous or earlier. India had separated from
Africa in the Late Cretaceous and was making its mad dash across the Indian Ocean, eventually to collide with the belly of Asia to form the Himalayas. Australia had begun to separate from Antarctica, but the con-

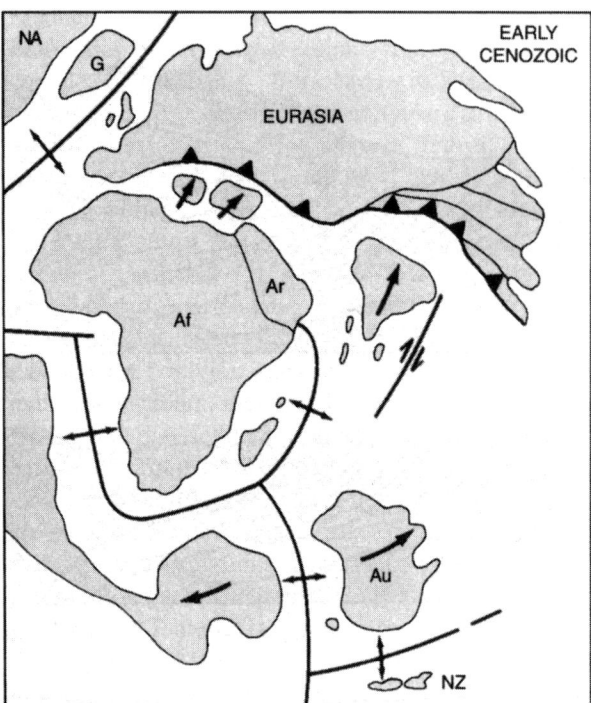

Figure 3.2.
Paleogeographic map of the
continents in the early
Cenozoic. After Prothero
and Dott 2003.

nection was still there. More importantly, there was still a huge tropical
seaway that ran from Gibraltar to Indonesia. Known as the Tethys Seaway, it had once been the habitat of large reefs made of the cone-shaped
rudistid clams and was still a realm of shallow, limey seas forming chalks.
Likewise, much of western Europe was still under shallow seas as well.

North America continued the trends that had begun in the Late Cretaceous. The Rocky Mountains, which had begun to rise in the latest Cretaceous with the Laramide Orogeny, continued to rise during the Paleocene
and Eocene. The shallow interior seaway that once connected the Gulf of
Mexico with the Arctic Ocean retreated in the great Late Cretaceous sea
level lowering, but there were still remnants known as the Cannonball Sea
in Montana, North Dakota, and Alberta. The Atlantic and Gulf Coastal
plains continued to subside and accumulate sediment, although the sea
level drop at the end of the Cretaceous meant that they were more exposed and emergent than they had been in the past, or would be when sea
level rose again in the Eocene.

In the midst of all these continental movements and sea level changes,
only certain regions deposited and preserved a good record of the Paleocene. For the terrestrial record, the best known sequences occur in the
Laramide basins of the Rocky Mountain region, especially those in the
Fort Union Group rocks of Montana, Wyoming, and North Dakota (such
as the Bighorn Basin, Crazy Mountain Basin, and Williston Basin) and the
Nacimiento Formation in the San Juan Basin in northwestern New Mex-

ico. For the marine record, the shallow seas of Europe and the Tethys are relatively well studied, as are the deep-sea records of the Paleocene recovered from numerous cores drilled all over the oceans of the world. These regions are the basis for our understanding of Paleocene climate and life.

Of these regions, only a handful of sections preserve the details of the Paleocene recovery. The early Paleocene is known as the Danian Stage, because some of the best lower Paleocene rocks are preserved along the coastal cliffs in Denmark (fig. 3.1). One particular section at Nye Klov in Denmark has an extraordinarily detailed record of the earliest Paleocene aftermath of the K/T event (Birkelund and Hakansson 1982). In this section, the basal Danian chalks are dominated by one species of crinoid, or sea lily, with almost all the other groups that were common in the latest Cretaceous chalks (bryozoans, brachiopods, and clams) absent. This crinoid, *Bourgeticrinus,* is abundant only at this level and is absent from later in the Danian. Birkelund and Hakansson (1982) suggest that it was a pioneer or weed-like species that took over a recently vacated sea bottom and established roots as the oceans returned to normal. Further up the Nye Klov section, life on the chalky seafloor returned to normal, with abundant bryozoans, clams, sea urchins, and brachiopods, and few crinoids.

Another important sequence occurs in the coastal plain sediments of New Jersey (Gallagher 2002). In a sand quarry known as the Inversand Pit, the Late Cretaceous shell beds are diverse, with about twenty-six species of invertebrates (mostly mollusks). The lowest Danian shell bed in the middle part of the Hornerstown Formation is very different, with only six species of mollusks preserved, and these are all dwarfed. Mollusks are also less abundant than sponge fragments, brachiopods, and solitary corals, but only one species of each of these groups is represented. Gallagher (2002) argues that all of these opportunistic survivors are minimalists that could live in a world with reduced planktonic food supply, and possibly even with lower oxygen levels. In the middle Paleocene shell beds of the Vincentown Formation (Gallagher 1993), the diversity rebounds to the pre-K/T levels, and a giant species of the clam *Cucullaea* (which was dwarfed in the early Danian) also occurs. Clearly, by the middle Paleocene the world had recovered to its pre-K/T conditions in many ways.

Deep-sea cores reveal the driving force behind this low diversity of benthic invertebrates. According to Olsson and Liu (1993), Olsson et al. (1999), Huber et al. (2002), and most other micropaleontologists, only three species of planktonic foraminifera (*Guembelitria cretacea, Hedbergella monmouthensis,* and *Heterohelix globulosa;* fig. 3.3a) are believed to have survived the K/T event into the Danian, so that the diversity and volume of plankton that makes up these early Danian chalks are drastically reduced. Similarly, only ten of about a hundred Cretaceous species of coccolithophorids (the calcareous planktonic algae that make up chalk; fig. 2.4) are thought to have survived into the Danian, at which time about thirty new species evolved (MacLeod et al. 1997). Clearly, the ex-

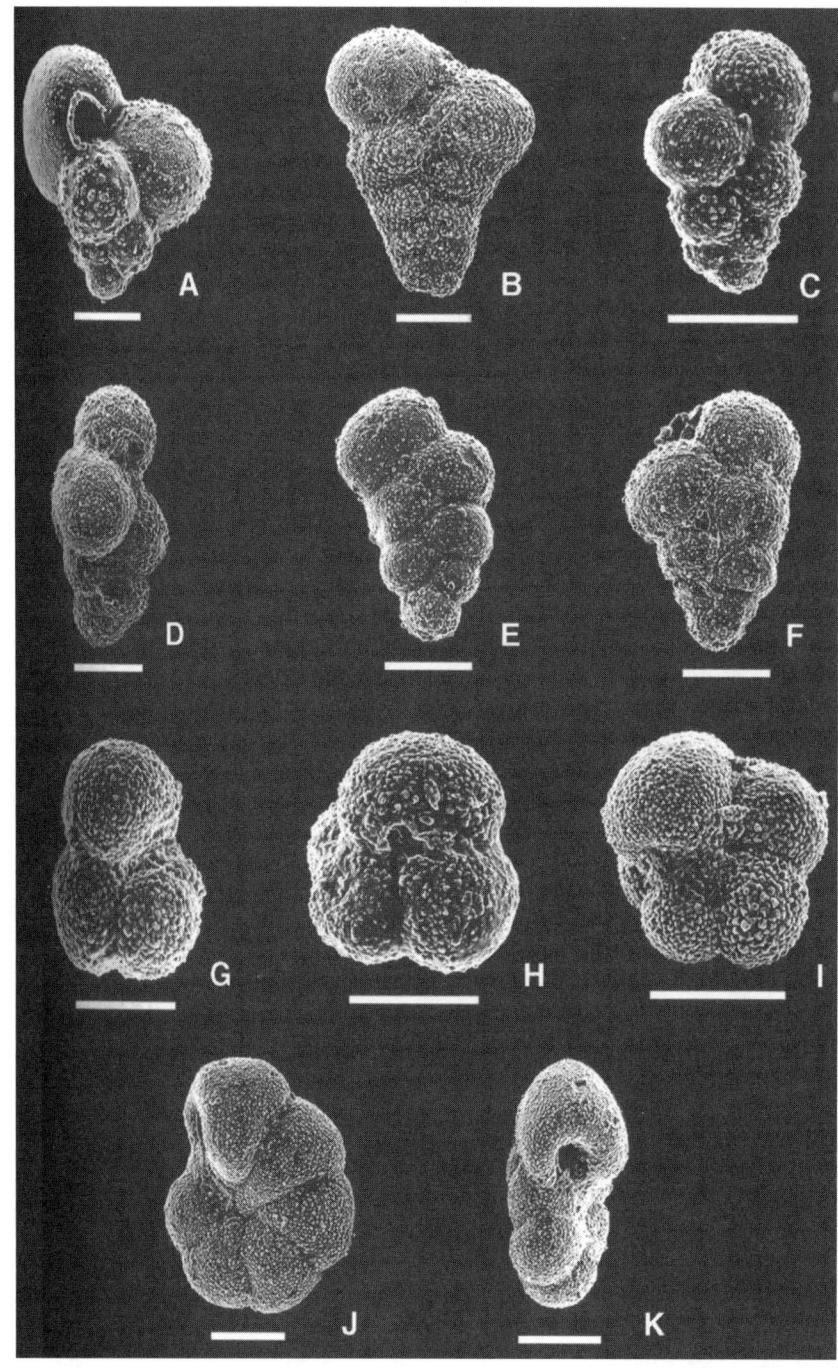

A

Figure 3.3. Paleocene foraminifera. (A) Lower Paleocene foraminifera from the western Atlantic. After Huber et al. 2002, fig. 7. (B) Evolutionary radiation of the Paleocene planktonic foraminifera from three survivors of the K/T extinctions. After Olsson et al. 1999.

Cretac.	early Paleocene				late Paleocene			Eocene	Morphologic Radiation	Genus
Maastr.	PO	Pα	P1 a·b·c	P2	P3 a·b	P4 a·b·c	P5	P6		
								coalingensis →	nonspinose, cancellate wall, no keel, strongly muricate	*Acarinia*
						nitida		*subsphaerica** ····		
						mckanni		*soldadoensis*		
						strabocella				
						acutispira			nonspinose, conical chambers, muricate wall, two lineages: 1. heavy muricate keel; 2. weak keel	*Morozovella*
								pasionensis		
								occlusa		
								velascoensis		
								acuta		
					conicotruncata					
					angulata					
						apanthesma				
								aequa		
								subbotinae		
								gracilis →		
					praeangulata					
								ovalis →	nonspinose, smooth wall, keel in advanced forms	*Globanomalina*
						imitata				
								australiformis		
					planocompressa			*pseudomenardii*		
					ehrenbergi					
				compressa				*chapmani*		
				archeocompressa			*planoconica* →		
Hedbergella holmdelensis									nonspinose, K/T survivors	Hedb.
H. monmouthensis										
		taurica							nonspinose, cancellate wall, no keel.	Praemuria
			pseudoinconstans							
			inconstans							
			uncinata							
					tadjiki-stanensis →				nonspinose, thick & crusted wall, keel in advanced form	Igorina
					pusilla					
					albeari					
		aff. pseudobulloides							spinose, cancellate wall, low trochospiral	Parasubbotina
			pseudobulloides ····							
					varianta					
				variospira						
			spiralis						spinose, cancellate wall, elevated coil	Eoglobig.
			edita							
		eobulloides								
			trivialis					*triangularis*	spinose, cancellate wall, reduction in chamber number	*Subbotina*
					cancellata					
					triloculinoides					
								velascoensis →		

49

tinction of so much of the surface plankton severely affected the bottom-dwelling invertebrates that fed upon them. Although the bottom-dwelling groups did not experience the catastrophic extinction that was once claimed, their diversity for the first part of the Danian was very reduced, and most of the species that thrived were opportunistic weed-like forms that could survive in a world with reduced food supply from the plankton, and possibly harsh temperature conditions as well. Soon thereafter there was an evolutionary explosion of planktonic foraminifera and coccolithophorids, with dozens of new species appearing by the late Danian. Many of these new species of planktonic foraminifera are very spiny, an adaptation that is thought to enhance the ability of the amoeba-like foraminiferans to float in the water column and trap more food (Olsson et al. 1999).

In the deepest ocean, similar patterns can be seen in the benthic foraminifera. For example, Coccioni and Galeotti (1994) documented an excellent deep-marine sequence now exposed in Caravaca, Spain. The earliest Danian is thought to represent only a few hundred years, and only the foraminiferan species that lived within the sediment, feeding on detritus and surviving reduced food and oxygen conditions, are found. After about 600 years, some benthic foraminifera that lived upon the seafloor appeared, although they were still adapted to low-oxygen conditions. About 1,500 years after the K/T event, the normal benthic foraminiferal assemblages began to reappear, which indicates that normal oxygen conditions had been restored. Speijer and van der Zwaan (1996) reported a similar trend in benthic foraminifera from near the tropical Tethys Seaway in Tunisia. About 50% of the species from the Cretaceous disappeared, which resulted in an impoverished fauna tolerant of cooler, low-oxygen conditions, even in shallow water. But within a few meters of section, nearly all the benthic foraminiferal groups appear or reappear, which suggests that conditions had returned to normal—and that many of the benthic foraminiferal lineages were not casualties of extinction but instead were Lazarus taxa that went into hiding during the K/T event.

Only a few places preserve the record of the recovery of mollusks from the K/T event. As we saw in chapter 2, most mollusks (except the ammonites) survived the K/T event with relatively little extinction, except at the species level. According to Thor Hansen (1988; Hansen et al. 1993), the Brazos River sediments (fig. 3.4) in the Texas Gulf Coastal Plain preserve one of the best-studied early Danian marine sequences. For the first 200,000 years of the Danian, the Texas sections suggest that the environment was still stressed, with low species richness, low abundances of individuals, and high species turnover. Most of the mollusks are deposit feeders, which depend not upon the plankton or life in the water column directly but on the organic matter already entombed in the seafloor sediments. About 2 million years after the K/T event, the normal bottom community (fig. 3.5) of mollusks, including suspension feeders that depend on plankton, and diverse carnivores as well, was finally reestablished.

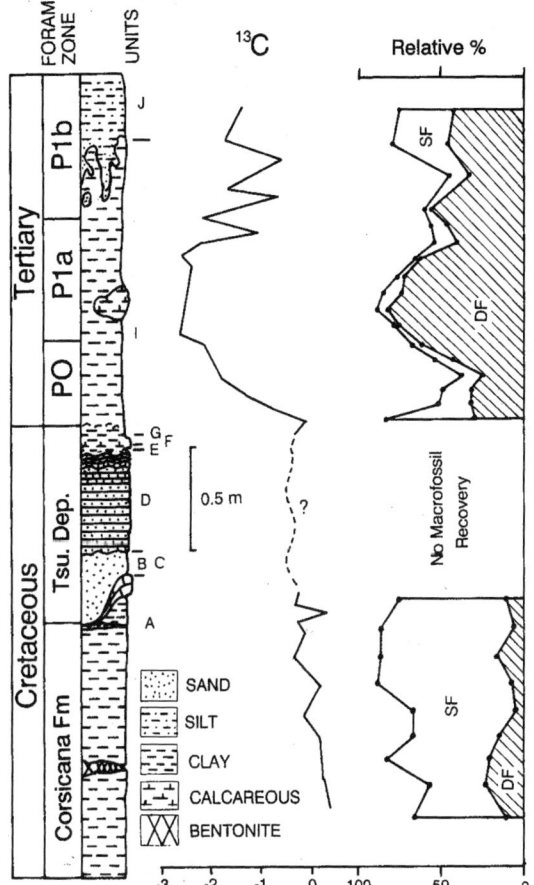

Figure 3.4. Early Paleocene section along the Brazos River, Texas. Tsu. Dep. refers to a terminal Cretaceous deposit interpreted as formed by an impact-induced tsunami across the Gulf of Mexico. The early Paleocene carbon isotope record shows an abrupt negative shift, indicating that carbon-12 was either upwelling from the deep or not being consumed by the plankton. The record of mollusks (right) shows a dominance of detritus feeders (DF) in the Paleocene, whereas surface feeders (SF) were much more common in the Cretaceous. Fm = Formation. After Hansen et al. 1987, fig. 6.

Another striking phenomenon of the Danian seafloor is the absence of large reefs. As we saw in chapter 2, most of the tropical Cretaceous reefs were made up of the cone-shaped rudistid clams, with a minor contribution from corals. Rudistids vanished in the later Cretaceous, and corals too were declining through the Late Cretaceous as well (Rosen and Turnsek 1989). According to Rosen (2000), only a single genus of coral is recorded from the early Danian, a Lazarus survivor, *Siderastraea,* from the Cretaceous. Clearly, the suppression of the plankton strongly affected the filter-feeding corals. By the later Danian, at least three coral genera are recorded, and by the late Paleocene, corals were diverse again (although many are apparently Lazarus taxa that must have been alive but not preserved in the early Danian). However, corals did not yet form the huge reefs that Cretaceous rudistids once formed.

Yet another proxy of the changes in the early Paleocene oceans is the carbon isotope system. About 98.89% of the earth's carbon is carbon-12, with six protons and six neutrons. A bit more than 1% of the carbon, however, is the rare isotope carbon-13, which has an extra neutron. Both

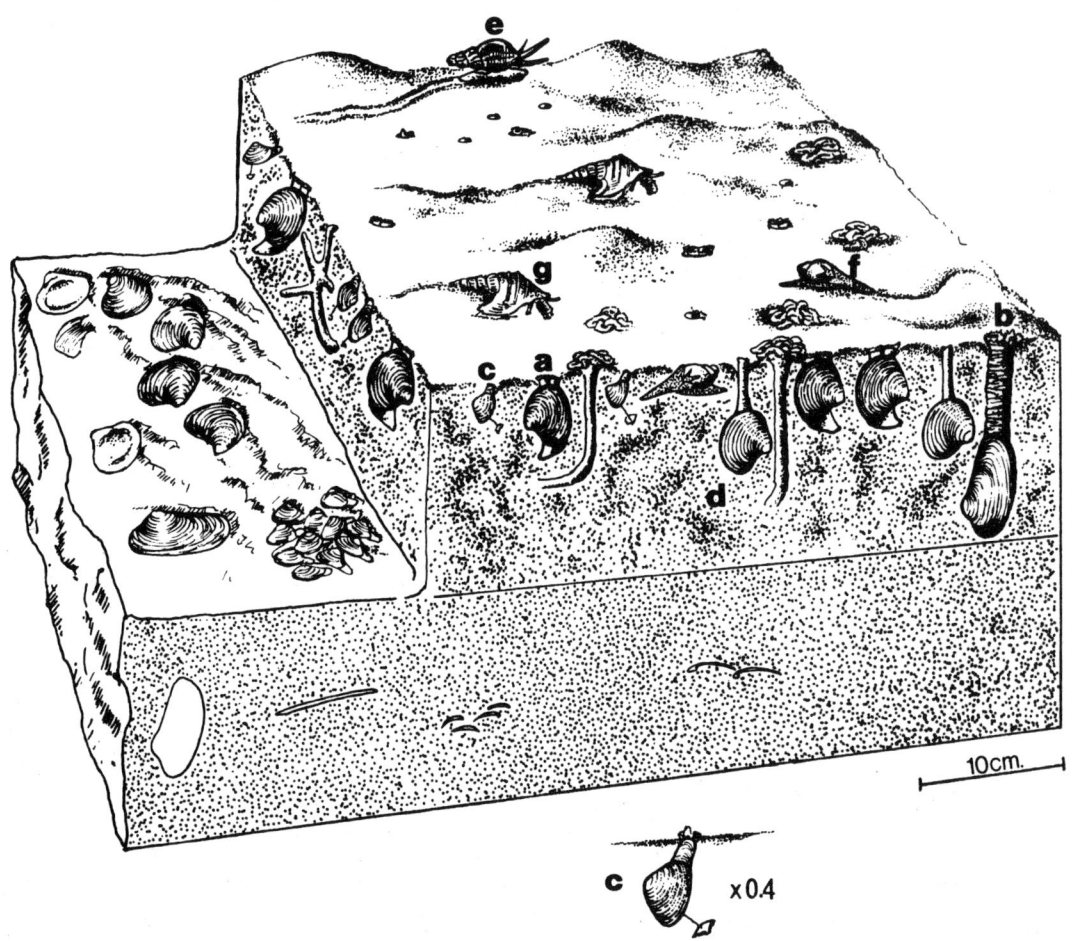

Figure 3.5. Seafloor life after the K/T extinctions. Life on the seafloor recovered after the extinctions, but mollusks and echinoids of a typical Paleocene marine sand community were not as diverse as they had been in the Late Cretaceous. There were many burrowing bivalves (a–d) but few oysters or surface dwellers. The surviving gastropods (e–g) were mostly smaller forms from archaic groups. After McKerrow 1978, fig. 108.

isotopes of carbon cycle freely through the world's oceans and atmospheres at a predictable ratio of about 99:1. Both are trapped in the shells of calcite-secreting organisms in approximately the same ratio as they were in the seawater at the time the organism lived. Thus, we can take tiny samples of calcite from the shells of foraminifera or mollusks and measure the carbon isotope ratio of the seawater in which they lived. Although the differences are tiny, modern mass spectrometers can measure the difference in ratios of carbon isotopes in parts per thousand, or parts per mil. This ratio is indicated by the symbol $\delta^{13}C$, pronounced "delta C-13."

Most organic material tends to be high in carbon-12, so that the deep sea bottom, which traps lots of organic matter, is enriched in carbon-12.

Conversely, during normal times, the shallow-marine realm is enriched in carbon-13, so the $\delta^{13}C$ is more positive. However, when oceanographic events recycle trapped organic matter through upwelling of deep ocean waters and stronger oceanic circulation, the balance of the ocean can shift away from carbon-13 and become enriched in carbon-12, and $\delta^{13}C$ becomes more negative. In addition, the normal ocean is enriched in carbon-13 because the plankton preferentially take up carbon-12 in their shells, making $\delta^{13}C$ more positive.

In many marine sections (Hsu et al. 1982; Zachos et al. 1989; Barrera and Keller 1990; Hansen et al. 1993) around the world, the earliest Paleocene (figs. 3.4, 3.6) shows a striking shift toward negative carbon isotope ratios, indicating enrichment in carbon-12. Geologists (Hsu et al. 1982; Zachos et al. 1989; Barrera and Keller 1990) interpret this shift as evidence of the breakdown of primary productivity in the calcareous planktonic foraminifera and coccolithophorids, presumably because so many became extinct. In the Brazos River section (fig. 3.4), the carbon isotope ratios had returned to their normal, more positive (more carbon-13-rich) values within about 400,000 years after the K/T event, indicating that oceanic circulation had returned to normal and that planktonic productivity had resumed its pre-K/T level.

In summary, the earliest Paleocene sections in many places in the world paint a similar picture of the post-K/T oceans: a crash in the diversity and abundance of plankton led to a seafloor dominated by a few opportunistic weedy species that could survive in the low-productivity, possibly low-oxygen conditions that persisted in the aftermath of the K/T event. Most of the surviving species (whether benthic foraminifera or mollusks) were adapted to living off detritus on the sea bottom, rather than depending upon planktonic productivity. Within a few thousand years after the K/T event, however, the seafloor was beginning to return to normal, with abundant planktonic productivity supporting a more normal assemblage of mollusks, echinoderms, corals, brachiopods, bryozoans, and benthic foraminifera.

Evolutionary Explosion on Land

Although the K/T event temporarily suppressed the productivity of the plankton and the diversity of seafloor life—and wiped out a few dominant Mesozoic groups, like the ammonites—it did not fundamentally rearrange the ecological relationships of seafloor life. To the untrained eye, a Paleocene shell bed (fig. 3.5) looks much like one from the Cretaceous (minus the rudistids, inoceramids, and ammonites). The plankton soon returned, to feed the diverse fauna of mollusks, echinoderms, corals, worms, crustaceans, bryozoans, fish, and the like. This rebound is not true on the land. Land plants remained at the base of the terrestrial food pyramid, but nearly all the large terrestrial herbivores and carnivores were non-avian dinosaurs. When these vanished, they left a world devoid of large verte-

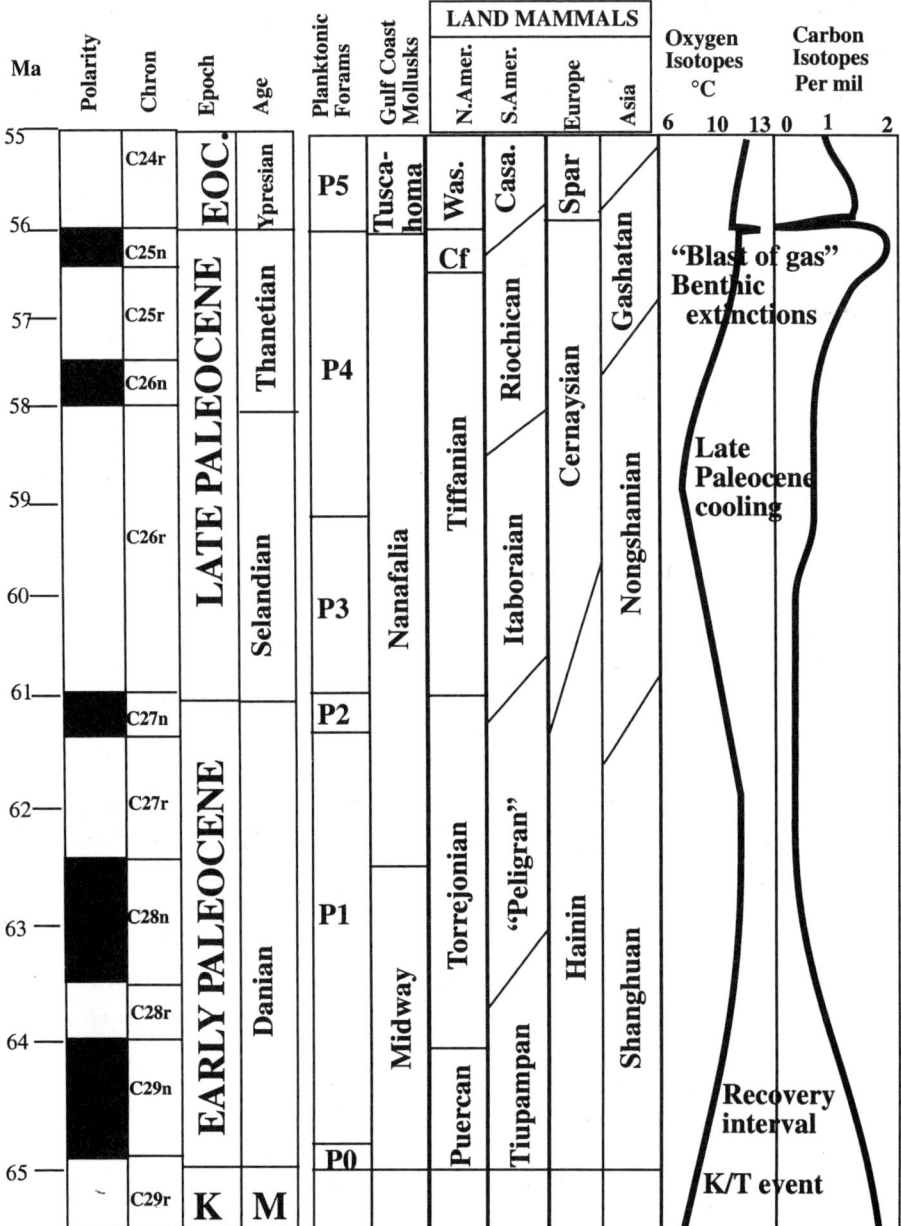

Figure 3.6. The Paleocene timescale (left columns), as recently calibrated by Berggren et al. (1995). The North American land mammal record is after Prothero (1995) and Woodburne and Swisher (1995); the South American land mammal ages are after Flynn and Swisher (1995); the European record is after Schmidt-Kittler (1997); and the Asian record is after Wang et al. (1998) and Meng and McKenna (1998). Note that the boundaries between North American land mammal ages are precisely calibrated, so they are drawn as synchronous horizontal lines; whereas those of South America, Asia, and Europe are less well defined, so they are drawn as time-transgressive diagonal boundaries. The isotopic records are after Zachos et al. (2001), but they are generalized and smoothed out. Abbreviations: Cf = Clarkforkian; EOC. = Eocene; K = Cretaceous; M = Maastrichtian; Spar = Sparnacian; Was. = Wasatchian; Casa. = Casamayoran.

brates. Into this void evolved the mammals, in one of the most spectacular evolutionary radiations ever documented.

And what a spectacular radiation it was! In the Late Cretaceous, mammals were tiny, shrew-sized to rat-sized creatures, forced to hide in the darkness and undergrowth and crevices from the non-avian dinosaurs that ruled the daytime. Although the first marsupials (pouched mammals) and placentals (mammals that have a placenta for nourishment and a membrane to protect their embryos) had evolved in the Early Cretaceous, over 120 Ma, they had still not made much progress by 65 Ma. In the Late Cretaceous faunas of the Hell Creek Formation (fig. 2.7) of Montana (Archibald 1982, 1996; Archibald and Bryant 1990), for example, there were ten species of marsupials, all opossum-like in their overall aspect. In addition, there were eleven species of multituberculates (fig. 3.7). These primitive squirrel-like mammals were most closely related to the living egg-laying monotremes (platypus and echidna). They had chisel-like front teeth resembling those of rodents; large molars with multiple bumps, or tubercles (hence the name); and their most distinctive feature, a large blade-like fourth lower premolar that may have helped shear and husk the fruits, nuts, and seeds that they ate. Multituberculates arose early in the evolution of mammals (members of a Late Triassic group known as haramyids may be their earliest representatives) and had a long and successful history through the entire Mesozoic, whereas more-advanced marsupials and placentals evolved to replace the other archaic Mesozoic mammal groups.

In contrast to the diverse marsupials and multituberculates, only six species of placental mammals are known from the Late Cretaceous. Although they would all look similar to shrews to untrained eyes, the details of their teeth show that they included some of the earliest hoofed mammals (*Protungulatum*) and even the earliest members of our own order Primates, a creature known as *Purgatorius* (the Dantesque name refers to their discovery in the Purgatory Hill locality in Hell Creek, Montana). This dominance of marsupials over placentals had changed rapidly by the early Paleocene. Only a handful of multituberculate species and one species of marsupial are found in the earliest Paleocene in Montana, whereas there were already thirteen species of placental mammals, including nine species of archaic hoofed mammals belonging to at least three families (Archibald 1982, 1996; Archibald and Bryant 1990). In contrast to the measly eighteen genera of mammals in the Late Cretaceous, by the middle Paleocene, there were eighty-four genera of mammals in North America alone; and by the late Paleocene, there were a hundred genera (Archibald 1993). Cutting to the chase, by the end of the Paleocene and the beginning of the early Eocene, just 10–12 million years later, these few Cretaceous placentals had given rise to all the orders of mammals (fig. 3.8), from shrews and rodents to giant whales and flying bats. That indeed is an impressive rate of increase and ecological diversification!

The earliest details of the incredible evolutionary radiation of placental

Figure 3.7. A multituberculate reconstructed as a squirrel-like creature with a prehensile tail. Courtesy of the American Association for the Advancement of Science.

mammals are recorded in the Hell Creek region of Montana (in a Paleocene unit known as the Tullock Formation), but the rest of the Paleocene is better known from places such as the Bighorn Basin of Wyoming and Montana, the Williston Basin of North Dakota and Montana, and the San Juan Basin of New Mexico (fig. 3.9). In these basins there is an excellent record of the details of the Paleocene in North America; the record is much less well known on other continents. The succession of land mammal ages (fig. 3.6) through this interval are (1) the Puercan, named for the Rio Puerco in the San Juan Basin, which spans the interval from 65 to

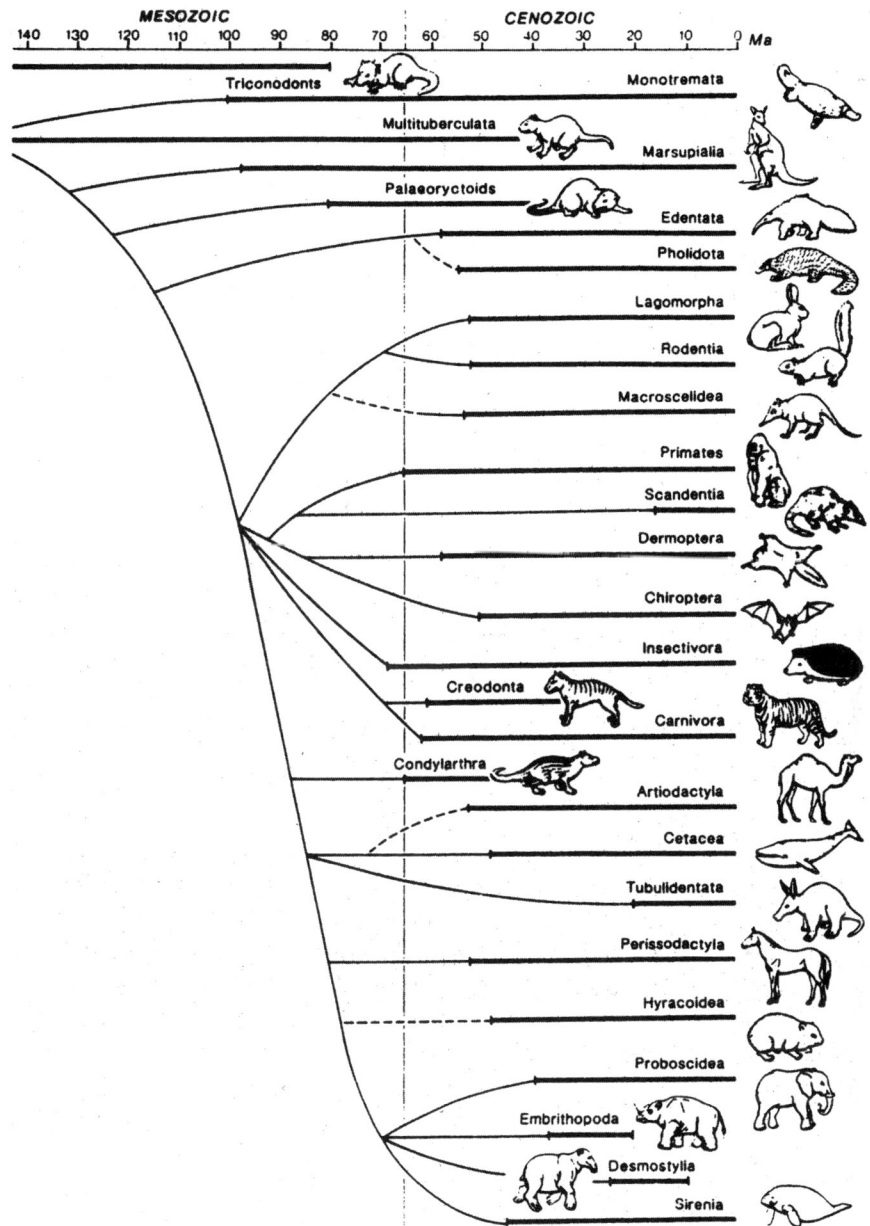

Figure 3.8. Evolutionary radiation of the major orders of mammals. After Novacek 1994.

63.5 Ma; (2) the Torrejonian, named for Torreon Wash in the San Juan Basin, which runs from 63.5 to 61 Ma; (3) the Tiffanian, named for the Tiffany fauna from Mason Pocket on the northwestern rim of the San Juan Basin in Colorado, which spans the interval from 61 to 56 Ma; and finally, in the latest Paleocene, the (4) Clarkforkian, named for the Clark's Fork of the Yellowstone River in the Bighorn Basin of Wyoming, which

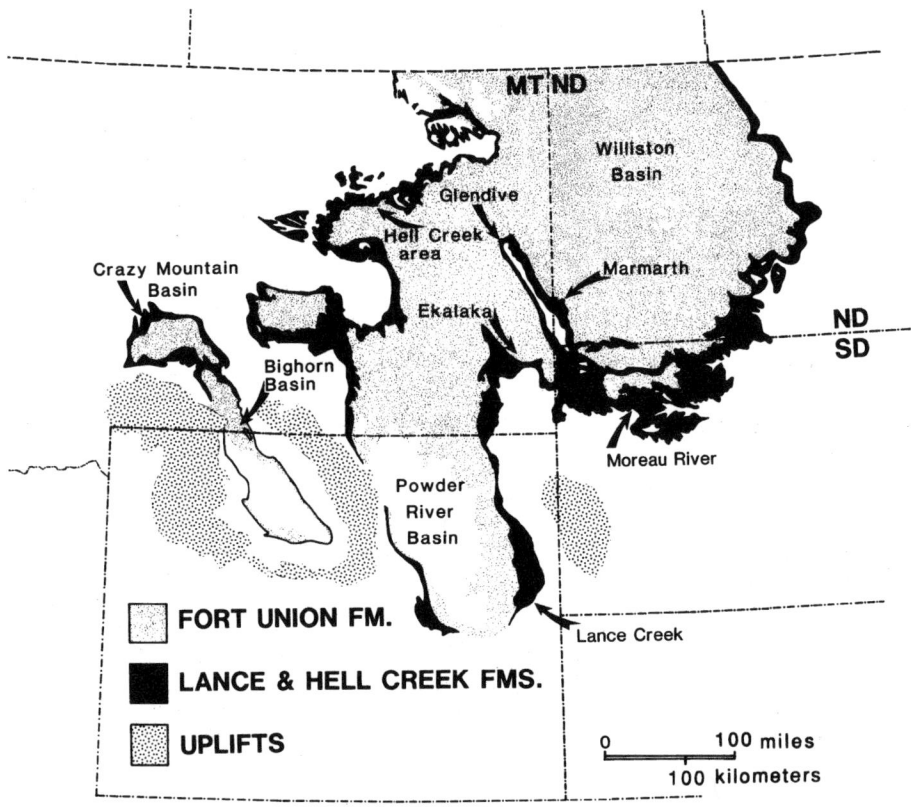

Figure 3.9. Map of major Paleocene basins in the northern Rockies and Plains, showing exposures of the Cretaceous Lance and Hell Creek formations and the Paleocene Fort Union Formation. After Johnson and Hickey 1990.

runs from 56 to 55 Ma. Each of these four land mammal ages is further subdivided into, for example, Puercan 1, Puercan 2, and so on (Archibald et al. 1987; Lofgren et al. 2004), recognized by distinctive assemblages of mammals, so that the 10 million years of the Paleocene (55 to 65 Ma) are subdivided into at least fifteen mammal zones, some with durations as short as 250,000 years. Thus, we can obtain a fine-scale detailed record of the rapid change in mammalian faunas in the Paleocene, and we can track the details of the great evolutionary explosion of mammals.

Who were the players in this immense radiation of Paleocene mammals? As mentioned above, only a handful of tiny shrew-like placentals lived in the Late Cretaceous, but by the Paleocene they had diverse teeth characteristics and distinct diets and ecology, even though they were still not much bigger than a cat. The most diverse group was the earliest hoofed mammals, or the archaic ungulates (once placed in the wastebasket group "Condylarthra"). Although they are related to modern hoofed mammals, you would never guess that their descendants would be cows, pigs, or horses. They ranged from the omnivorous arctocyonids (fig. 3.10A, B), which resembled raccoons or coatimundis in size and proportions, to the more stocky periptychids, with distinctive crenulated ridges on their tooth enamel, to the abundant meniscotheres (fig. 3.10C), to the dachshund-like hyopsodonts (fig. 4.24A), to the sheep-like phenacodonts (fig. 4.24B). The archaic ungulates even included a predatory group, the wolf-like mesonychids (fig. 3.11), which fed on meat and carrion, even though they were related to herbivorous hoofed mammals. By the early Eocene, the mesonychids gave rise to the earliest whales (see chapter 4). This immense radiation of archaic ungulates (at least fifteen to twenty genera in nearly every zone in the North American Paleocene) were the dominant mammals roaming the ground in the Paleocene forests, mostly eating leaves and fruits with their low-crowned teeth.

Besides the archaic hoofed mammals, there were other peculiar beasts living in the Paleocene forests of North America. The largest were the pantodonts, which vaguely resembled sheep; they had stocky legs and bodies, and their teeth were characterized by distinctive V-shaped crests suited for chopping up leaves (fig. 3.12A, B). Pantodonts started out cat-sized in the early Paleocene, but by the early Eocene, they became had become cow-sized herbivores, the largest land mammals of their time. Then they abruptly died out by the middle Eocene. Along with pantodonts were two other extinct zoological mysteries, the taeniodonts and the tillodonts (fig. 3.13). Although they were herbivorous, they were not related to hoofed mammals, and many had claws, possibly for digging up roots and tubers. Some taeniodonts had chisel-like front teeth like those of an enormous rodent, and both groups tended to have simple, low-crowned molars with short crests, presumably for an omnivorous diet. The heavy wear on their teeth suggests that they may have ingested a lot of grit with their food. Whatever pantodonts, taeniodonts, and tillodonts ate and looked like, they remain a zoological mystery—all three were extinct by the middle Eocene, and none left living descendants, or even close relatives among the living mammalian fauna.

If the ground was populated by archaic hoofed mammals, plus pantodonts, taeniodonts, and tillodonts (pl. 1), the dense forest canopy of the Paleocene was ruled by a variety of arboreal leaf and fruit eaters. Foremost among these were the multituberculates, those squirrel-like, egg-laying holdovers from the Mesozoic (fig. 3.7). They apparently occupied part of the squirrel niche in the trees before rodents evolved. Other multituberculates, such as *Taeniolabis,* attained the size of beavers and probably lived more like ground squirrels. Next in abundance is another tree-dwelling group: the early primates (fig. 3.14). Although they are distantly related to us and were members of our order, they looked more like the most primitive living primates, the lemurs. They had well-developed tails

C

A

B

Figure 3.10. (A) Reconstruction of the coatimundi-like arctocyonid Chriacus. *After Rose 1990, fig. 10, drawing by Elaine Kasmer, courtesy of the Geological Society of America. (B) Reconstruction of the bear-sized arctocyonid* Arctocyon. *After Agusti and Anton 2002. (C) Reconstruction of the common archaic ungulate* Meniscotherium. *Painting by B. Horsfall, in Scott 1913.*

Figure 3.11. Reconstruction of the wolf-like hoofed mammal Mesonyx, *which was ancestral to the whales. Painting by Charles R. Knight, negative no. 35777 courtesy of the Department of Library Services, American Museum of Natural History.*

(sometimes prehensile), dog-like snouts rather than the short, flat faces of most monkeys, and relatively short limbs. Some of them had proportions more like those of a woodchuck, and they may have lived partially on the ground as well. Others had rodent-like chisel-shaped front teeth, so they also occupied the rodent niche (shared with the multituberculates) before true rodents appeared.

The trees and underbrush may have been dominated by squirrel-like herbivorous fruit- and nut-eating multituberculates and primates, but they also harbored a variety of insectivorous mammals (fig. 3.15). At least seven families and over a dozen genera of Paleocene insectivorous mammals are known, although to the nonspecialist they resemble shrews. One wastebasket group, the palaeoryctids, was very shrew-like in most of its features. Another group, known as the leptictids, had a long snout like

A

B

Figure 3.12. (A) Reconstruction of the dog-sized early Paleocene pantodont
Pantolambda. *Painting by B. Horsfall, in Scott 1913. (B) Reconstruction of the sheep-
sized late Paleocene pantodont* Barylambda. *From Scheele 1955.*

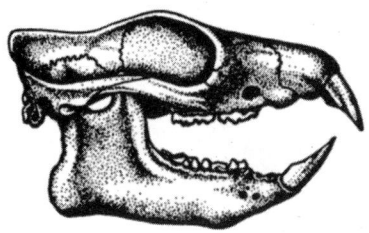 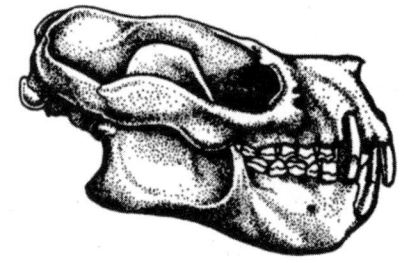

Figure 3.13. Tillodonts (left) and taeniodonts (right). These rare mammals had long, chisel-like incisors; relatively simple, low-crowned teeth; and many other peculiar features. Their relationships to other mammals are still unknown. Courtesy of R. M. Schoch.

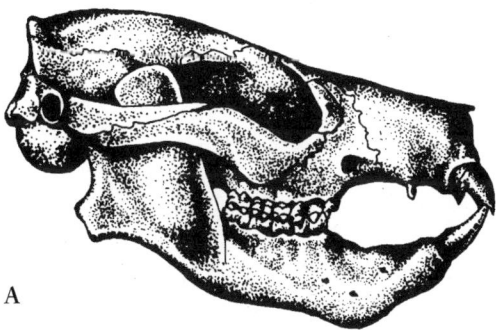

Figure 3.14. Plesiadapids were primitive relatives of primates with a rodent-like skull (including chisel-like incisors) and a skeleton resembling that of a ground squirrel. After Gingerich 1976, courtesy of the University of Michigan Museum of Paleontology.

A

B

63

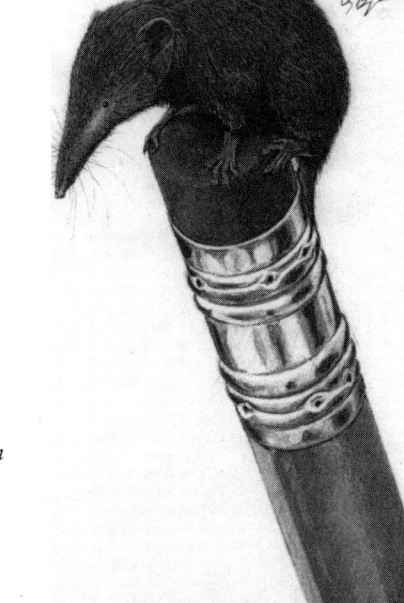

Figure 3.15. Restorations of early Cenozoic insectivores. (A) The tiny Batonoides vanhouteni, *the smallest mammal fossil ever discovered. Drawing by Doug Boyer, courtesy of J. Bloch. (B) The long-snouted running* leptictid Leptictidium *from the Eocene of Messel (see chapter 4). After Agusti and Anton 2002.*

A

B

that of an elephant shrew for probing the undergrowth in search of insects and worms, and kangaroo-like limb proportions, so they were excellent hoppers. A third group, the pantolestids, apparently lived like otters, with powerful forelimbs for swimming and burrowing. Yet another group, the apatemyids, had long curved front teeth like those of rodents (and multituberculates and primates), with scissor-like premolars for cutting bark, and long fingers—all features that suggest a habit of prying up bark and fishing for grubs. Finally, there were true insectivorans (members of the modern order Insectivora), including archaic members of the shrew and hedgehog lineages as well.

So far, we have mentioned only the herbivorous and insectivorous

Figure 3.16. Oxyaena, a late Paleocene–early Eocene creodont about the size of a jaguar. Creodonts were archaic carnivorous mammals with relatively simple shearing dentitions, robust limbs, and smaller brains compared to the members of the living order Carnivora. Painting by B. Horsfall, in Scott 1913.

mammals in the Paleocene fauna. Naturally, there were predators, too. The most common predators were an archaic group of carnivorous mammals, the creodonts (fig. 3.16). Although they look vaguely wolf-like or wolverine-like to the untrained eye, they were much more primitive than wolves or any other living carnivorous mammal. They were relatively short-limbed, inefficient runners, with stocky bodies, small brains, and teeth that had only a rudimentary shear mechanism for cutting flesh. They probably relied on ambushing their prey from the dense undergrowth, and did not require great intelligence to outwit their small-brained prey. In addition to the archaic creodonts, predators from the living order Carnivora were also present, represented by small weasel-like forms (fig. 3.17) once placed in the wastebasket group "Miacidae." These "miacids" were not members of the specialized living families of the order Carnivora, such as the cat, dog, bear, weasel, hyena, mongoose, and raccoon families—they were too primitive to belong to any living family. However, they were truly members of the order Carnivora, as shown by the fact that they had developed enlarged shearing teeth, the carnassials, between the fourth lower premolar and the first upper molar. These carnassial teeth

Figure 3.17. One of the weasel-like miacids. Miacids were the earliest members of the living order Carnivora, but most were about the size and shape of weasels or raccoons. From Scheele 1955.

are found in all living carnivorans and are used for cutting large tendons and breaking bones. If you watch a dog eat, you will notice that it often turns its snout to cut with the side of its mouth, bringing these strong carnassial shearing teeth into play.

Mammals became the most diverse group of land vertebrates once the non-avian dinosaurs had vanished, but there were many relicts from the Mesozoic that still haunted the Paleocene forests. No mammalian predator in the Paleocene was larger than a modern dog, so the role of largest predator was still occupied by Mesozoic holdovers. Among these were the crocodilians, which were unaffected by the K/T event, and the crocodile-like champsosaurs. The crocodiles included *Asiatosuchus* in Europe, which was related to the modern Nile crocodile and reached 4 meters in length; smaller (1.5 meters long) caiman-like alligators like *Diplocynodon;* the crocodilian *Pristichampsus,* with serrated blade-like teeth; and the small alligator *Allognathosuchus,* which had both sharp teeth and blunt, knob-like teeth for crushing mollusk shells. There were also the common alligator *Leidyosuchus* and the horned crocodile *Ceratosuchus.* This high diversity of crocodilians was accompanied by dozens of species of mostly aquatic turtles. Both are strong evidence that the climate was wet and mild even in Montana, because these animals do not tolerate temperatures much below 10°C and prefer subtropical to tropical conditions.

Figure 3.18. Diatryma, *a huge predatory flightless bird, over 2 meters tall, that could prey on nearly any mammal of the Paleocene or Eocene in North America. Witmer and Rose 1991.*

Although the non-avian dinosaurs had died out at the end of the Cretaceous, as I pointed out in chapter 2, the birds (avian dinosaurs) survived and continued to diversify in the Paleocene. Perhaps the most impressive of these were the 2-meter-tall "terror cranes," represented by *Diatryma* in North America and *Gastornis* in Europe (fig. 3.18). These huge birds had heavy, sharp beaks for crushing their small mammalian prey and ripping flesh, and thick legs with long claws for running and grabbing their prey. Like other large ground-dwelling birds, they had reduced wings and were flightless, but in the dense Paleocene forests there was no advantage to flight—they were better served by being able to run through the undergrowth and grab their prey with their beaks. These were the largest terrestrial predators through most of the Paleocene and into the middle Eocene, so in a real sense, the dinosaurs continued to rule the earth long after the Cretaceous ended.

The Swamps of Wyoming

The large number of leaf-eating and arboreal Paleocene mammals, plus crocodilians and aquatic turtles, suggests that the landscape of the Paleocene was tropical, wet, and heavily forested. This inference is confirmed by plant fossils. In fact, the fossil evidence suggests that much of Eurasia and North America was covered by dense forests through most of the Paleocene (pl. 1). In some places, such as the Powder River Basin of Wyoming, the habitat was so lush and swampy that huge coal beds over 100 meters thick accumulated (fig. 3.19). These coal beds are now the major source of coal in North America, since they are extremely thick and close to the surface. By contrast, most coal beds from the Appalachians

Figure 3.19. Exposure of a black coal seam over 70 meters thick in the Powder River coal beds of northeastern Wyoming. Photo by the author.

and Illinois are only a few meters thick and are found deep below the surface, so huge strip mines are used to recover the coal; some beds are so deeply buried that they require underground shaft mining. Not only are these Powder River Basin coals thick and easy to reach, but they are also low in sulfur, so they do not cause much acid rain when they burn. In fact, most of the Paleocene deposits of North America (such as the Fort Union Group rocks of Theodore Roosevelt National Park in the badlands of North Dakota; fig. 3.9) are full of fossil plants. Some of these show that dense forests ranged as far north as the shores of the Arctic Ocean (Sweet and Braman 1992).

Although these forests looked superficially like our modern tropics, in most ways they were not (Wing 1998; Graham 1999). The swampy regions were dominated by gymnosperms, such as the bald cypresses (fig. 3.20A) and the living fossils *Ginkgo* and *Metasequoia* (fig. 3.20B). There were abundant citrus plants, figs, cinnamon trees, magnolias, laurels (including relatives of the avocado, sassafras, and the oriental camphor trees), the cashew family (including pistachios and mangos), and the tropical *Annona* trees (fig. 3.20C) including fruit familiar in Central America as pawpaws, cherimoya, custard apple, soursop, and sugar apples. Any-

A

B

C

Figure 3.20. Tropical and subtropical plants that made up much of the forests of northern continents in the Paleocene. (A) The bald cypress (Taxodium, *which forms dense tropical swamps today). (B) The living fossil known as the dawn redwood (*Metasequoia), *as depicted on a Chinese postage stamp. (C) Annona trees (pawpaws and their tropical kin). Photos courtesy of B. Tiffney.*

one familiar with fruit such as avocadoes, mangos, pawpaws, and citrus knows that these plants are tropical or subtropical and cannot tolerate freezing. Further evidence of the dense tropical multistory vegetation is the presence of vines, including those of the moonseed and icacina families. The trees also included many members of the oak, birch, and witch hazel family as well. The shrubs were mostly primitive ferns and horsetails, with relatively few flowering plants compared to today's underbrush. In areas such as Colorado and New Mexico, there were abundant palm trees, with fewer conifers and bald cypresses.

On the basis of the temperature preferences of the modern descendants of these well-known plants, the paleotemperature history of the Paleocene can be estimated. In the Bighorn Basin of Wyoming, the early Paleocene (Puercan and Torrejonian) temperatures were mild, averaging about 10°C (50°F), according to Wing et al. (1991, 1995). But by the late Paleocene, the temperatures began to climb, averaging 14°C by the latest Paleocene in the Bighorn Basin. For the late Paleocene floras of the Golden Valley Formation in North Dakota, Hickey (1977) estimated a mean annual temperature of 18°C (65°F), with winter temperatures no colder than 13°C (55°F), since so many plants were intolerant of freezing. Down on the Gulf Coast, temperatures were much warmer, with mean annual temperatures estimated to be as high as 27°C (81°F). These data are consistent with the trend from the deep oceans, which indicates that oceanic temperatures were cooler in the early Paleocene after the warm Cretaceous, but warmed up again through the later Paleocene (fig. 3.6).

How do all these trends compare to one another? Wing et al. (1995) showed that the steady climb in temperatures is paralleled in some ways by the diversification of mammals in the later Paleocene and early Eocene, and both of these trends accelerated after a plateau in the Tiffanian. In addition, the mammalian curve also shows the explosive diversification of mammals in the early Paleocene (especially the Torrejonian) as they began to take over the terrestrial habitats vacated by the non-avian dinosaurs. By contrast, the curve for plant species increases steadily but slowly through the entire Paleocene, peaking at the end of the Tiffanian, then undergoing a slight dip before another peak in the early Eocene. This pattern is not surprising, considering that plants were only slightly affected by the K/T event; most were well established in their habitats and had only limited space to diversify and expand. Mammals, by contrast, moved into the niches left vacant by the non-avian dinosaurs, so there was enormous room for diversification and expansion throughout the Paleocene.

Land Bridges and Isolation

So far, we have focused on North America. How did other terrestrial ecologies compare with North America in the Paleocene? The best-studied region is western Europe, where the fossil records of numerous land mammal and plant assemblages span most of the Paleocene (Agusti and Anton

2002). Although the record is not as rich or continuous as that found in the bone beds of North America, nevertheless there are important localities in Spain, Germany, England, Romania, and Belgium and in the famous locality of Cernay in France. Overall, the western European Paleocene mammal faunas are much like those in North America, which suggests that some kind of land bridge (possibly across the North Atlantic between Greenland and Iceland) existed (fig. 4.8). Although the species in Europe are generally different from those in North America, many genera are shared, and most of the families and orders of mammals are the same. As in North America, in Europe the most common and diverse animals were the archaic hoofed mammals, including the wolf-like *Arctocyon* (fig. 3.10B). The rodent-like primates and multituberculates were also common, and the predators included the bear-sized *Dissacus* (one of the large hoofed mesonychids), creodonts and the huge predatory bird *Gastornis*, and many of the same crocodiles and turtles. However, there were important differences as well. There were no true "miacid" carnivorans in Europe. Among large mammals, there were few pantodonts and no taeniodonts or tillodonts, even though these groups are common in North America.

Asia at this time was isolated from western Europe by the Obik Sea (fig. 4.8), which cut north-south across Siberia and separated most of Asia from Europe. At times, there were openings across the Obik Sea when the Turgai Strait closed, so occasionally we see Asian mammals in Europe. Apparently there was also some interchange with North America across the Bering Strait as well, because North America and Asia share a number of taxa in the Paleocene (Hooker and Dashzeveg 2003).

Until recently, the only well-studied Paleocene mammalian fauna from Asia was the late Paleocene Gashato fauna of Mongolia, first described by Matthew and Granger (1925) from specimens collected in the 1920s by the famous Central Asiatic Expeditions of the American Museum of Natural History. Before 1960, no fossil mammals were published from the Paleocene of China. Beginning in the 1970s and 1980s, when Chinese and non-Chinese paleontologists were free to travel and study at the end of the Cultural Revolution, there was an explosion of research in China. Now the Chinese Paleocene record is known in great detail (Russell and Zhai 1987; Ting 1998; Wang et al. 1998), with many faunas divided into three successive land mammal ages (fig. 3.6): the Shanghuan (equivalent to the Puercan and Torrejonian in North America); the Nongshanian (equivalent to the Tiffanian), and the Gashatan (equivalent to the Clarkforkian). One glance at the Asian faunas and it is immediately apparent how different they are from those in Europe and North America at the same time. The most abundant large mammals are the pantodonts (fig. 3.12), along with a few tillodonts (but no taeniodonts, which were restricted to North America). There were only a handful of archaic ungulates ("condylarths") plus the mesonychids, the hoofed predators that gave rise to whales. In the early Paleocene (Shanghuan), the small mammal fauna was dominated by groups such as the anagalids, which were ancestral to the rabbits and rodents. There were only a few primates and insectivores, and none of the abundant multituberculates that are so common in the Puercan and Torrejonian. In the Nongshanian (Ting 1998), pantodonts were even more dominant, along with rare tillodonts, archaic ungulates, and mesonychids. In addition, we see the first primitive members of the uintatheres, which we shall discuss further in chapter 4. Small-mammal assemblages continued to be dominated by the anagalids and other groups closely related to rabbits, but a few primates and even a multituberculate are known. There is also a group known as the arctostylopids, which previously were known only from South America, and a sloth-like beast known as *Ernanodon*, which may be related to the edentates (sloths, armadillos, and anteaters and their kin) that were found mainly in South America through most of the Cenozoic.

Not only were the abundance patterns of pantodonts, mesonychids, and anagalids in Asia strikingly different from the patterns on other continents, but the scarcity of other groups that dominated in North America and Europe was also remarkable. Primates were by far the most common small mammals on both those continents in the Paleocene, followed closely by multituberculates. In fact, the zonation of the later part of the North American Paleocene is largely based on primates. Yet primates are absent from the Shanghuan and Nongshanian in Asia, and only one possible species of multituberculate is known from the Nongshanian. Likewise, the archaic hoofed mammals were incredibly abundant in both Europe and North America during the entire Paleocene, yet they are relatively rare throughout the Paleocene of China, their place being taken by pantodonts. What this pattern suggests about the ecology of the Asian Paleocene is still unclear. The scarcity of the arboreal primates and multituberculates, and the abundance of ground-dwelling ancestors of the rodents and rabbits, might suggest that the habitat in Asia was much more open, with few dense forests to support the arboreal forms (Meng et al. 1998; Meng and McKenna 1998). Unfortunately, there are still no detailed published studies of the floras or pollen from the Paleocene of China to corroborate this interpretation, or to provide more details of the vegetation. Mehrotra (2003) reported that the Paleocene plants from the Danian of India suggest a warm, tropical climate with a mixture of tropical evergreens and moist deciduous forests. However, India was still an island continent moving rapidly across the Indian Ocean in the Danian, so its vegetation may not be representative of China, which was part of the Asian mainland.

Finally, during the Gashatan, the fauna became more complex. Pantodonts were still abundant, but they shared the large mammalian niches with uintatheres, arctostylopids, and creodont and mesonychid predators. The anagalids and their kin were still the most common small mammals, but multituberculates were slightly more common than they had been earlier; the scarce primates, however, had vanished (yet were still dominant in Europe and North America). In the Gashatan, we also see the earliest rodents (which migrated out of Asia to Europe and North America late in the Paleocene) and the earliest relatives of the perissodactyls (odd-toed hoofed mammals, today including horses, rhinos, and tapirs). The latter include an animal known as *Radinskya* (McKenna et al. 1989), which may be the most primitive relative of the perissodactyls known. Finally, the occurrence of *Minchenella*, one of the earliest relatives of the elephants and their kin, suggests that many important mammalian groups had their origin in Asia in the Paleocene. This possibility is known as the East of Eden hypothesis (Beard 1998). As Beard (1998) points out, the evidence is strong that many higher groups of mammals (perissodactyls, artiodactyls, elephants, whales, uintatheres, tillodonts, arctostylopids, pantodonts, rodents, rabbits, hyaenodont creodonts, and advanced primates, or anthropoids) all originated in Asia in the Paleocene. Many of these migrated elsewhere during the Paleocene (for example, pantodonts, tillodonts, uintatheres, arctostylopids), and the wave of Eocene immigrants (rodents, rabbits, perissodactyls, and artiodactyls) fundamentally changed the nature of land mammal faunas in Europe and North America.

In contrast to the abundant fossil records of Europe, Asia, and North America in the Paleocene, the records in most of the rest of the continents are still poorly known. There are still no unquestioned Paleocene terrestrial faunas in Australia or Antarctica. Africa is known only from a handful of Paleocene localities (mostly in North Africa), which up to now have produced a few fragments of insectivores, creodonts, and carnivorans, plus one of the earliest relatives of the elephants, *Paschatherium* (Capetta et al. 1978; Gheerbrant 1990; Gheerbrant et al. 1996).

At one time, the South American Paleocene was known from only one peculiar fauna from the Rio Chico Formation of Argentina (Simpson 1940). Since that time, a much more diverse Paleocene fauna has been discovered, and divided into four sequential land mammal ages: Tiupampan, Peligran, Itaboraian, and Ricochican (Flynn and Swisher 1995). These fossils show that South America was already isolated from the rest of the continents, and developing its own endemic fauna. Unlike the placental-dominated faunas of all the northern continents, the main mammalian predators in South America were marsupials, including many opossum-like forms (which still live there today), as well as the borhyaenids, a group of marsupial predators that paralleled wolves and hyenas and other placental carnivores. Instead of the families of archaic hoofed mammals found in the Northern Hemisphere, South America had its own native radiation of endemic hoofed mammals, including the notoungulates, which eventually evolved into forms that looked like hippos and rhinos; the litopterns, which evolved into forms that paralleled the camels and horses; the peculiar *Carodnia*, which bore many resemblances to uintatheres of the Northern Hemisphere; and a group called the didolodonts, which many paleontologists have tried to link to the archaic hoofed mammals known as hyopsodonts (fig. 4.24A) that were so common in North America. In addition to the marsupials and native hoofed mammal groups, the third major component of South America's endemic fauna was the edentates, which today are represented by the sloths, armadillos, and anteaters (all three of which are still common in South America). This group is the most primitive order of placental mammals known, and most of their evolution took place in South America. For a long time, it was thought that the edentates had been restricted to South America through most of the Cenozoic, but now we have the peculiar form *Ernanodon* from the middle Paleocene of China, and the anteater *Eurotamandua* from the middle Eocene of Messel, Germany, so somehow edentates were traveling to other continents. Add to this the peculiar arctostylopids of Asia, which also appear in South America, and it is apparent that South America was not as isolated as was long believed. Nevertheless, most of its native mammalian fauna was highly peculiar and endemic, and would remain so until the Pliocene, when the Panamanian land bridge opened the floodgates to North American immigrants (see chapter 7).

A Blast of Gas from the Past

As we saw from the land plant record, temperatures on the North American continent were undergoing a warming trend from the cool conditions of the early Paleocene (fig. 3.6). How does the land plant record compare with the marine record? The best proxy for marine temperature is the oxygen isotope system. Like carbon, oxygen also occurs in two principal isotopes in natural systems. The common form, oxygen-16, has eight protons and eight neutrons and makes up about 99.756% of the oxygen in the atmosphere and oceans. The heavier isotope, oxygen-18, has eight protons but ten neutrons and makes up only 0.205% of the oxygen in the globe. Like carbon isotopes, these two oxygen isotopes have a predictable ratio in nature, which is preserved in the calcite shells of mollusks, foraminiferans, and anything else that traps oxygen from the atmosphere or seawater at the time it lives. Since the 1950s, oxygen isotopes have been the principal system for determining paleotemperature, particularly in the world's oceans. Today, it is routine to run hundreds of analyses of oxygen isotopes through an automated mass spectrometer and get detailed records on a millimeter-by-millimeter basis. Like carbon isotope differences, oxygen isotope differences are measured in parts per thousand, or per mil, and are represented by the expression $\delta^{18}O$, pronounced "delta O-18."

How does the system work? In organisms, the ratio of oxygen isotopes taken up by the shell is directly proportional to the temperature, so that for each 4.2°C increase in temperature, the $\delta^{18}O$ decreases (becomes more negative) by 1 per mil. However, since the 1970s, scientists have come to realize that the size of the ice caps has an even greater effect than temperature (fig. 3.21). Because oxygen-16 is lighter and therefore evaporates

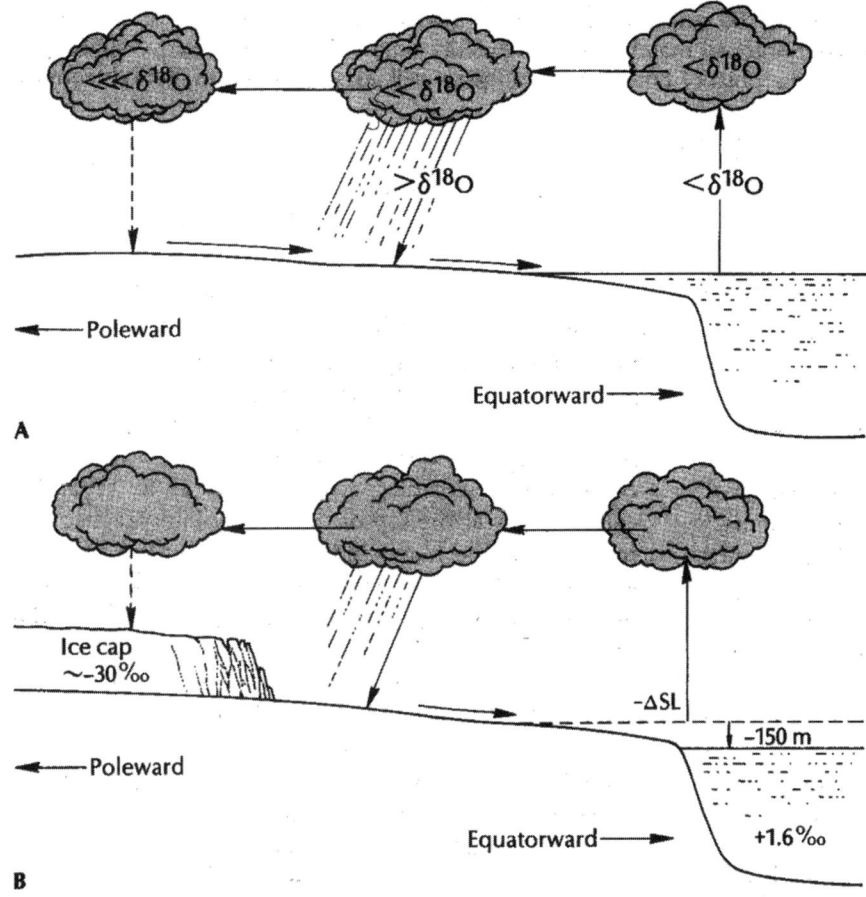

Figure 3.21. Relationship between oxygen isotopes and global ice volume. During normal times, light-oxygen-rich (^{16}O-rich) water evaporates more easily and makes up most of the water vapor that rains onto the land and returns to the sea via the rivers. However, during glacial times, light-oxygen-rich water is trapped in the ice caps, drawing down sea level by as much as 150 meters and enriching the oceans in the heavier but rarer isotope, ^{18}O. Modified from Matthews 1984.

more readily than heavier oxygen-18, it tends to be found in clouds and rainfall more abundantly than oxygen-18. The clouds, full of light oxygen-16, rain their water down on the land, where it normally returns to the oceans via the rivers and streams of the world, and the ratio remains unchanged. However, during glacial times, most of this light water ($\delta^{18}O$ is a −30 to −55 per mil) is locked up on the continents in the polar ice caps, leaving the oceans enriched (positive $\delta^{18}O$ values) by 3–4 per mil in oxygen-18. Sea levels are also lower, because of all the water locked up in ice. Thus, when we measure the oxygen isotope ratios in foraminiferal shells from deep-sea cores, a shift toward oxygen-16 (more negative $\delta^{18}O$) indicates warming, whereas a shift to oxygen-18 (more positive $\delta^{18}O$) indicates cooling, glaciation, or both.

Detailed records of oxygen and carbon isotopes thus can be analyzed from thousands of foraminiferal shells from hundreds of deep-sea cores all over the world's oceans, and a few marine sections now uplifted onto the land as well (Miller et al. 1987; Zachos et al. 1994, 2001). They all show a similar trend (figs. 3.6, 3.22): early Paleocene oxygen isotopes of the deep oceans indicate a slight warming, from about 8°C (46°F) to about 10°C for mean annual temperature, followed by slight cooling after the middle Paleocene. (Remember that these are sea-bottom temperatures, which are stable, fluctuating less than a tenth of a degree during the seasonal cycles; so a difference of a few degrees means much more dramatic temperature changes at the sea surface, and especially on land.) In the late Paleocene, the warming trend returned, with late Paleocene global average temperatures as high as 11°C (52°F). Likewise, the carbon isotopes show a parallel trend (fig. 3.6) from the more negative (lighter) values when the post-K/T loss of plankton meant that less carbon-12 was being consumed, to a middle and late Paleocene trend of more positive (heavier) average carbon isotope values of the oceans as the plankton diversified and consumed more and more of the surface water carbon-12, enriching the ocean in carbon-13.

Parallel tendencies can be seen in the Paleocene macrofossils. The late Paleocene in Europe is known as the Thanetian Stage, named after the Thanet Sands exposed near the mouth of the Thames River in England. This unit produces a rich molluscan fauna, as well as land plants and mammals. In both Europe and North America, the diversification of the plankton and the gradually warming conditions apparently triggered diversification in the mollusks as well. For example, Dockery (1986, 1998) reports an interesting trend in the diversity of mollusks from the Gulf Coastal Plain of Alabama and Mississippi. After a moderately diverse Danian molluscan fauna in the Clayton and Porters Creek formations, diversity drops to low in the middle Paleocene Naheola and Nanafalia formations, and then nearly doubles in the upper Paleocene Tuscahoma Formation. This trend nicely parallels that of the oxygen isotopes (figs. 3.6, 3.22), with the middle Paleocene cooling reflected in lower molluscan diversity, and the late Paleocene warming triggering a doubling in molluscan diversity.

Although the benthic macrofossil record for the Pacific Coast of the United States is not as long or rich as that of the Gulf Coast, some trends are apparent. Late Paleocene ("Martinez Stage") faunas are known widely from northern California to Mexico, and they include a variety of solitary corals as well as tropical forms such as cowries, sundial shells, fig shells, olive shells, pen shells, and the thick-shelled cockles known as *Venericardia* (a taxon restricted to the warm Paleocene and Eocene). There are also large numbers of the ubiquitous high-spired snail *Turritella*, which favored (and still lives in) temperate and tropical waters throughout the Cretaceous and Cenozoic. It is so abundant and speciose that its distinctive shells are one of the main index fossils for shallow-

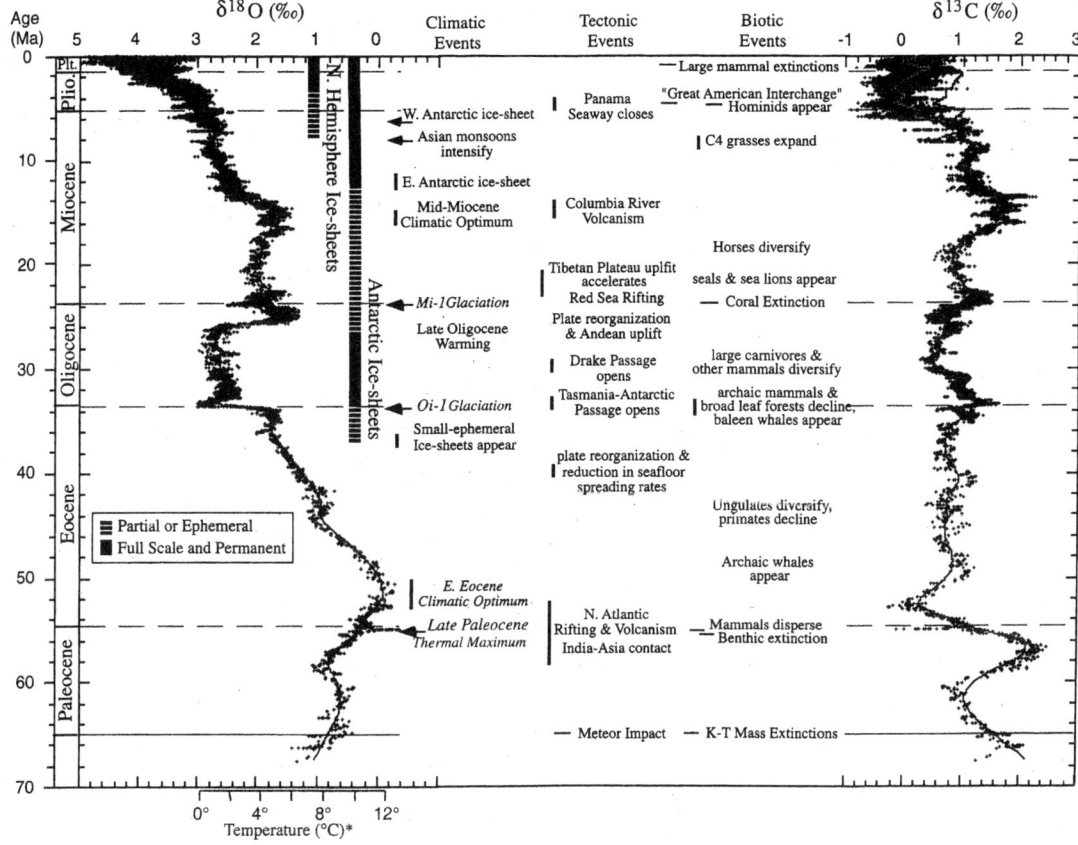

Figure 3.22. Summary of the Cenozoic record of oxygen and carbon isotopes, with major climatic, tectonic, and biotic events. From Zachos et al. 2001.

marine Cenozoic rocks around the world. Unlike most snails, which creep along the bottom and find food, it has a distinctive way of feeding; it sits on the bottom and uses its sticky foot to trap floating food particles. Together, all of these subtropical taxa as far north as northern California suggest that the Pacific Coast was warm and mild, with average water temperatures of 20°C (68°F), much warmer than today (Durham 1950).

This gradual warming and diversification through the later Paleocene culminated in a dramatic event at the end of the Paleocene. It was originally named the Late Paleocene Thermal Maximum (LPTM) by Zachos et al. (1993), but recently the name has been changed to the Initial Eocene Thermal Maximum (IETM), because it may have peaked in the early Eocene (Schmitz 2000), or to the noncommittal Paleocene-Eocene Thermal Maximum (PETM) (various papers cited in Wing et al. 2003). It has been the subject of intense research and many publications since it was first documented in detail in the late 1980s and 1990s. Originally, it was recognized by a dramatic mass extinction in the benthic foraminifera, with more than 50% of the species going extinct (Tjalsma and Lohmann 1983; E. Thomas 1998). This is the largest extinction in the history of the benthic foraminifera, and it contrasts with the lack of extinction among the benthic foraminifera during the K/T event, and most other mass extinction events in the Mesozoic and Cenozoic. The bottom of the ocean is a stable, unchanging environment, so most events that occur in the shallow seas or on land are nearly invisible to the quiet realm of the deeper ocean. By contrast, the end of the Paleocene saw a diversification in most of the planktonic foraminifera and coccolithophorids, as well as the other plankton such as radiolarians and dinoflagellates. In fact, the dinoflagellate bloom is huge (comparable to the bloom that occurs when the red tide floods the waters with their tiny cells), suggesting that something unusual took place in the nutrient supplies in the ocean surface (Crouch et al. 2001). Even the molluscan diversity in places like the Gulf Coast of Alabama and Mississippi doubled (Dockery 1986, 1998).

So what could have caused massive extinction in the deep-water benthos but favorably affected the surface plankton and the shallow-marine benthic mollusks? Clearly, it had to be an event that changed some aspect of the deep-ocean currents but had less of an effect on the shallow-marine system. Initially, speculation centered on some sort of gradual change in continents that triggered new oceanic circulation patterns. For example, Kennett and Stott (1990) proposed a model whereby the deep oceans of the Paleocene were supplied by warm, salty waters that formed in the tropical Tethys Seaway by evaporation. As these waters became saltier, they also became denser, and so sank to the ocean bottom and spread from the Tethys to the polar regions. Kennett and Stott (1990) suggested that the collision of India with Asia at the end of the Paleocene may have gradually broken up the Tethys circulation pattern and triggered extinction in the Paleocene benthic foraminifera that were accustomed to warm, oxygenated waters in the deep ocean.

But as these ideas were being published, new high-resolution isotopic studies were done on deep-sea cores that spanned the Paleocene-Eocene boundary (Kennett and Stott 1991). These studies showed a remarkable detail that had been missed in the lower-resolution analyses: the isotopic shift was too abrupt and extreme to be have been caused by gradual processes such as the shifting of oceanic currents due to plate motions. Detailed studies of a number of cores indicate that the initial isotopic shift took less than 10,000 years, after which the PETM gradually tapered off over an interval of no more than 200,000 years (Röhl et al. 2000). Yet the oxygen isotopes show a dramatic spike (fig. 3.6) of 3–5 per mil. This is equivalent to a global increase in deep-ocean mean annual temperature from 11°C to 15°C (52°F to 59°F), which is an extreme and rapid warming event. The carbon isotopes show an even more dramatic negative spike, suggesting that something happened to the carbon reservoirs in the world's oceans, or that the uptake of carbon by the plankton may have been interrupted. But what could have caused such an event? There was no K/T-style mass extinction in the plankton, and the temperatures were getting warmer, not cooler, which does not usually suppress planktonic uptake of carbon-12 from the oceans. In 1991, Kennett and Stott (1991) abandoned their salty Tethys deep-water model of just the year before and argued that a dramatic transient change in oceanic circulation must have taken place. This concept was further reinforced by Zachos et al. (1993, 1994), but it eventually ran into problems, because the magnitude of the carbon isotope shift was too great to be explained by subtle differences in oceanic circulation. Eldholm and Thomas (1993) suggested that carbon dioxide outgassed from the eruption of lavas in the North Atlantic might have driven this change, but mantle carbon dioxide is not sufficiently rich in light carbon-12. In addition, this model does not explain either the huge volumes of isotopically light carbon needed to change the world oceans or the rapidity of the event (since these North Atlantic eruptions had been going on for some time in the late Paleocene). Another suggestion was that a carbon-rich comet struck the earth (Deming 1999; Kent et al. 2001), but again such an impact would not have provided enough carbon to cause the isotopic results we see.

Then in the late 1990s, a new idea took hold. Marine geologists found that within the pore spaces between the sediments of the deep ocean today there is a great reservoir of carbon in the form of methane (CH_4, or natural gas) that is bound in a crystalline cage of frozen water molecules into a complex known as a methane hydrate, or methane clathrate (Kvenvolden 1993). Essentially, these clathrates are just natural gas and water bound together in a frozen state. They occur in huge volumes today along the continental margins, where the pressure produced by the overlying water and sediments stabilizes them in frozen form at 5°C (41°F). Today, this reservoir holds a staggering amount of carbon, perhaps as much as 11×10^{18} g, trapped in a semi-stable state! All that is needed is a slight rise in ocean temperatures above a critical threshold (well documented in the marine record, according to Thomas et al. 2002), and this unstable cage of water molecules will break down, rapidly releasing huge amounts of isotopically light (more negative ratios) carbon into the oceans (Dickens et al. 1995, 1997, 1998; Thomas and Shackleton 1996; Thomas et al. 2002). If all this gas were released, it would cause a tremendous extinction in the deep-sea benthos, which live in the same sediment. But as the methane reached the surface and then the atmosphere, it would act as a greenhouse gas, dramatically warming the globe in a "super-greenhouse" effect. The effect is so dramatic that even the carbon isotopes recorded in the ancient soils and teeth of land mammals show a dramatic carbon isotope shift (Koch et al. 1992, 1995). Dickens et al. (2003) showed that the deposition of barium sulfate (barite) at the Paleocene-Eocene boundary can also be explained by the release of so much methane.

What could have triggered the release of all this methane from the seafloor? Most marine geologists have pointed out that the temperatures were gradually warming during the late Thanetian, and when the temperature threshold was exceeded, the sudden melting of the methane hydrates occurred. Svensen et al. (2004) suggested that the intrusion of molten rock in the Norwegian Sea was the trigger that melted and released methane hydrates into the oceans and atmosphere. However, McKenna (2002) has suggested a novel idea. He points out that the Arctic Ocean had been sealed off from the rest of the world's oceans since the Cretaceous Western Interior Seaway across North America had dried up. All the evidence suggests that the Arctic was becoming a giant body of fresh water even larger than the Great Lakes. New plate tectonic evidence suggests that rifting between Greenland, Iceland, and Europe opened a major seaway at the end of the Paleocene. McKenna suggests that this rifting would have released a burst of fresh water from the Arctic lake, which could easily have changed the chemistry and circulation of the oceans and released all the methane hydrates frozen on the seafloor. This idea is so recent that the paleoceanographers and climate modelers have not tested it yet, but it is an intriguing suggestion.

After about 200,000 years, the various carbon reservoirs in the oceans absorbed much of the methane, and the "blast of gas" episode was over. The incredibly warm methane-induced greenhouse conditions, however, had just begun, making the beginning of the Eocene the warmest time in the Cenozoic.

For Further Reading

Agusti, J., and M. Anton. 2002. *Mammoths, Sabertooths, and Hominids: 65 Million Years of Mammalian Evolution in Europe.* New York: Columbia University Press.

Aubry, M.-P., S. G. Lucas, and W. A. Berggren, eds. 1998. *Late Paleocene–Early Eocene Climatic and Biotic Events in the Marine and Terrestrial Records.* New York: Columbia University Press.

Dickens, G. R., J. R. O'Neill, D. K. Rea, and R. M. Owen. 1995. Dissociation of oceanic methane hydrate as a cause of the carbon isotope excursion at the end of the Paleocene. *Paleoceanography* 10:965–971.

Thomas, D. J., J. C. Zachos, T. J. Bralower, E. Thomas, and S. Bohaty. 2002. Warming the fuel for the fire: Evidence for thermal dissociation of methane hydrate during the Paleocene-Eocene thermal maximum. *Geology* 30:1067–1070.

Wing, S. L., P. D. Gingerich, B. Schmitz, and E. Thomas, eds. 2003. Causes and consequences of globally warm climates in the early Paleogene. *Geological Society of America Special Paper* 369.

Color Plate 1. The early Paleocene landscape. As far north as Montana, the earth was dominated by tropical jungles, inhabited by a variety of archaic leaf-eating and tree-dwelling mammals. In the right foreground are the primitive pantodonts *Pantolambda* (left) and *Barylambda* (right). In front of them is the archaic hoofed mammal *Tetraclaenodon*. Gliding from the trees are the dermopterans *Planetetherium*. Large reptiles, such as the boa constrictor, the soft-shelled turtle, and the alligator, also dominated the Paleocene forests. From a mural by R. Zallinger, courtesy Yale Peabody Museum of Natural History.

Color Plate 2. Diorama of vegetation and animals during the Eocene of Wyoming or Montana. The left half of the panel represents the early Eocene, the right half depicts the middle Eocene. Dense jungles dominated by palms and other tropical plants were inhabited by the pantodont *Coryphodon* (in the left foreground with tusks), archaic hoofed *Phenacodus* (running in the left background), the archaic creodont *Oxyaena* (on rock at left), the earliest horse *Protorohippus* (center foreground), the early brontothere *Palaeosyops* (just behind and to the right of the horse), and small insectivores and lemur-like primates (in the bush, left foreground). The middle Eocene forms in the right half of the mural include the huge horned uintathere *Eobasileus*, the predatory bird *Diatryma*, the wolf-like hoofed mammal *Mesonyx* (snarling in front of the uintathere), and the predatory creodont *Tritemnodon* (to the left of *Diatryma*). From a mural by R. Zallinger, courtesy Yale Peabody Museum of Natural History.

Color Plate 3. Diorama of life in the Big Badlands during the latest Eocene and Oligocene. Huge brontotheres move in the right background, along with the primitive rhinoceros *Subhyracodon*. To the left are the rhinoceros *Hyracodon* and a single primitive tapir, *Protapirus*. Far right is a herd of the oreodont *Merycoidodon*. Nearby two *Hyaenodon* fight over a lizard, *Glyptosaurus*. In the right foreground, the sabertoothed nimravid *Hoplophoneus* chases the tiny deer-like *Leptomeryx* and *Hypertragulus*, while the even tinier *Hypisodus* hides in front of the nearest bush. At the edge of the dry wash stand a herd of the camel *Poebrotherium* and, to their left, the horned *Protoceras*. Battling within the ravine are two large enteledonts, *Archaeotherium*. In the left foreground is the anthracothere *Bothriodon*, and behind it a herd of the three-toed horse *Mesohippus*. Painting by J. Matternes, courtesy of the National Museum of Natural History, Smithsonian Institution.

Color Plate 4. *Above:* Diorama of life near Agate Springs, Nebraska, during the early Miocene. A herd of three-toed horses *Parahippus* (center) faces a herd of the oreodont *Merychyus* (lower left). The large clawed herbivores in the middle left are the chalicotheres *Moropus*, defending themselves against *Daphoenodon* beardogs. The paired-horn rhino *Menoceras* lies under the shade of the tree in the center background, and a herd of the gazelle-like camel *Stenomylus* stand before them. The pig-like oreodonts *Promerycochoerus* wallow in the mud of the river, and the huge pig-like enterodont *Daeodon* walks to the right. Painting by J. Matternes, courtesy of the National Museum of Natural History, Smithsonian Institution.

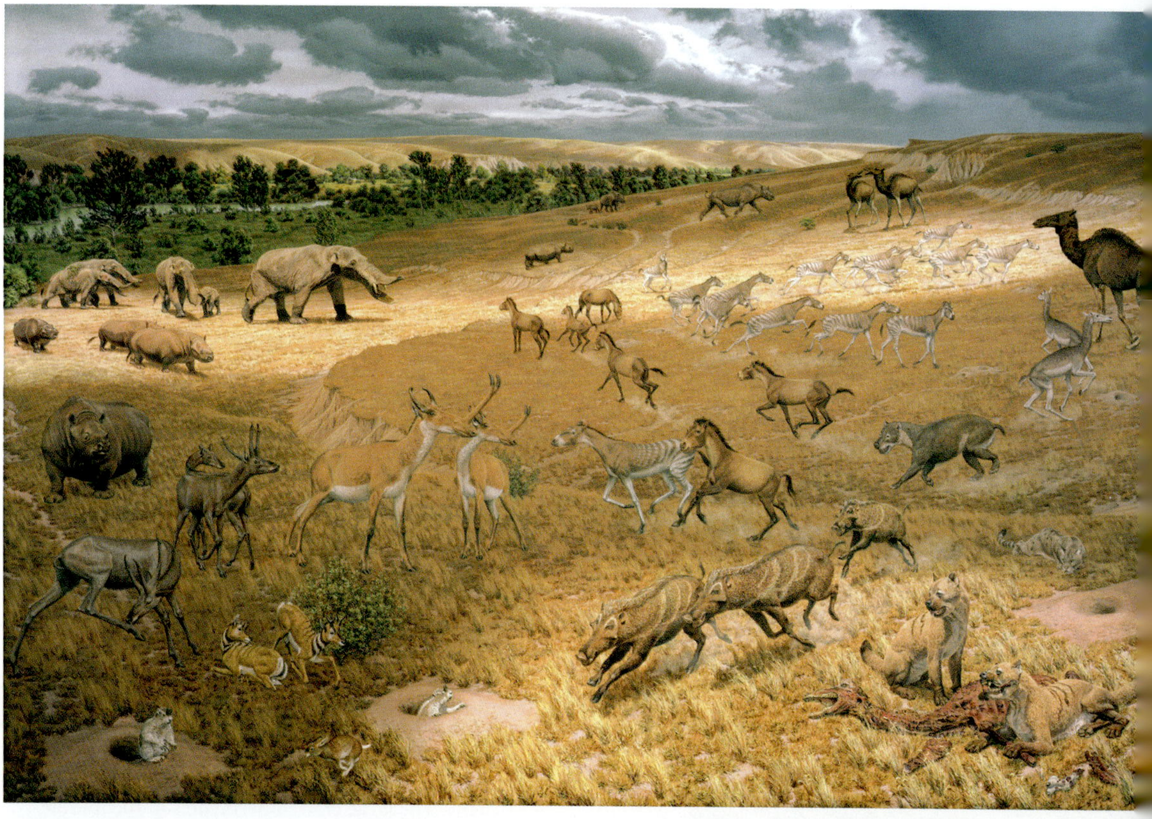

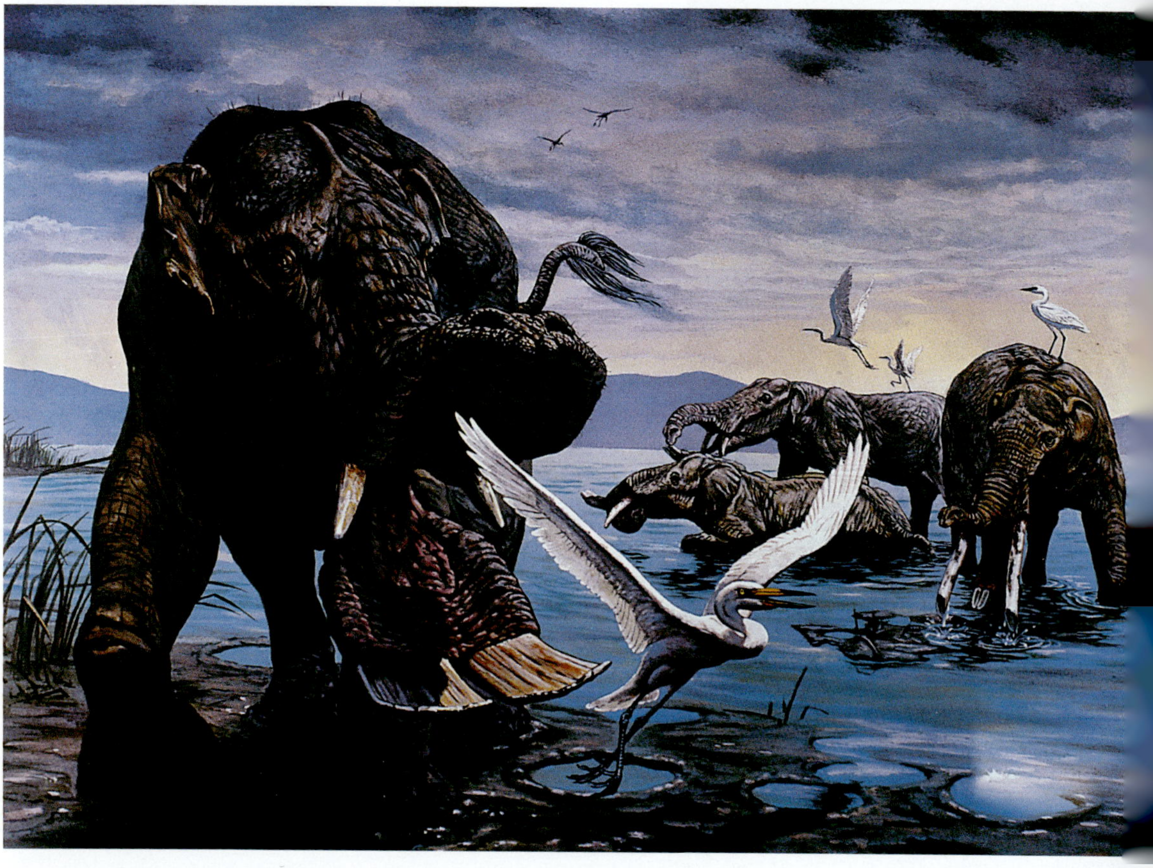

Color Plate 5. *Above left:* Reconstruction of the great American savanna during the late Miocene. Many different species of one-toed and three-toed horses (*Neohipparion* and *Pliohippus*) and pig-like peccaries (foreground) run around, chased by the bone-crushing borophagine dog *Osteoborus* (lower right). The "slingshot beast" *Synthetoceras* (a protoceratid) stands to the lower left, next to the three-horned dromomerycid *Cranioceras* and the gazelle-like pronghorn *Merycodus*. The hippo-like rhinoceros *Teleoceras* walks in the left distance, and the long-legged aceratherine rhinoceros *Aphelops* runs in the middle distance. Giant camels (*Megatylopus*) are visible in the right distance. The horned mylagaulid rodent *Epigaulus* emerges from burrows in the foreground. Painting by J. Matternes, courtesy of the National Museum of Natural History, Smithsonian Institution.

Color Plate 6. *Below left:* A scene from the Miocene swamps of North America. In the foreground, the shovel-tusked mastodont *Torynobelodon* scoops up vegetation, while other primitive masto-donts (from left to right) *Trilophodon*, *Amebelodon*, and *Ocalientius* bathe nearby. Painting by Mark Hallett.

Color Plate 7. *Above:* During the late Miocene (8 million years ago) in Nebraska, scenes such as this must have been common. Here, the wolf-sized borophagine dog *Epicyon* attacks the "slingshot beast" protoceratid *Synthetoceras*. In the background, the fox-like borophagine *Eucyon* attacks the peccary *Dyseohyus*. Painting by Mark Hallett.

Color Plate 8. Diorama of the landscape in North America during Pliocene time, representing the fossils found around the lake deposits at Hagerman Fossil Beds National Monument in Idaho. In the center are the common horses at Hagerman, *Equus simplicidens* (related to zebras), and in front of them, the peccary *Platygonus pearcei*. Behind the horses are the

extinct llama *Tanupolama* and the short-faced bear *Arctodus*. In the right center is the pronghorn *Ceratomeryx prenticei*. In the foreground, the sabertoothed *Machairodus hesperus* attacks the beaver *Castor californicus*. In the far distance are mastodonts (*Mammut*). Painting by J. Matternes, courtesy of the Smithsonian Institution.

Color Plate 9. Reconstruction of the late Pleistocene landscape at the La Brea Tar Pits about 40,000 years ago. Bison and horses run from the ice age lion, while Columbian mammoths walk in the background. A giant short-faced bear stands on the right, and giant vultures (*Teratornis*) fly overhead. Painting by Mark Hallett.

Color Plate 10. Reconstruction of the Eurasian landscape at the peak of the last glacial advance about 20,000 years ago. The gigantic cave bear and ice age lions fight in the foreground, while on the left the sabertoothed *Homotherium* eats its prey as hyenas look on. In the background (left to right) roam woolly mammoths, the "Irish elk" *Megaloceros,* horses, musk oxen, caribou, and a dueling pair of woolly rhinos. Painting by M. Hallett.

Figure 4.1. Badlands cliffs in the Bighorn Basin of Wyoming and Montana made of sandstones and mudstones deposited on floodplains and rivers during the early Eocene. Full of the jaws and teeth of fossil mammals, they preserve one of the best records of the Paleocene and early Eocene. Photo by the author.

4

Dawn of the Recent: The Eocene

The great mass of gypsum maybe considered as a purely fresh-water deposit, containing land and fluviatile shells, together with fragments of palm-wood, and great numbers of skeletons of quadrupeds and birds, an assemblage of organic remains which has given great celebrity to the Paris Basin. The bones of fresh-water fish, and of crocodiles, and many land and fluviatile reptiles occur in this rock. . . . The heat of European latitudes during the Eocene period . . . seem[s] . . . equal to that now experienced between the tropics.

Charles Lyell, *Principles of Geology,* 1833

Hot! Hot! Hot!

The methane blast at the end of the Paleocene produced a dramatically warmer world in the early Eocene. In fact, the entire globe was in a true greenhouse state, with temperatures warmer than at any other time in the Cenozoic, and as warm as the warmest temperatures in the Cretaceous. Oxygen isotopes suggest that even the deep-water benthic foraminifera were experiencing temperatures around 14°C (57°F), and average tropical sea-surface temperatures were in the 28–32°C (82–89°F) range (Zachos et al. 1994; Pearson et al. 2001). However, the tropics had been warm since the Cretaceous and had not cooled that much during the Paleocene. What is striking about the early Eocene greenhouse earth is how dramatically the temperate and polar regions warmed. Fossil plants in Wyoming, North Dakota, and Montana (Hickey 1977; Wing and Greenwood 1993; Wing 1998; Graham 1999) show that mean annual temperatures were as high as 21°C (70°F), and even the mean annual cold-month temperature was only 13°C (55°F), because most of the plants are intolerant of freezing. These plants also suggest that the climate was very wet, with mean annual rainfall exceeding 150 centimeters (60 inches), so the region was well watered. By contrast, western North Dakota today has a steppe climate. The mean annual temperature is only 5°C (41°F), and the spread between daily extremes ranges over 33°C (over 90°F). In North Dakota and eastern Montana, it is not at all unusual for the temperature on a hot spring or fall day to start out above 32°C (90°F) and then drop below freezing in a matter of hours as an Arctic cold front moves in.

From the evidence of floras in the Bighorn Basin of Wyoming and the Williston Basin of Montana and North Dakota, we can visualize a dense tropical forest much like that found in modern Panama (pl. 2). Tall trees formed a dense canopy, with vines and lianas growing all around them. The fossil plants include many of the tropical groups described in the previous chapter, including citrus, avocado, cashew, and pawpaw trees (fig. 3.20C). Many of the plant genera are found today only in the jungles of Southeast Asia or tropical Central America. In addition to the direct evidence of the plant fossils, there is information in the striking color bands that stripe the badlands slopes (fig. 4.1). Each band represents an ancient soil horizon, and in many places there are hundreds of bands stacked on top of one another, representing millions of years of the early Eocene. Each represents another episode of floodplain mud deposition, followed by the development of plants and a soil horizon and then by another episode of flooding that buried the old soil. According to Tom Bown and Mary Kraus (1981, 1987), these ancient soils were deposited on broad floodplains bordering meandering rivers, much like those of the modern Amazon.

The same pattern can be seen in other temperate and tropical regions throughout the early Eocene. On the Gulf Coast of Alabama and Mississippi (from the lower Eocene Wilcox Group), dense tropical rain forests developed, with extensive mangrove swamps (Frederiksen 1991; Wing and Greenwood 1993). Temperatures averaged around 27°C (81°F), consistent with estimates provided by the vegetation (Graham 1999). Even the Pacific Northwest and southern Alaska were relatively warm (25°C) and wet, blanketed with broad-leaved evergreen forests, with abundant vines and lianas, and many plants of tropical Asian affinities (Wolfe 1977, 1985).

From the London Clay in the basements of London comes an important early Eocene flora. As in Montana, there are mostly tropical trees and shrubs, including cinnamon, figs, magnolias, palms, laurels, citrus, pawpaw (fig. 3.20C), cashews, and laurels, as well as vines and lianas such as moonseed, icacina, and grapes. Collinson (1983) and Collinson and Hooker (1987) showed that 92% of these plants have living relatives in the jungles of Southeast Asia. Fringing the coasts of the tropical jungles were mangrove swamps full of *Nypa* palms (fig. 4.2A), also restricted to Southeast Asia today. From this evidence, the average temperatures in London were about 25°C (77°F), compared to the modern average of 10°C (50°F). Instead of the cold, foggy London of Sherlock Holmes, London was as warm and tropical as Singapore.

A

B

Figure 4.2. (A) Nypa *palms in a coastal swamp. Photo courtesy of B. Tiffney. (B) The Norfolk Island pine,* Araucaria, *found in the cool-temperate and polar regions in the Eocene. Photo by C. R. Prothero.*

Anywhere we find fossil floras of early Eocene age in temperate or tropical regions around the world, we encounter a similar story. Floras from China (Guo 1985), Siberia (Budantsev 1992), India (Mehrotra 2003), and southern South America (Romero 1986) all show the same tropical-subtropical pattern, even though many of these regions were at fairly high latitudes and inland locations. Naturally, the few floras known from tropical regions, such as Panama, show that conditions were hot and wet there in the early Eocene, as they are today (Graham 1999).

Even more striking are the flora and fauna of the polar regions in the early Eocene. Although these fossils come from above the Arctic and Antarctic circles, and must have experienced six months of darkness, they look nothing like the floras of the Arctic tundra and Antarctic ice caps that are found there today. Instead, there were broad-leaved evergreens, including palm trees and cycads, above 61° north latitude, in Alaska, indicating average temperatures around 18°C (65°F). There were broad-leaved deciduous forests, and even rich coal beds, indicating dense forest vegetation. Spitsbergen produced a flora that could not have tolerated freezing (Schweitzer 1980). Ellesmere Island in the Canadian Arctic, which today lies at 78° north latitude and was not far from that latitude in the early Eocene, produced similar fossil plants. It also yields fossil alligators, pond turtles, land tortoises, and monitor lizards, as well as garfish and bowfin fish (Estes and Hutchison 1980). Most of these animals are typical of subtropical climates today, and none can tolerate freezing for long. Alligators are limited by a mean coldest-month temperature of 10°C (50°F).

In the high latitudes of the Southern Hemisphere, the story was the same. Australia (Kemp 1978; Greenwood et al. 2003) was cloaked in a subtropical-temperate forest dominated by typical Southern Hemisphere conifers, such as *Araucaria* (Norfolk Island pine [fig. 4.2B] and monkey puzzle tree) and podocarps, along with flowering plants such as the Proteaceae (including the familiar flowering shrub *Banksia*), and laurels. In the swampy areas, *Nypa* palms (fig. 4.2A) and tree ferns were common, and coals formed. According to Greenwood et al. (2003), the mean annual temperature for this region was 16–22°C (up to 72°F), with the coldest mean temperature no lower than 10°C (50°F), and the mean annual precipitation was more than 150 centimeters per year. Crouch and Visscher (2003) report a similar pattern for New Zealand as well. A multistory coniferous warm-temperate rain forest grew there in the early Eocene, dominated by *Araucaria* pines and podocarps, with abundant *Nypa* palms in the wet regions. Finally, even Antarctica itself was ice-free and vegetated at this time. The continent was cloaked in cool-temperate forests composed mostly of southern beech (*Nothofagus*), podocarps, *Araucaria*, and abundant ferns (Case 1988; Askin 1992; Francis and Poole 2002; Hunt and Poole 2003). Mean annual temperatures, even in this region of six months of darkness, were in the 10–15°C (50–59°F) range, and mean annual precipitation was between 200 and 400 centimeters of rainfall!

How could the polar regions have been so balmy, even though they experienced six months of darkness? When the evidence from Ellesmere Island and other polar localities was first discovered in the 1980s, paleoclimatologists struggled to answer this question. Paleomagnetic pole positions for Eocene rocks of the high Arctic show that they have not drifted significantly northward since they formed, so we can rule that possibility out (McKenna 1980, 1983). Ellesmere Island, for example, was at 75° north latitude in the Eocene, not significantly different from its present latitude.

Wolfe (1980) and others suggested that the tilt of the earth's axis might have been less in the Eocene, so the Arctic Circle, and regions exposed to the six months of darkness, would have been smaller. However, geophysicists have shown that such an extreme change in the earth's tilt would have had much more dramatic effects than are shown by the geological record (W. Ward 1982; Harris and Ward 1982). Vanyo and Aramwik (1982) examined the history of seasonality as preserved in Precambrian cyanobacterial mats and found no evidence that the earth's spin axis had changed over the late Precambrian and Phanerozoic. Barron's (1984) model of the climate of the earth with less axial tilt produced polar cooling, not warming. Creber and Chaloner (1984) re-examined the botanical evidence cited by Wolfe (1980) and concluded that even with six months of darkness, the polar regions received sufficient sunlight for a seasonally productive forest. The main limiting factor was temperature. As long as it did not get too cold, the forests could grow during their midnight sun summers and remain dormant during their six months of darkness. The modern dominance of tundra vegetation in the Arctic is dictated by the cold, dry conditions, not by the darkness. In addition, the Eocene fossil wood from Spitsbergen, Greenland, and Ellesmere Island shows pronounced growth rings, indicating a strongly seasonal climate. This seasonality would not have happened if the earth's tilt had decreased.

So all we need is to keep the poles relatively warm, and they can host crocodiles and temperate plants. Here we run into some interesting debates. At one time, the prevailing idea was that the early Eocene greenhouse conditions were due to excess carbon dioxide, with levels as high as eight times the present values (Berner et al. 1983), and some people still advocate this model (Sloan and Rea 1995; DeConto and Pollard 2003). However, it is contradicted by the evidence from stomata, the tiny holes on the undersides of leaves that allow gases such as carbon dioxide and water vapor to enter and leave the leaf's interior. Most plants show fewer stomata under conditions of high carbon dioxide, because the plants can still get enough carbon dioxide with fewer stomata, and they lose less water vapor in the process. Conversely, low carbon dioxide conditions force leaves to have more stomata to allow more carbon dioxide in. Royer et al. (2001) and Royer (2003) found that the number of stomata on such living fossil plants as *Metasequoia* (fig. 3.20B) and *Ginkgo* have been fairly constant since the Cretaceous (with possibly one high value in the early Eocene), and this evidence suggests that carbon dioxide levels have remained within present limits throughout this time. Retallack (2001a) examined plant cuticles and determined that early Eocene carbon dioxide levels were only slightly higher than today's levels. In addition, Pearson and Palmer (1999, 2000) have shown that the carbon dioxide balance and pH profile of the Eocene oceans are consistent with an atmospheric carbon dioxide level at modern values, or only slightly higher. Clearly, there were greenhouse conditions in the early Eocene, but if the paleobotanical and geochemical evidence is correct, it is not clear that carbon dioxide levels were that elevated.

Some geologists have advocated the idea that ocean currents need to bring tropical and subtropical waters to polar regions to decrease the temperature gradient between the poles and Equator (Barron 1987). However, the various computer models of earth's climate do not produce mild temperatures at the poles by ocean currents alone (Crowley 1991). More-recent models (Huber et al. 2003) produce not a vigorously circulating early Eocene ocean but rather one in which there is about as much oceanic heat transport as today.

Currently, the preferred gas for explaining the Eocene greenhouse conditions is methane (Sloan et al. 1992; Dickens et al. 1995, 1997; Thomas et al. 2002), which was apparently released during the huge methane burp that terminated the Paleocene. Unlike carbon dioxide, methane would have warmed the poles without making the tropics noticeably warmer. Atmospheric methane promotes the formation of optically thick polar stratospheric ice clouds, which would have trapped heat and prevented the severe winter cooling of the polar regions that occurs today with clear skies over the poles (Sloan et al. 1992).

The Eocene Global Village

As we saw in the previous chapter, the early Eocene mammals of the Northern Hemisphere were already quite diverse, and most were adapted to the dense forests that formed in the late Paleocene. We also saw, however, that the northern continents exhibited some significant differences in their land mammal faunas. North America was dominated by archaic hoofed mammals, pantodonts, and taeniodonts on the ground and by multituberculates, primates, and diverse insectivores in the trees. Europe, in contrast, lacked pantodonts, tillodonts, and taeniodonts in the Paleocene and had no "miacid" carnivorans either. Each region had a handful of genera in common, but some isolation was clearly operating as well, because many groups were not shared, and most genera of the two continents were different.

Eastern Asia (mostly China and Mongolia) was yet another situation altogether. In the late Paleocene (Gashatan), it was the home of diverse pantodonts, plus tillodonts and many archaic ungulates. Yet it had almost no primates or multituberculates, and the smaller herbivore niche was filled entirely by the anagalids, a group related to the rodents and rabbits. But as we saw in the previous chapter, other groups were appearing in the late Paleocene of Asia as well. Among the new groups that evolved there, according to the East of Eden hypothesis of Beard (1998), were the first rodents, represented by the primitive squirrel-like form *Tribosphenomys*. They may have evolved into their familiar lifestyle (with their chisel-like front teeth) in Asia because there was no competition from the rodent-like primates or multituberculates found in North America and Europe. Also appearing in Asia in the late Paleocene were the first odd-toed hoofed mammals, or perissodactyls. These included not only *Radinskya* mentioned previously but also the earliest representatives of the horse-like and tapir-like groups as well. There were also the strange horned beasts known as uintatheres, which we shall discuss later in this chapter. And there was *Minchenella*, one of the earliest relatives of the lineage that led to elephants.

But as the Paleocene came to an end, this distinction between Europe, Asia, and North America was eliminated as immigrants moved freely between the continents. The first wave came at the very end of the Paleocene (fig. 3.6), in the Clarkforkian land mammal age in North America (at about 56 Ma), when a number of Asian groups appeared suddenly in North America for the first time. These included the first rodents, plus the tillodonts and the cow-sized pantodont *Coryphodon* (pl. 2). Given that most of these taxa were Asian in origin, the likeliest route is across the Bering Strait.

The floodgates really burst open at about 55 Ma, at the beginning of the Eocene (Ypresian or Sparnacian in Europe, Wasatchian in North America, Bumbanian in Asia). North American land mammal faunas were radically rearranged as numerous immigrant groups arrived from elsewhere (Woodburne and Swisher 1995). Most notable among these are the odd-toed perissodactyls, including the earliest members of the lineages that include horses, tapirs, and rhinos. The popular literature has long called these earliest North American horses *Eohippus* or *Hyracotherium* (fig. 4.3), but recent research shows that neither of these names is valid for most of these horses. Hooker (1989) showed that the original English fossils that were the basis of the name *Hyracotherium* are actually related to an endemic European horse-like group, the palaeotheres. Froehlich (2002) analyzed the North American fossils in detail and found that the old name *Eohippus* is applicable to only one of the species, *E. angustidens*. Instead, many of these early Eocene horses belong to *Protorohippus*, and others are assigned to a variety of genera, including previously proposed names such as *Xenicohippus*, *Systemodon*, and *Pliolophus*, as well as to new genera such as *Sifrhippus*, *Minippus*, and *Arenahippus*. The old days when all early Eocene horses could be lumped into one genus (whether *Eohippus* or *Hyracotherium*) are long gone.

Next in importance is the other major living order of hoofed mammals, the even-toed Artiodactyla (fig. 4.4). Today, artiodactyls include pigs, hippos, camels, sheep, goats, deer, antelopes, and cattle, but their earliest relatives, the dichobunids, looked nothing like these animals. In-

Figure 4.3. The early Eocene horse Protorohippus *(formerly known as* Hyracotherium *or Eohippus),* which had four toes on the front feet and three toes on the hind feet and was about the size of a beagle. Painting by B. Horsfall, in Scott 1913.

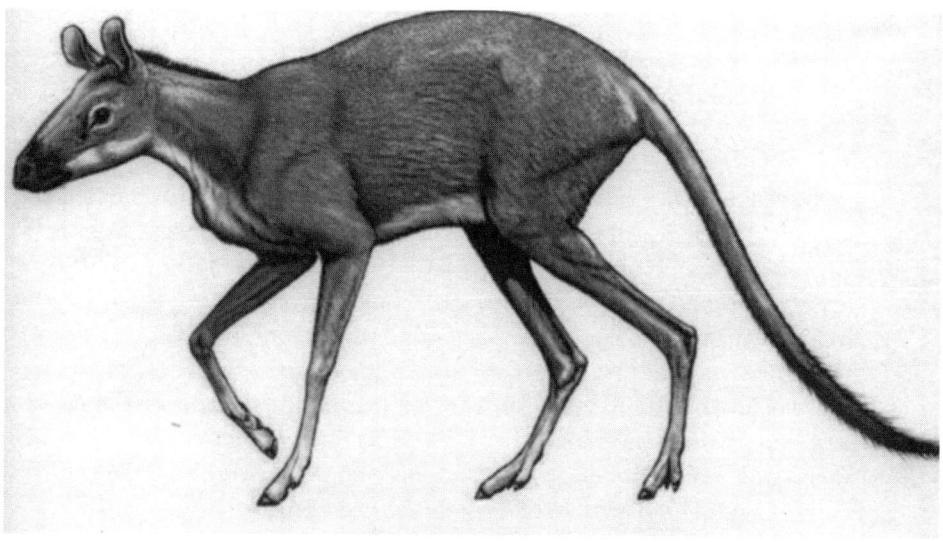

Figure 4.4. The tiny deer-like Diacodexis, *one of the earliest artiodactyls, which had long hind limbs for running and hopping. Painting courtesy Carl Buell.*

Figure 4.5. The tarsier-like omomyid primate Tetonius, *one of the more common animals of the early Eocene forests. Drawing by Lydia Kibiuk, courtesy of K. Rose.*

stead, they were the size of tiny deer, with long delicate legs. Some had such long hind legs that they were apparently adapted for rapid hopping. These two new groups of hoofed mammals (artiodactyls and perisso-dactyls) rapidly displaced many of the archaic ungulates that had domi-nated the Paleocene in North America. Joining these immigrants to North America were two advanced groups of lemur-like primates, the Omomyi-dae and the Adapidae (fig. 4.5), which rapidly displaced the more rodent-like Plesiadapidae (fig. 3.15) that had dominated the Paleocene. Finally, there were the first advanced creodonts, known as hyaenodonts, plus a bizarre Asian group known as didymoconids, a number of new kinds of insectivorous mammals, and opossum-like marsupials that reappeared after disappearing in the early Paleocene.

Europe, too, saw an astonishing wave of immigration from Asia and North America at this time (Agusti and Anton 2002). Pantodonts, which were largely Asian in origin, and tillodonts both appeared for the first time, and the big pantodont *Coryphodon* (pl. 2) is particularly striking in its cosmopolitan distribution (Lucas 1998). Adapid and omomyid pri-mates (fig. 4.5) also reached Europe for the first time, as did the rodents; all of these groups rapidly displaced the rodent-like plesiadapid primates and the few surviving multituberculates. Finally, the perissodactyls and artiodactyls also arrived, pushing out the archaic hoofed mammals, just as in North America. Europe was home of *Hyracotherium*, a member of an endemic lineage of horse-like perissodactyls known as palaeotheres. Eu-rope also gained the small deer-like dichobunid artiodactyls, which were common in North America too. The beginning of the Eocene in Europe saw the first true "miacid" carnivorans (fig. 3.16), which began to replace the archaic creodont (fig. 4.6) and mesonychid flesh-eaters that had been dominant in the Paleocene.

Figure 4.6. *The archaic creodont* Patriofelis, *a leopard-sized predator from the early and middle Eocene. Painting by B. Horsfall, in Scott 1913.*

Finally, the early Eocene (Bumbanian) in Asia was also marked by faunal change (Ting 1998). Rodents quickly came to dominate the small-mammal fauna, pushing out their anagalid ancestors. Perissodactyls, too, became common, reducing the role of their archaic hoofed mammal relatives. Other archaic hoofed mammals, such as the carnivorous mesonychids, were becoming scarce, as were the pantodonts, uintatheres, and arctostylopids. Most striking is the complete absence of artiodactyls, which were already abundant in Europe and North America. If not from China or Mongolia, where did they come from?

Kraus and Maas (1990) suggest that at least some of the immigrant groups (such as artiodactyls and perhaps some advanced primates) might have originated on the Indian continent while it was drifting away from Africa in the Paleocene (fig. 4.7). During the Early Cretaceous, India was still attached to Africa, Madagascar, and the rest of the Gondwana land masses; and the limited fossil evidence from the Cretaceous of India suggests that it shared primitive placental mammals with the rest of the continents. Unfortunately, we have no Paleocene mammal fossils from India, but Krause and Maas (1990) suggest that India served as a great Noah's ark during this time, allowing groups such as the artiodactyls to evolve in isolation. Then, when India began its collision with Asia in the early Eocene, they were able to hop off the ark and spread widely through the Northern Hemisphere. This hypothesis is corroborated by the occurrence

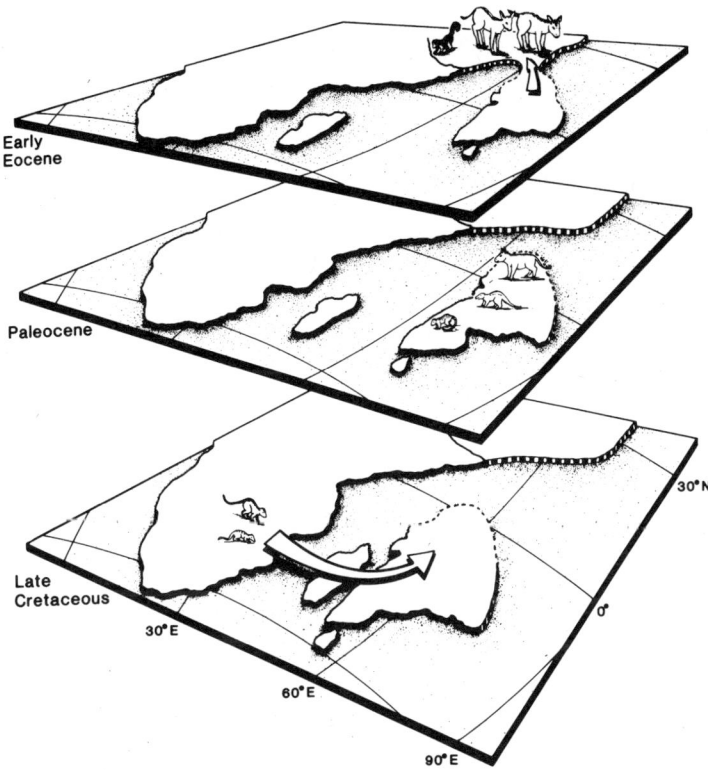

Figure 4.7. Noah's ark. India split from the other Gondwana continents in the Cretaceous and may have acted as a Noah's ark, docking with Asia in the early Eocene to release mammals that had evolved in isolation (including perissodactyls, artiodactyls, several groups of primates, and hyaenodont creodonts). From Krause and Maas 1990; by permission of the Geological Society of America.

of the most primitive artiodactyl known, *Diacodexis pakistanensis*, from the early Eocene beds of Pakistan (Thewissen et al. 1983). The oldest faunas that come from India and Pakistan are early Eocene, and they consist of elements common worldwide: artiodactyls, perissodactyls (including primitive tapirs and rhinos), omomyid primates, anthracobunids (relatives of the elephant), rodents, and carnivorans (Russell and Zhai 1987). Clearly, if India had harbored an endemic fauna on the ark during the Paleocene, by the early Eocene, the ark was fully docked and had received mammals from the rest of Asia. Apparently, however, some barrier prevented artiodactyls from reaching China in the early Eocene, and they must have reached North America by way of Europe and the North Atlantic land bridge.

The beginning of the Eocene marks a worldwide migration event that made most faunas in the Northern Hemisphere similar to one another. Clearly, there was migration between Asia and North America via the

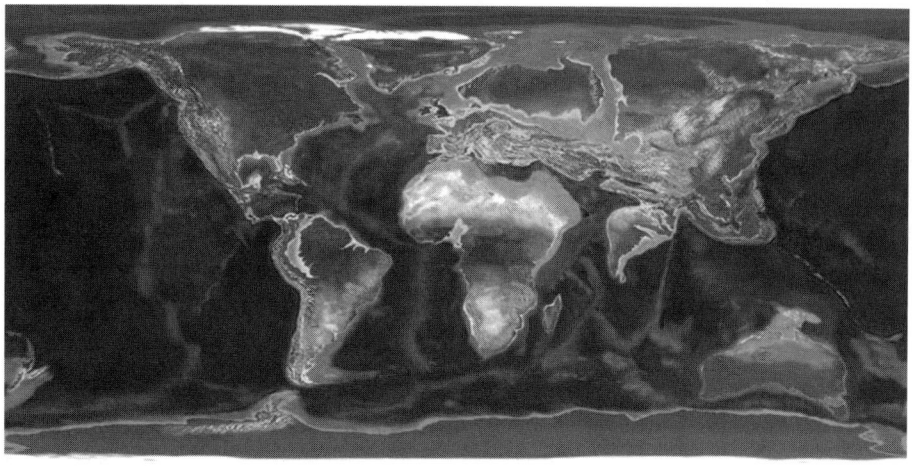

Figure 4.8. Paleogeographic map of the continents and oceans in the middle Eocene. Note that India has just begun its collision with Asia and that the Tethys Seaway is beginning to be fragmented. Europe was an archipelago, and the Obik Sea cut Siberia in half. There were several possible land routes between the northern continents during the early Eocene, including two potential North Atlantic corridors and the Bering land bridge. The great similarity between the early Eocene faunas of North America and Europe suggests that the North Atlantic route was effective, as does the presence of these same animals on Ellesmere Island near Greenland. However, by the middle Eocene, this North Atlantic connections was severed, and Europe began to develop its own endemic faunas. Europe was also isolated from Asia by the Obik Sea, and from Africa by the Tethys-Mediterranean Sea. The Bering route was used throughout the Cenozoic, and there were many periods of faunal interchange between Asia and North America. Courtesy of R. Blakey.

Bering Strait. In addition, there also was migration across the North Atlantic because the warm climatic conditions enabled many mammals to walk from Europe to Iceland to Greenland to North America and back again (McKenna 1983). Indeed, the mammals of Ellesmere Island in the Canadian Arctic (McKenna 1980) are evidence for how easily mammals traveled between Europe and North America. Even though they were living at 78° north latitude in six months of darkness and midnight sun summers, Ellesmere Island supported a sparse fauna of primates, rodents, multituberculates, tapir-like perissodactyls, and the ubiquitous pantodont *Coryphodon* (pl. 2). Finally, the free migration of Asian forms suggests that the Bering Strait was not the only road from China. Instead, the collision of India with Asia in the early Eocene may have caused the Turgai Strait across the Obik Sea to dry up (fig. 4.8), allowing free migration across Eurasia. Such free migration between the continents would disappear by the middle Eocene, when the North Atlantic corridor disappeared; and endemism would return to Europe, which became an archipelago of isolated islands with many peculiar mammals. Not until the Pliocene would there be such sharing of mammal faunas around the continents again.

Good Times in Warm Oceans

Belgium is famous for being the country with the most battlefields, from wars fought mainly by non-Belgian armies, because it is a low country sandwiched between frequently warring nations such as France, England, and Germany. These battles ranged from the many medieval conflicts, to the power grabs by Louis XIV and various Spanish and Austrian nobles, to Napoleon's final defeat at Waterloo in 1815, to the Battle of the Bulge in December 1944. Today the town of Ypres (pronounced "EE-pruh" in French; it is spelled Ieper and pronounced "YAY-per" by the Flemish) is a quiet town in the Flanders region of Belgium, just inland of the port of Ostend and a few miles from the French border. Once famous for its linen and lace industry, in medieval times it was as large as London, and it was also renowned for its beautiful medieval buildings as well. But students of military history will recognize Ypres as the site of some of the bloodiest fighting in World War I. The British Tommies bitterly mispronounced it "Wipers," and indeed many brave men were wiped out there. When the German juggernaut swept through the rest of Belgium at the beginning of the war in August 1914, they took Ypres briefly but were thrown back and blocked from taking the coastal port of Ostend (where they could dock their U-boats), and from invading the coast of France around Calais. The Ypres salient became a constant source of anger in the German High Command, and a symbol of resistance by the Allies, so it was attacked and defended fiercely. During the course of four years of war, Ypres saw five major bloody battles, with more than a million German, French, and British casualties. Today, Ypres is dotted with war memorials that cover the mass graves of thousands of brave men.

But the name Ypres has a different association for geologists. In 1849, Dumont proposed the Ypresian Stage for the earliest stage of the Eocene, on the basis of the *argiles de Flandres*, clays in the Flanders region near Ypres (fig. 4.9). In modern timescales (Berggren 1971), the Ypresian Stage has long been known as the earliest stage of the Eocene. It is the warmest period in the entire Cenozoic, and the time when life was at its most diverse. Military historians associate Ypres with death, but geologists associate it with warm, tropical greenhouse conditions and a profusion of life. There is, however, one geological battle going on. The beginning of the Ypresian Stage, which is historically accepted as the beginning of the Eocene, is about 800,000 years younger than the carbon isotope event and the big climatic and faunal turnover event (Aubry et al. 1998). Many

Figure 4.9. Exposures of the type Ypresian (lower Eocene) Ypres clays from Egem Quarry in the Flanders region of northwest Belgium. Photo courtesy of S. Van Simaeys.

geologists prefer to draw the Paleocene-Eocene boundary at the global carbon signal and the big faunal event, whereas others prefer to follow historical precedent. In 2003, the international committees decided to place the boundary at the climatic event (the base of the Sparnacian mammalian stage in Europe) and to ignore the historical precedent for the Ypresian Stage.

As we have already seen, the temperate and polar land regions were unusually warm and wet and tropical in the early Eocene, more so than at any other time since the Cretaceous. So too were the oceans. The tropical shallow oceans (Pearson et al. 2001) warmed to mean annual temperatures as high as 28–32°C (82–89°F), warmer even than today (the modern tropical sea surface is typically 23–27°C). But even the waters of the North Sea Basin (near freezing most of the year now) were as warm as 17°C (63°F) in the winter and 22°C (82°F) in the summer, so even Denmark was subtropical then (Schmitz 2000). At higher latitudes, the mean annual temperatures were as high as 15°C (59°F), whereas today those same oceans are at or below freezing (Zachos et al. 1994). Not only were the oceans warm, but there was a small difference in temperatures between the poles and Equator, so oceanic circulation had to be different as well. Warm salty waters formed within the Tethys Seaway (fig. 4.9). According to Boersma et al. (1987), not only was the entire world ocean warmer on average, but tropical surface waters extended closer to the poles than at any other time in the past, and the difference in temperature between surface and bottom waters was not very great either. Recent climatic models (Huber et al. 2003) predict that warm (12–15°C) salty bottom waters formed in the Tethys and North Atlantic and sank to the bottom. This process made the entire world ocean warmer and more stable, with a temperature inversion effect, where surface waters in most regions formed a lid that failed to sink to the bottom because the warm salty deep water beneath it was denser than the surface waters. In contrast, today the oceans are vigorously mixed as cold polar surface waters sink to the bottom and bring their oxygen with them.

These warm tropical seas were home to a wide variety of organisms. The planktonic algae, or coccolithophorids, were at a peak in diversity (Aubry 1998). Planktonic and benthic foraminifera became incredibly diverse (Boersma et al. 1987, 1998), with tropical surface plankton extending to middle and high latitudes and even to the Antarctic. By the end of the early Eocene, their diversity was further increased by the evolution of a number of new planktonic foraminiferans (figs. 4.10A, B), including the tiny globigerinathekids, the spiny hantkeninids, and the first true globigerinids, the most abundant group in the oceans today. The warm shallow bottom of the Tethys Seaway was covered by an incredible density of benthic foraminiferans. One group, the nummulitids (fig. 4.10C) secreted a disk-shaped shell made of tiny chambers in a flat spiral. Even though the organism was like a single-celled amoeba, the shells it secreted sometimes reached several centimeters in diameter—an immense construction for a single-celled organism! Paleontologists, long puzzled over how such a tiny cell could secrete so much calcite, have concluded that like many living large tropical benthic foraminifera, nummulitids probably incorporated symbiotic algae in their tissues to allow them to grow so large. The waters of the Tethys were not only warm but clear and shallow, like parts of the modern South Pacific and Caribbean, so they promoted the growth of plant-animal symbioses like benthic foraminifera and corals.

The volume of these nummulitids is truly staggering. Many limestones in the region are made entirely of them, with enough shells to represent trillions of individuals, so these rocks resemble a bunch of coins glued together (fig. 4.10C). Indeed, they get their name from the Latin word *nummus,* a kind of ancient coin. In older French books, some stratigraphers called the Eocene the Nummulitic Period. Thick nummulitic limestones are the major building material in the Gizeh Plateau of Egypt, and most of the blocks in the Great Pyramids are made of them (fig. 4.10D). When the Greek historian Herodotus visited the pyramids (which were already ancient when Herodotus saw them around 450 B.C.), he noticed all the stone disks on the ground and thought they were petrified lentils from the lunches of the slaves who built the pyramids. In the Paris Basin, the same nummulitids gave the middle Eocene limestones the name *banc royal.* However, they were not restricted to the tropics (Durham 1950); in the early Eocene, large nummulitic benthic foraminifera were found wherever

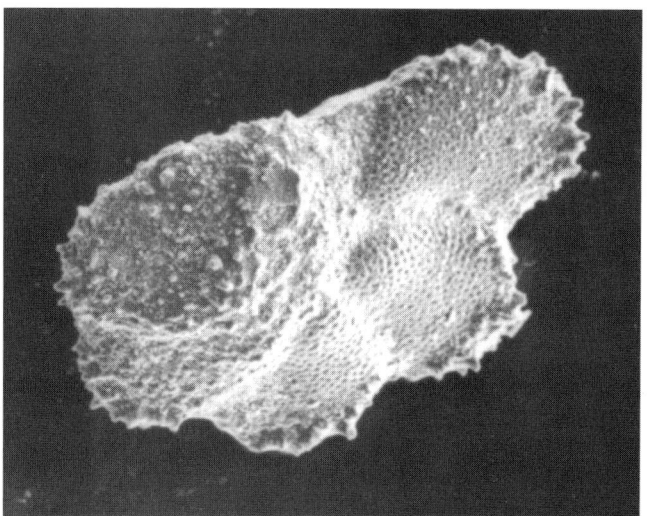

A

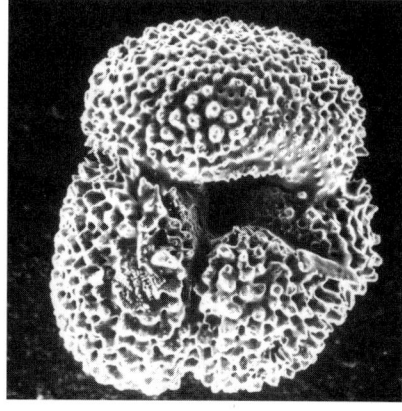

B

Figure 4.10. Warm-water foraminifera that died out at the end of the middle Eocene. Planktonic foraminifera include (A) Morozovella lehneri and (B) Trucorotaloides rohri. Benthic foraminifera include (C) the coin-sized nummulitids, which were so abundant in the tropical Tethys Seaway that (D) nummulitic limestones make up the Gizeh Plateau in Egypt and were used to build the pyramids. Photos A and B courtesy of G. Keller; photo C by the author; photo D courtesy of R. Schoch.

C

D

the warm-water conditions prevailed, as far north as Vancouver Island and even England and northwest Germany.

The tropical Tethys and Caribbean were also home to extensive reef corals, with a new peak in diversity of coral genera and species (Rosen 2000). Both reef-forming and solitary corals were also common on the Gulf Coast of the United States; and on the Pacific Coast, their abundance suggests a minimum water temperature of 25°C (77°F) as far north as British Columbia (Durham 1950).

The mollusks were as happy as clams (fig. 4.11). Nearly every marine unit sets a new record of molluscan diversity in the early Eocene. Dockery (1986) reports the highest molluscan diversity yet in the Cenozoic from the lower Eocene Hatchetigbee Formation of Mississippi and Alabama. On the Pacific Coast of North America, there were a record number of molluscan species and genera in the "Capay" and "Meganos" Stages (Durham 1950; Squires 2003). Most of these were genera that live in the subtropics or tropics today, including olive, fig, cowry, pen, sundial, and cone shells, as well as the thick-shelled Eocene clam *Venericardia*, and the ubiquitous snail *Turritella* (fig. 4.11). Conditions were so hospitable across the world that as many as 30 of the 141 Pacific Coast genera of mollusks are shared with the Paris Basin or Tethys (Squires 2003); a similar number were shared between the Gulf Coast of Mississippi and Alabama and the Paris Basin (Dockery and Lozouet 2003). Clearly, planktonic molluscan larvae found it easy to float around the tropics from Europe through the Tethys to the Pacific. Ivany et al. (2004) analyzed the annual change in oxygen isotopes trapped in the growth lines of *Venericardia* shells from the lower Eocene Hatchetigbee Formation in Alabama and found that these organisms lived in a mean water temperature of 26°C (79°F) and that temperatures fluctuated only from a summer high of 27°C (81°F) to 25°C (77°F) in the winter, so the conditions were mild and non-seasonal.

After mollusks, the next most common shelled marine invertebrates were the echinoids (sea urchins and their relatives). Early and middle Eocene echinoids were richly diverse, and again they were mostly warm-water species. In the early Eocene, the first sand dollars evolved from more-spherical ancestors, probably in the Tethys near Africa or India (fig. 4.12). These modified sea urchins have flattened bodies for rapid burrowing, and they live in shallow, shifting nearshore sands, tilted like a shingle with one edge stuck in the sand and their mouths (on the bottom of the shell) facing into the current to trap food particles. By the middle Eocene, sand dollars had spread out from the Tethys and had a worldwide distribution.

Fish were also successful. In a few exceptional localities like Monte Bolca in Italy (fig. 4.13A), an incredible diversity of fishes is preserved. Although the fossils mostly represent extinct species, their bodies are shaped like many of the living fishes that are found in coral reefs and tropical waters around the world today. Meanwhile, a tremendous diversification of both marine and freshwater fishes took place. Many of the living fish fam-

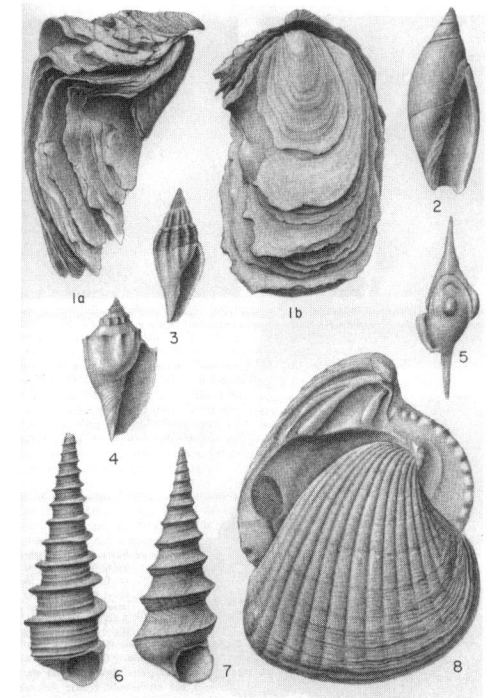

A

B

Figure 4.11. (A) Typical mollusks of the Eocene: 1a,b. the oyster Ostrea sellaeformis; *2. the olive snail* Oliva alabamiensis; *3.* Plejona rugata; *4. the volute snail* Volutilithes sayana; *5. the peculiar snail* Calyptrophorus trinodiferus; *6 and 7. two common species of the high-spired snail,* Turritella praecincta *and* Turritella mortoni; *8. the thick-shelled clam that is an index fossil of the Eocene,* Venericardia planicosta. *From Dunbar 1966. (B) Thick concretion of* Venericardia bashiplata *from the Bashi Formation, Tombigbee River, Alabama. Photo courtesy of L. C. Ivany.*

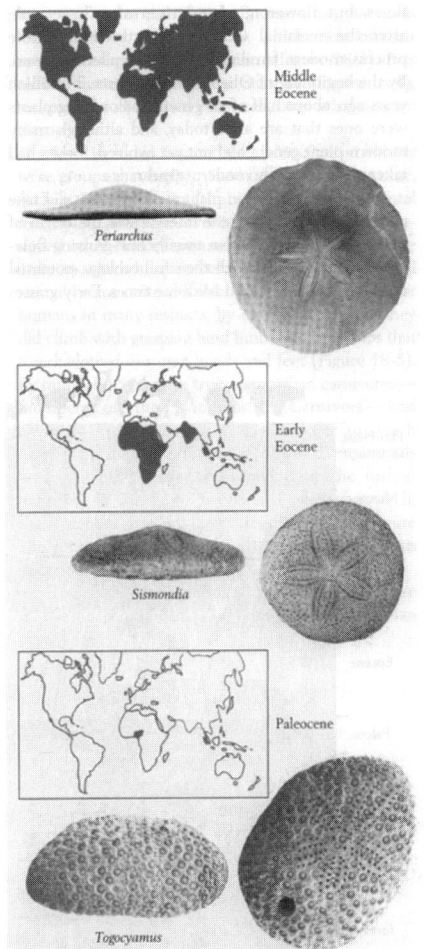

Figure 4.12. Sand dollar evolution. After Kier 1975.

ilies arose in the early and middle Eocene (Sepkoski 1982), including eight families of eels and their relatives (Angulliformes), four families of salmon and their relatives (Salmoniformes), two families of catfish (Siluriformes), four families of lizardfishes (Aulopiformes), all three families of anglerfish (Lophiiformes), five families of codfish and their relatives (Gadiformes), three families of silversides and flying fish (Atheriniformes), six families of pipefish and seahorses (Syngnathiformes), and seven families of scorpionfishes and sculpins (Scorpaeniformes). Most impressive of all was the huge diversification of the Perciformes, or perches, and their relatives. Thirty families appeared in the Eocene, including perches, surgeonfishes, sand lances, cardinalfishes, blennies, dragonets, pompanos, snooks, bandfishes, surf perches, batfishes, gobies, snappers, breams, whipfishes, damselfishes, bluefishes, bigeyes, parrotfishes, barracudas, archerfishes, weeverfishes, stargazers, and cichlids (so popular in fish tanks today). In addition, the Eocene saw the appearance of numerous other fish families,

A

Figure 4.13. (A) Typical fish fossils from the Eocene locality in Monte Bolca, Italy. Photo courtesy of the Field Museum of Natural History. (B) Eocene marine diorama representing the shallow-marine beds of the London Clay. A wide variety of modern-looking snails and clams (plus fish and one nautiloid) inhabit the sea bottom while the sand shark Odontaspis swims above them. However, many of the mollusks are archaic species that are now extinct, and the seascape lacks most of the modern groups of fishes, plus whales, seals, and sea lions, all of which evolved later. From McKerrow 1978.

B

including five families of flatfish and flounders (soles had already appeared in the Paleocene) and six families of puffers, filefishes, and triggerfishes. Although many of these fishes are known from both marine and freshwater habitats, the radiation of freshwater fishes on land was the most profound (Cavender 1986). Impressive numbers of fossils of these new fish groups are found fossilized in Eocene lake deposits, such as the famous Green River shales of the Rocky Mountains.

Clearly, the warm tropical Tethyan reefs and shallow nummulitic sandy shoals were havens for fish and mollusks and sand dollars, like the Great Barrier Reef of Australia today. In addition to bony fish, there was a diversity of sharks and rays comparable to that of today (fig. 4.13B).

A Whale of a Tale

Marine life was at an all-time high in diversity in the early Eocene, but still some niches were vacant. The role of a giant marine predator had gone unoccupied since the extinction of the marine reptiles in the Late Cretaceous, and the Paleocene oceans saw nothing larger than small sharks as top predators. But the large diversity of fish and squid in the seas was an important resource that would not go unexploited forever. Since reptiles no longer dominated the large vertebrate niches, it would fall to mammals to evolve into marine predators, and they would do so multiple times during the Cenozoic. The first such invasion of the water took place in the early Eocene, and this evolutionary transition is one of the most spectacular and best-documented in the fossil record: when whales left the land and returned to the water.

Ever since people realized that whales and dolphins were mammals, they have speculated about how they might have evolved from land-dwelling mammals, and from which group of mammals they originated. By the 1830s and 1840s, specimens of huge primitive whales known as archaeocetes were being discovered in middle Eocene beds of Alabama, but these specimens were fully aquatic, with flippers and tail flukes and a sinuous body that was 24 meters (80 feet) long (fig. 4.14). Clearly, whales must have originated before the middle Eocene, but nothing was known of their fossil record prior to that time. In 1966, Leigh van Valen and others showed that the skulls and teeth of primitive whales looked much like the predatory archaic hoofed mammals known as mesonychids (fig. 3.11), which we encountered in the previous chapter. Even though mesonychids were land mammals with hooves, there were many similarities in the skull and skeleton (especially the large, serrated, triangular blade-like teeth) that suggested a close relationship with archaeocete whales. Yet for over a century there were no transitional fossils known between mesonychids and archaeocetes.

The breakthrough occurred when scientists began to examine the lower Eocene beds of Pakistan. Deposited on shorelines and shallow seas along the north shore of the Tethys Seaway (fig. 4.9), these beds have since been uplifted by the collision of India with Asia, which also began in

Figure 4.14. Reconstruction of the 50-foot-long archaeocete whale Basilosaurus. *After Burian.*

the early Eocene. Gingerich et al. (1983) described *Pakicetus* on the basis of a skull with an archaeocete braincase, but lacking ears that were capable of echolocation, and with teeth intermediate between those of mesonychids and archaeocetes (fig. 4.15A). *Pakicetus* came from river sediments bordering shallow seaways, which suggests that it might have been a semi-aquatic predator that waded in rivers part-time to find food. The skeleton of *Pakicetus* is quite wolf-like, with long slender limbs and a tail, so it still resembles a mesonychid in most features.

The next development occurred a few years later, when Gingerich et al. (1990) described new specimens of the archaeocete *Basilosaurus* from the middle and upper Eocene deposits of Egypt. Although it was like other archaeocetes in being fully aquatic, these new specimens had something never previously preserved: the hind limbs. In most whales, there are no

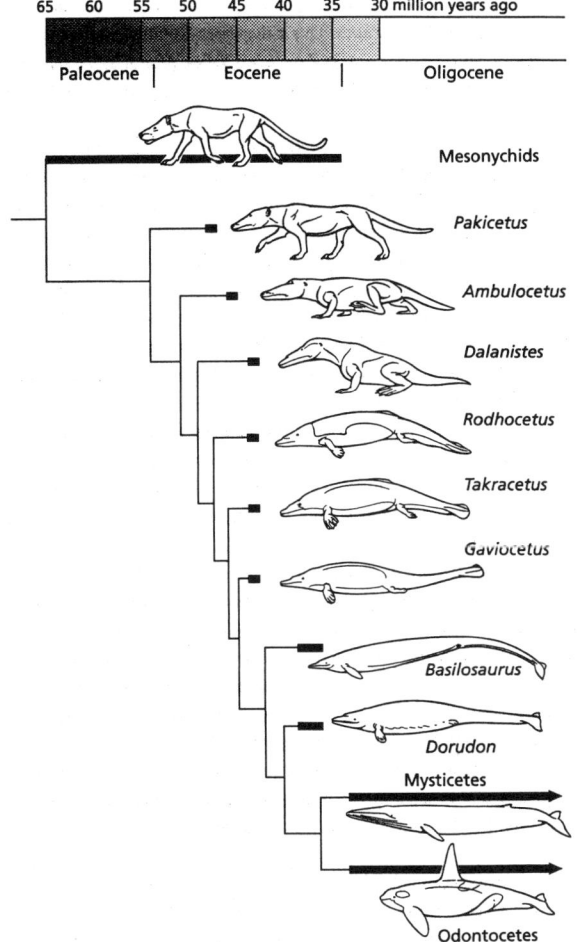

65 60 55 50 45 40 35 30 million years ago

Paleocene | Eocene | Oligocene

Mesonychids

Pakicetus

Ambulocetus

Dalanistes

Rodhocetus

Takracetus

Gaviocetus

Basilosaurus

Dorudon

Mysticetes

Odontocetes   A

Figure 4.15. (A) Family tree of early whales, showing the remarkable transitional sequence from terrestrial mesonychids through Pakicetus, Ambulocetus, Dalanistes, Rodhocetus *and archaeocetes to modern whales. From Zimmer 1999. Reprinted with the permission of The Free Press, a division of Simon & Schuster Adult Publishing Group, from* At the Water's Edge *by Carl Zimmer. © 1998 by Carl Zimmer. All Rights Reserved. (B) Reconstruction of* Ambulocetus natans *attacking an Eocene phenacodont. Painting by Carl Buell.*

B

external hind limbs, and the remnants of the hip and thigh bones are buried in muscles along the spine halfway down the body. These specimens, however, had tiny hind limbs (about as large as a human arm on a 24-meter-long body), which clearly did not function for locomotion. Like the vestigial hind limbs of modern whales buried inside the body, these tiny limbs were functionless relics of the day when "whales" did walk on land. Since this discovery, other archaeocetes, such as *Takracetus* and *Gaviocetus,* have been found to retain vestigial hind limbs.

The most important discovery occurred when Thewissen et al. (1994) described *Ambulocetus natans,* whose name means literally "walking swimming whale." Found in middle Eocene marine beds of Pakistan, it was about the size of a sea lion (fig. 4.15B), with functional flippers on both its forefeet and huge hind feet (which had vestigial hooves as well). Its skull and teeth, however, were still like those of mesonychids. On the basis of its highly flexible vertebrae, Thewissen et al. (1994) suggested that *Ambulocetus* swam with an up-and-down flexure of its body, similar to the swimming motion of an otter, rather than paddling with its feet like a penguin or seal or wriggling side-to-side like a fish. This motion is a precursor to the up-and-down motion of a whale's tail flukes as it swims through the water.

Further discoveries (mostly in the middle Eocene of Pakistan) followed one after another. *Dalanistes,* for example, had fully functional front and hind limbs with webbed feet and a long tail but was much more whale-like, with a longer snout. *Rodhocetus* was more like a dolphin yet still retained functional hind limbs. As the years go by, more and more transitional whales are being discovered, so that by now the amazing transformation from land mesonychid to whale is one of the best examples of evolutionary transitions in the fossil record (fig. 4.15A). This may not make creationists happy, but the fossils cannot be denied.

The final nail in the coffin came from a variety of lines of evidence. For years, molecular biologists had been saying that whales seemed to be most closely related to the even-toed artiodactyls among the living mammals. Some went so far as to put the whales *within* the Artiodactyla, and derived them from hippopotami, not mesonychids! Then in 2001, both Gingerich et al. (2001) and Thewissen et al. (2001) independently described the ankle regions of several primitive whales, which turn out to have the characteristic double-pulley astragalus bone that is the signature of the Artiodactyla. Now the once-outrageous notion that whales are artiodactyls is being seriously entertained, with mesonychids considered the close relative of both groups.

The discovery that whales originated in the late Paleocene and early Eocene around the Tethys Seaway is consistent with the idea that many other mammal groups originated at this time and place as well. We saw that Krause and Maas (1990) have argued that artiodactyls originated when India docked with Asia and closed the Tethys, and indeed the oldest and most primitive known artiodactyl, *Diacodexis pakistanensis,* comes from the early Eocene of Pakistan (Thewissen et al. 1983). This informa-

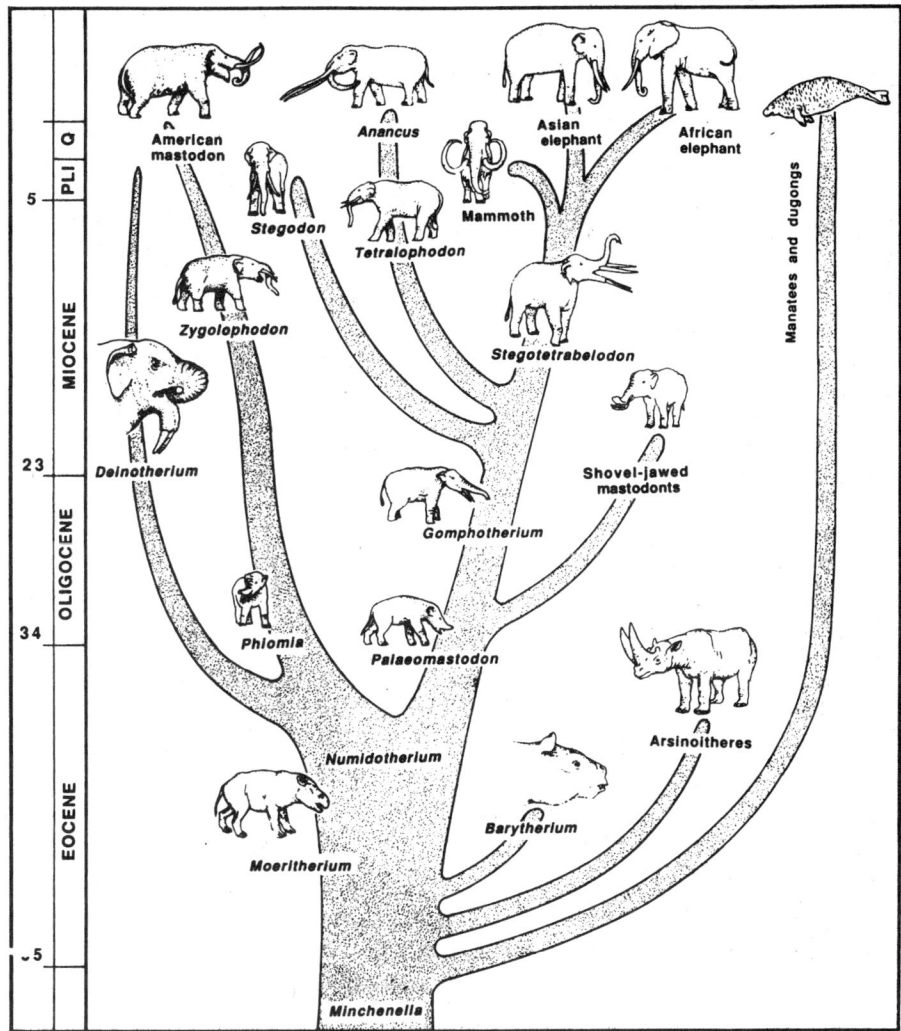

Figure 4.16. The radiation of tethytheres from primitive groups in the Eocene to sirenians, arsinoitheres, and the huge diversity of trunked proboscideans. Drawn by C. R. Prothero.

tion fits well with the idea that whales are artiodactyls and first appeared in the early Eocene of Pakistan.

In addition to whales and artiodactyls, the earliest known members of the sea cows and manatees (order Sirenia) also occurred in the Tethys. Recently, Daryl Domning (2001) reported the discovery of a fossil sea cow from the Eocene of Jamaica. Known as *Pezosiren*, it still had functional hands and feet, rather than flippers, yet its teeth and skull and thick ribs were identical to those of early sirenians. As McKenna (1975) showed, sirenians are part of a great radiation he called the Tethytheria (fig. 4.16), which also includes elephants and their kin (the order Proboscidea). The earliest known tethytheres were *Paschatherium* from the late Paleocene of Morocco (Gheerbrant et al. 1996) and *Minchenella* from the late Paleocene of China. By the early Eocene, there were many primitive anthracobunid proboscideans like *Minchenella* in Pakistan and Africa, and by the middle and late Eocene, the earliest and most primitive mastodonts with short tusks but no trunks (*Numidotherium, Barytherium, Moeritherium*) had appeared in North Africa. Recent molecular studies have confirmed that proboscideans and sirenians are indeed closely related as tethytheres, and these studies have added a variety of other animals, such as hyraxes, tenrecs, golden moles, aardvarks and elephant shrews (all found in Africa), to a group that molecular biologists call Afrotheria.

From Greenhouse to Icehouse

The early Eocene world was the warmest period since the Cretaceous. By the early Oligocene, however, glaciers had returned to Antarctica, and the global "icehouse" began. With this transformation came major changes in global vegetation, and in the animals that lived in it, and transformations of the life in the ocean. How did these changes take place? Was there an abrupt crash, comparable to the K/T impact or the abrupt beginning of the early Eocene warmth? In the 1980s, geologists were talking about a "Terminal Eocene Event" (TEE) and promoting it as similar to the K/T event. Or did the changes take place over a longer time, as a long, protracted climatic deterioration?

Fortunately, we have excellent, detailed records of the 12 million years of the middle Eocene (49–37 Ma) and the 3 million years of the late Eocene (37–34 Ma) in many places in the world, so we can examine the story in detail. Putting this record in context, the 12 million years of the middle Eocene alone are longer than the 10 million years of the Paleocene, or the 10 million years of the Oligocene, or twice the duration of the Pliocene and Quaternary (5 million years) put together—and these years were just as eventful. Not surprisingly, the answer is a complex one that requires a lot of careful painstaking work on many stratigraphic sections and many groups of organisms, so only the highlights can be presented here. Those interested in further details can consult the volumes edited by Prothero and Berggren (1992) and Prothero et al. (2003) for a more complete account.

We start with the most complete source of data, the oxygen isotope record of the deep sea (figs. 3.22, 4.17). No matter which version of the oxygen isotope curve we use (e.g., Miller et al. 1987; Zachos et al. 1994, 2001), the pattern is similar: a peak of warmth (lighter, more negative, more oxygen-16-rich values) in the early Eocene, followed by a steady drop in global temperature (heavier, more positive, more oxygen-18-rich values) through the middle and late Eocene. Occasionally the curve shows some steps and reversals caused by short-term warming events and more-rapid cooling pulses, but the overall trend is clear. From the early Eocene, when average sea-bottom temperatures were around 13°C (55°F), the average temperature dropped steadily, so that by the early Oligocene, the deep oceans were close to freezing on a global basis, and ice caps formed in the Antarctic. To place this trend in context, each Pleistocene glacial-interglacial cycle changed the global temperature by only 2–3°C, yet in the course of the 15 million years of the middle and late Eocene, global temperature dropped almost 12°C. By contrast, the global carbon isotope curve (figs. 3.22, 4.17) is relatively stable, with no net change after the early Eocene methane spike disappears. This suggests that although the oceans and the globe were cooling, there was no major perturbation in the carbon reservoirs through most of this interval.

So why did the earth change from a greenhouse to an icehouse? This topic is hotly debated even now (see the review in Ruddiman 2001). One school of thought (e.g., Kennett 1977; Berggren and Hollister 1977) points to evidence that major changes in oceanic circulation occurred in the Eocene and Oligocene (discussed in chapter 5), which rearranged the distribution of heat on the globe and triggered the growth of Antarctic ice caps. But critics of this school of thought argue that it explains only short-term changes in local climate, and not the long-term cooling of the earth over 55 million years. In addition, this model does not explain how the greenhouse gases (methane and possibly excess carbon dioxide) were removed from the atmosphere and locked up in the earth. Another school of thought (e.g., Ruddiman and Kutzbach 1991; Raymo and Ruddiman 1992) argues that the rapid uplift of new mountain ranges (especially the Himalayas) increased the rate of crustal weathering, which removes carbon dioxide from the atmosphere as this gas combines with the newly formed minerals in soils. This hypothesis may work well for the last 15 million years, but it does not explain the Eocene cooling, when the Himalayan uplift had just begun. Finally, there is the spreading-rate hypothesis (e.g., Berner et al. 1983), which argues that greenhouse conditions occurred when there were high rates of seafloor spreading, which released carbon dioxide from the mantle (e.g., during the Cretaceous). By contrast, slow rates of seafloor spreading produce less carbon dioxide, so this gas is gradually absorbed by weathering in soils and is pulled out of the atmosphere. Indeed, there is evidence that spreading rates were at their highest in the early and earliest middle Eocene; after plate rearrangement events at 54 and again at 44 Ma, spreading slowed down in parallel with the long-term cooling trend. However, for the past 15 million years, spreading rates have been increasing, yet climates have continued to cool; so this model does not work well for the late Cenozoic. In addition, many scientists doubt that the amount of carbon dioxide and methane taken up by weathering in either the crustal-uplift model or the seafloor-spreading model is sufficient to account for the loss of these greenhouse gases in the atmosphere. They argue that other reservoirs, such as limestones or coals (which are relatively rare in the middle and late Eocene), are needed, or possibly more methane hydrates on the seafloor (although there is no way to test this hypothesis in ancient sediments).

Like many complex questions, this one probably not does not have a simple answer. It seems likely that the slow decline in spreading rates was important for the long-term cooling trend in the Eocene, amplified and modified by important changes in oceanic circulation discussed later in this chapter. But it is equally likely that Miocene and later cooling was partially driven by the uplift of the Himalayas, in spite of the increased rate of seafloor spreading, again modified by the changes in oceanic circulation. We will explore this topic further in chapters 6 and 7.

Whatever the ultimate cause, the cooling trend seen in the oxygen isotopes is a reality. Evidence of cooling can be seen everywhere, including the formerly ice-free Antarctic continent. By 49 Ma, there were small glaciers on the Antarctic Peninsula and West Antarctica (Birkenmajer 1987; Birkenmajer et al. 2005). By the late middle Eocene, icebergs were melting in the Pacific sector of the Southern Ocean and dropping ice-rafted sediment, which has been retrieved from deep-sea cores (Wei 1989). However, Antarctica still had not frozen over, because cool-temperate forests of southern beeches, araucarias, podocarps, and abundant ferns were still cloaking the continent through most of the middle and late Eocene (Kemp 1975; Case 1988; Mohr 1990). But major ice caps did develop in the earliest Oligocene, as we shall discuss in chapter 5.

Terminal Eotoadstoolian Event?

How did marine life respond to the long-term middle Eocene cooling trend? The story is a complicated one, with a mixture of signals. According to Boersma et al. (1987), planktonic foraminifera show that the beginning of the middle Eocene (49 Ma) was marked by several degrees of cooling at high latitudes and in bottom waters that flowed from high latitudes all the way to the mid-Atlantic and the Gulf of Mexico. There was little cooling in the near-surface waters, especially in the tropics. By contrast, equatorial surface waters expanded to the poles, as they had in the early Eocene, which triggered equatorial upwelling and ended the period of stagnant warm oceans of the earliest Eocene. At about 47 Ma, the Atlantic became more intensely stratified by temperature. Bottom waters and even temperate surface waters cooled by 3°C, creating a thermal gradient between surface and bottom. Old tropical species persisted along with newly evolved cold-tolerant species, leading to a new high in planktonic foraminiferal diversity. Warm waters bathed the North Sea, allowing Tethyan foraminifera to spread much further north. This vigorous circulation subdivided the ocean into smaller biogeographic provinces, and foraminifera show their highest provincialism at this time. The increased circulation also brought up nutrients trapped in deep waters, leading to large-scale phytoplankton blooms, especially of diatoms.

The most dramatic event, however, occurred at 37–38 Ma, at the end

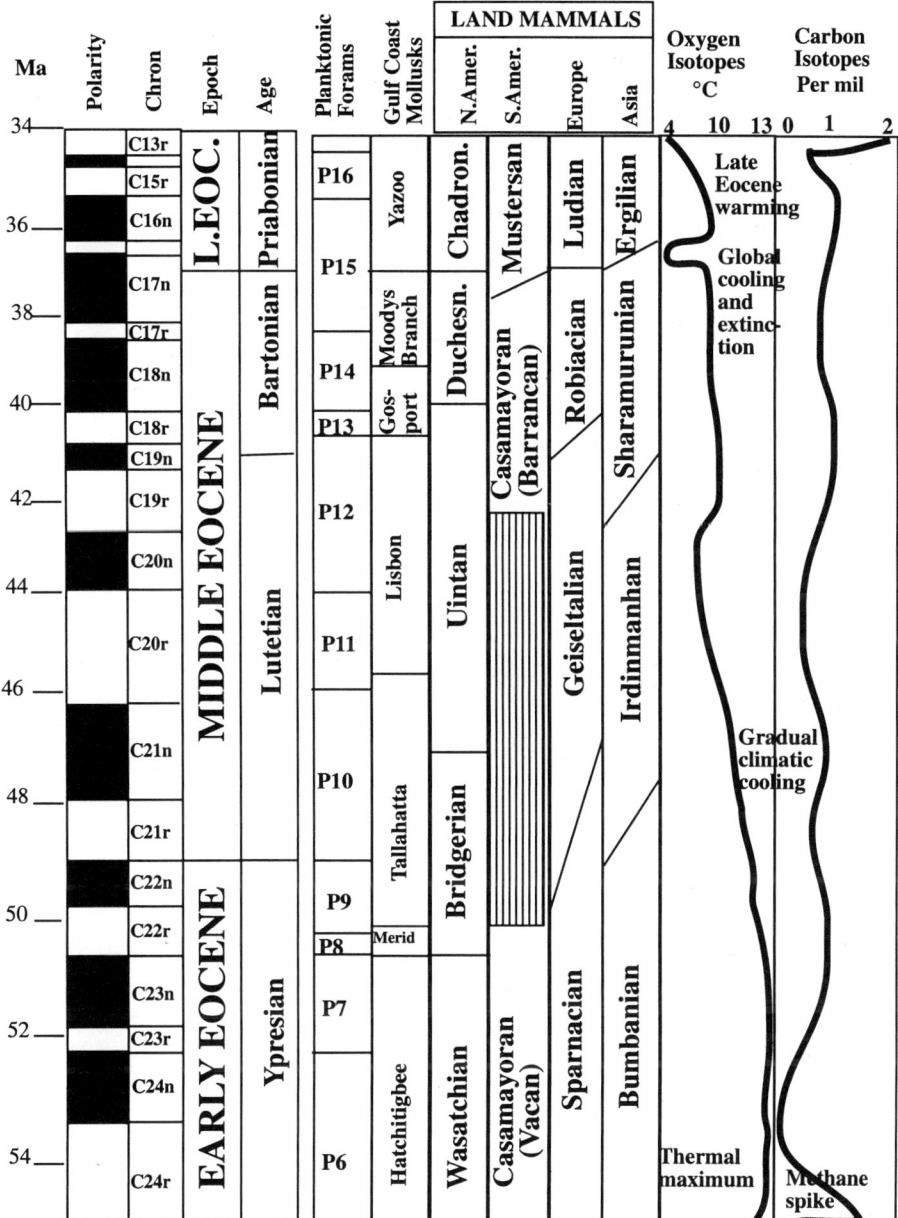

Figure 4.17. The Eocene timescale (left columns), as recently calibrated by Berggren et al. (1995). The North American land mammal record is after Prothero (1995) and Woodburne and Swisher (1995); the South American land mammal ages are after Flynn and Swisher (1995) and Kay et al. (1999); the European record is after Schmidt-Kittler (1997); and the Asian record is after Wang et al. (1998) and Meng and McKenna (1998). Note that the boundaries between the North American land mammal ages are precisely calibrated, so they are drawn as synchronous horizontal lines; whereas those of South America, Asia, and Europe are less well calibrated, so they are drawn as time-transgressive diagonal boundaries. The majority of the South American land mammal record is unknown, as indicated by the vertically ruled pattern. The isotopic records are after Zachos et al. 2001, but they are generalized and smoothed out. Abbreviation: Merid = Meridian.

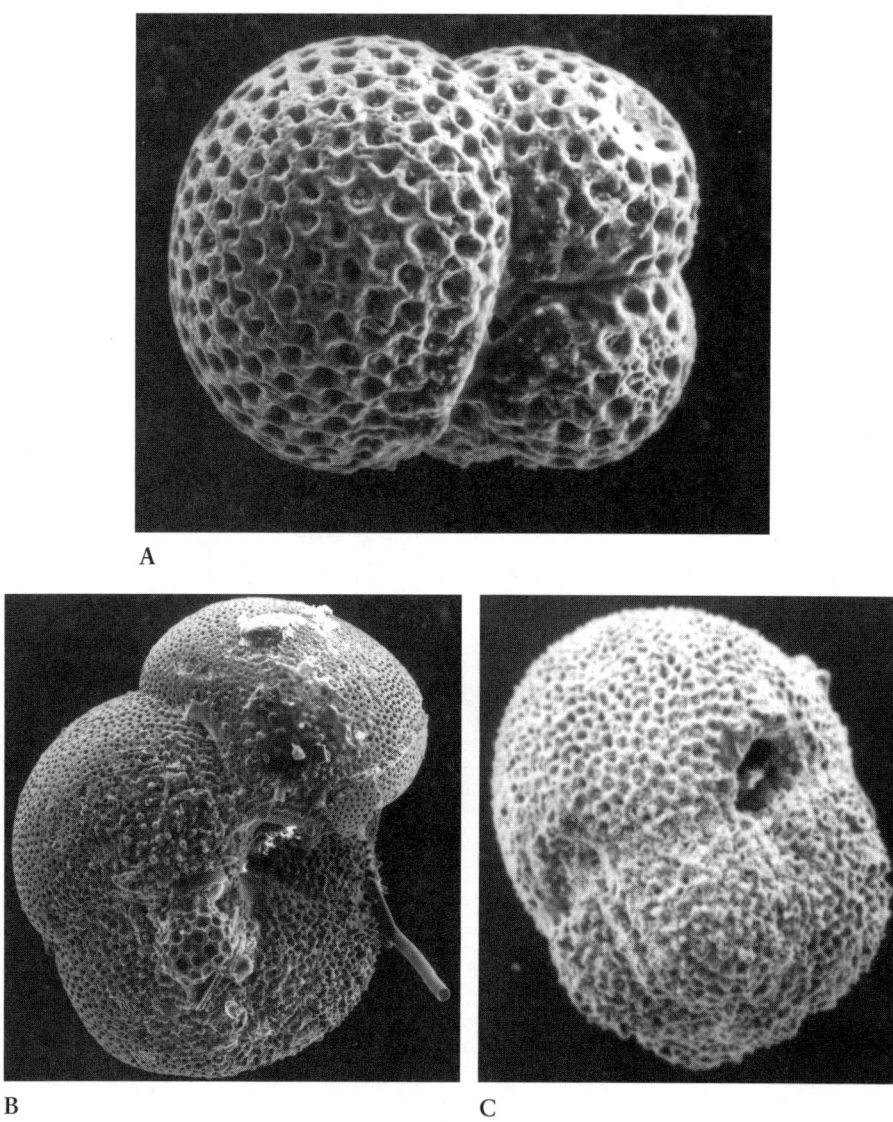

A

B   C

Figure 4.18. Typical planktonic foraminifera of the late Eocene, survivors of the great cooling event at the end of the middle Eocene: (A) Subbotina lineapertura; *(B)* Globigerinatheka semiinvoluta, *zonal indicator of the early late Eocene; and (C)* Turborotalia cerroazulensis, *zonal indicator of the latest Eocene. Photos courtesy of G. Keller.*

of the middle Eocene (fig. 4.17). Bottom waters became thermally decoupled from tropical surface waters, andwarm tropical waters were prevented from mixing with polar waters. The polar waters became trapped in the Antarctic region, increasing the temperature difference between Equator and poles. Although the tropics did not yet cool significantly, the temperate and polar regions did. As a result, there was a significant extinction of warm-water planktonic foraminiferans (fig. 4.18), with more than eighteen species going extinct (Boersma et al. 1987). In particular, the tropical morozovellids and acarininids were the major victims of the terminal middle Eocene event.

The end of the middle Eocene was a bad time for other tropical forms as well. Large nummulitids (fig. 4.10C) disappeared completely from the shrinking Tethys Seaway (now cut in half by the collision of India), and only smaller nummulitids managed to straggle through the end of the Eocene and into the Oligocene. Major extinctions occurred in the benthic foraminifera of the Caribbean (Robinson 2003) and the Gulf Coastal Plain of the United States (Fluegeman 2003). Aubry (1992) found that the tropical planktonic coccolithophorid protists also underwent a major extinction at this time, as the oceans became not only cooler but also more stratified between shallow and deep currents. Diatoms, in contrast, showed a large increase in diversity and abundance at the end of the middle Eocene, reflecting the more vigorous circulation and the increase in nutrients they require, such as silica (Jouse 1978; Fenner 1986; Baldauf 1992).

How about the marine mollusks? Clams and snails from the rich fossil beds of the Gulf Coastal Plain of the United States were hit hard at the beginning of the middle Eocene (faunas of the lower Lisbon and Tallahatta formations), particularly the high-spired snails of the family Turridae (Dockery 1986, 2003; Hansen 1988, 1992). Mollusks then recovered in the later middle Eocene and diversified to even higher levels, reaching their peak at the end of the middle Eocene (upper Lisbon, Cook Mountain, and Gosport formations). The Gosport Sand is particularly famous for its rich shell beds in Alabama and Mississippi (fig. 4.19). But the end of the middle Eocene (boundary between the Gosport and Moodys Branch formations) marked a severe decline in mollusks, with 89% of the snail species and 84% of the clam species dying out. Another lesser extinction occurred in the early late Eocene (between the Moodys Branch and Yazoo formations), which wiped out 72% of the species of snails and 63% of the clam species. Molluscan diversity then stayed low through the rest of the late Eocene (Yazoo Formation). A similar pattern is demonstrated in the Atlantic Coastal Plain of the Carolinas (Campbell and Campbell 2003). On the Pacific Coast of North America, 48% of the snail species went extinct at the end of the middle Eocene (Tejon Stage–Galvinian Stage boundary) according to Squires (2003). Nesbitt (2003) and Hickman (2003) noted a tremendous loss in molluscan species richness, especially in tropical taxa, at the end of the middle Eocene in the Pacific Northwest. Likewise, in Europe the late middle Eocene Bartonian Stage of the Paris Basin contains some 1,300 to 1,400 species of mollusks, but the late Eocene (Priabonian) retains only 800 species (Cavelier 1979; Lozouet 1997).

The other well-studied shelled marine invertebrate group is the echinoids. McKinney et al. (1992) showed that echinoid diversity peaked in the early and early middle Eocene, after which the numbers declined. This decline is particularly apparent at the boundary between the middle and

Figure 4.19. (A) Rich shell beds of the upper middle Eocene Gosport Sand, Tombigbee River, upstream from St. Stephens Quarry, Alabama, represent the last warm-adapted, high-diversity middle Eocene faunas before the cooling of the late Eocene. The spindle-shaped snail shell is Calyptraphorus velatus. *Photo courtesy of L. C. Ivany. (B) Typical assemblage of fossils from the lower middle Eocene Lisbon Group in Alabama and Mississippi (from the upper left in a clockwise spiral):* Flabellum cuneiforme *(solitary coral),* Distortio septemdentata *(snail),* Endopachys maclurii *(solitary coral), lunulitid bryozoan,* Odontaspis *(sand shark tooth),* Chlamys wahtubbeana *(scallop),* Levifusus mortoniopsis *(snail),* Discotrochus orbignianus *(solitary coral),* Calyptraphorus velatus *(snail),* Venericardia rotunda *(clam),* Athleta haleanus *(snail),* Cochlospira engonata *(snail),* Pseudoliva vetusta *(snail),* Fusimitra polita *(snail),* Eosurcula moorei *(snail), and* Neverita sp. *(moon snail). Photo courtesy of L. C. Ivany.*

late Eocene on the Gulf Coastal Plain of North America, especially in Florida (Carter 2003).

Thus, there is strong evidence that the most severe extinction event of the Cenozoic, which wiped out a huge number of tropical and warm-water taxa, and fundamentally rearranged the marine faunas, occurred at the cooling event at the end of the middle Eocene, not at a Terminal Eocene Event as was suggested in the 1980s. This conclusion was reinforced over and over at the 1989 Penrose Conference on the Eocene-Oligocene Transition. Spencer Lucas of the New Mexico Museum of Nature and Science reminded the attendees that they needed to focus not on the alleged importance of the Terminal Eocene Event but instead on the middle–late Eocene extinction event. The most profound faunal event of the Cenozoic was the boundary between the middle and late Eocene, not the Eocene-Oligocene boundary or the Paleogene-Neogene division of the Cenozoic at the Oligocene-Miocene boundary. Lucas whimsically proposed that we call the late Eocene through Holocene the "Toadstoolian" (named after the Eocene-Oligocene badlands of Toadstool Park in northwestern Nebraska). Thus we could call the Paleocene through middle Eocene the "Eotoadstoolian"—allowing the TEE acronym to be retained for the terminal Eotoadstoolian event! Even though this proposal was made in jest, the underlying point is serious. The end of the middle Eocene was one of the most profound faunal turnovers in the Cenozoic history, yet it is usually neglected or misunderstood in the misguided focus on the Eocene-Oligocene boundary (which was a relatively minor event, as we shall soon see).

Middle Eocene Terrestrial Deterioration

The middle Eocene marine and isotopic records show evidence of 12 million years of steady cooling and extinction, terminated by the extinctions at the end of the middle Eocene, at about 37–38 Ma. What do we see in the terrestrial record? Again, the most detailed information comes from North American. Sections in Europe and Asia are less well studied, and we are only beginning to understand the middle Eocene climate in Africa and South America.

The floral record is our best proxy of temperature and climatic change on land. Paleobotanists such as Jack Wolfe (1971, 1978, 1990) use the shapes of leaves as a proxy for temperature (fig. 4.20A). Regardless of the taxonomic group, plants that live in tropical regions produce leaves that have smooth (entire) margins, and are often thick and nondeciduous, with drip tips for shedding rain. Plants that live in cooler, drier regions have jagged margins and are frequently smaller, thinner, and deciduous. Although this distinction seems simplistic, the percentage of entire margins versus jagged margins has proven to be a remarkably reliable paleothermometer (Wolfe 1978), and even when more-complex, multivariate methods of measuring leaf shape (the CLAMP model of Wolfe 1994) were em-

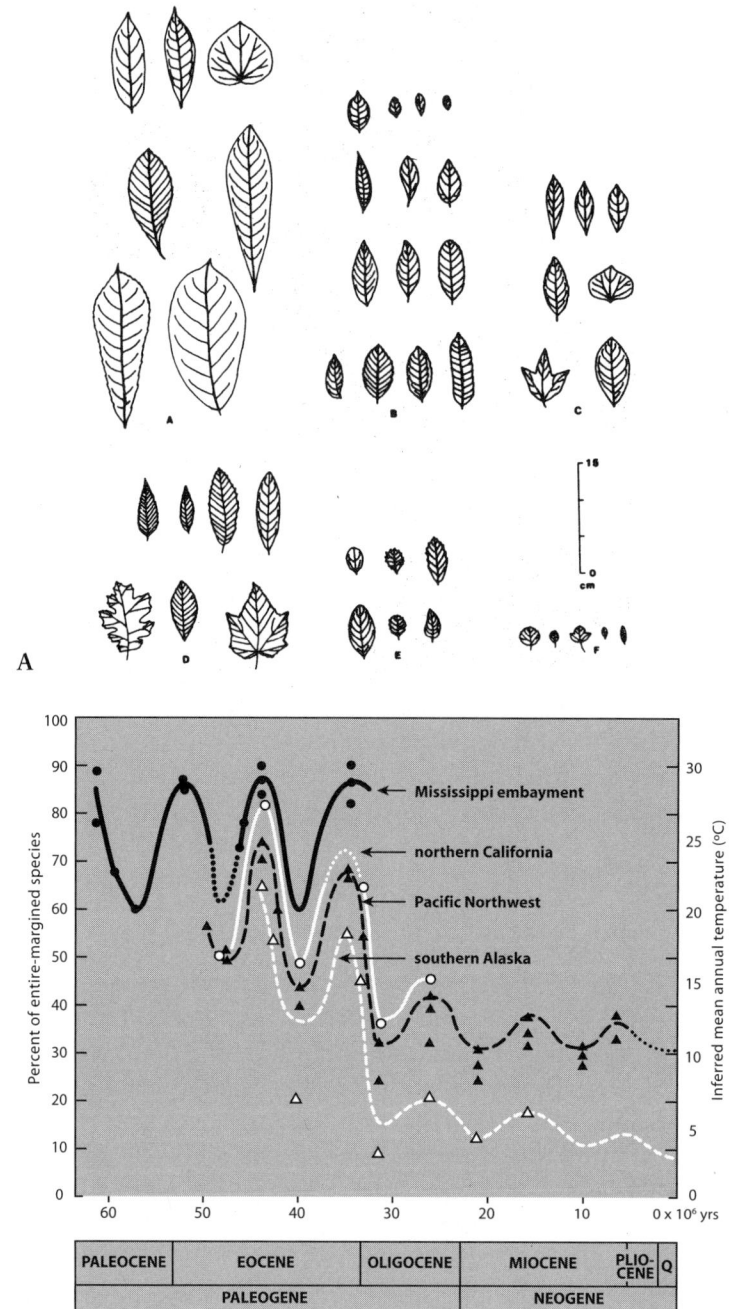

Figure 4.20. (A) The shapes of leaves are highly indicative of climate. The leaves in (a) are all large, entire-margined forms with drip tips typical of the tropical rain forest; the groups in (b) through (f), which have progressively smaller leaves with more and more jagged margins, grew in progressively cooler and drier climates. After Wolfe 1985. Reproduced by permission of American Geophysical Union. (B) The paleotemperature curve of Wolfe (1978) for the North American Cenozoic. Notice that the percentage entire margin species curve on the left is equated with a mean annual temperature curve on the right. The trends are the same, even though the ancient floras come from regions as different as the Gulf Coast (solid black line), California (upper white line), the Pacific Northwest (dashed black line), and Alaska (dashed white line). There is a big drop in temperature at the end of the middle Eocene, but the biggest drop occurs in the earliest Oligocene. After Wolfe 1978.

ployed, the results were the same. In every case, the fossil leaves of North America show that the Paleocene and Eocene were warm periods (as discussed already) and were followed by a cooling in the middle Eocene (fig. 4.20B), a short-term warming event in the late Eocene, and then a huge decline in the earliest Oligocene (what Wolfe 1978, originally called the Terminal Eocene Event and which has now be dated as earliest Oligocene). Both the middle Eocene and early Oligocene cooling events were found in floras from Alaska to the Gulf Coast of the United States, so the phenomenon occurs in all temperate and subpolar latitudes. The tropics would not be expected to cool this much, however. Each cooling episode saw a mean annual temperature decline from 30°C (86°F) to 21°C (70°F) on the Gulf Coast, and from 24°C (75°C) to 17°C (63°F) in northern California. These temperature swings of 7 to 11°C (12 to 16°F) in continental North America are much more extreme than the 4 to 5°C (7 to 9°F) changes in the oceans at the same time. Another difference is that the land floras rebounded with a warming event in the late Eocene, with mean annual temperatures as warm as 30°C, whereas no such rebound is seen in the marine isotope record (figs. 3.22, 4.17). This late Eocene rebound, as well as the overall shape and trend of the Eocene leaf margin–temperature curve of Wolfe (1978), has been corroborated by more-recent analyses (Smith et al. 1998; Myers 2003).

How are these floral fluctuations reflected in the plants themselves? Early middle Eocene floras show that seasonally dry climates were already occurring over much of North America, because the fossil plants include abundant deciduous taxa, including members of the legume, laurel, oak, and walnut families (Wolfe 1986). Early Eocene vegetation had been moist and tropical over most of North America, but early middle Eocene floras in the Rocky Mountains show evidence of cooling and drying, partially due to the uplift of the Rockies in the Laramide Orogeny (Wing 1987). Floras from the famous Green River lake beds of the Wyoming-Colorado-Utah border have a mixture of both deciduous open woodland forests and some subtropical to paratropical forests, with abundant palm trees (MacGinitie 1969; Wolfe 1985). In the northeastern corner of Yellowstone National Park are the spectacular fossil tree stumps of Specimen Ridge. Repeated volcanic mudflows buried these trees in upright position, with at least twenty-seven successive layers, each trapping another episode of forest growth (Dorf 1960). The fossil plants include a mixture of subtropical trees, such as figs, avocados, and bay laurels, along with warm-temperate magnolias, chestnuts, sycamores, walnuts, and many conifers, including pines and sequoias (Wing 1987). Thus, most of the middle Eocene floras contain mixtures of paratropical to temperate plants, depending upon location and how late they are in the cooling trend.

The late middle Eocene cooling event is particularly well preserved in the volcanic sediments of the Clarno Formation in central Oregon, and in numerous floras in Washington and Alaska (Wolfe 1971). These floras still contain elements of the paratropical rain forests, dominated by broad-leaved evergreen plants, with many drip tips. There were lush forests with lianas, palm trees, and many tropical tree families, such as the pawpaws (fig. 3.20), dipterocarps, and icacina and moonseed vines. According to Wolfe (1978, 1990), these late middle Eocene paratropical forests flourished at mean annual temperatures of 20–25°C (68–77°F), and seasonal shifts in temperature spanned only 5°C (9°F). By contrast, the late Eocene middle Clarno flora is a mixture of broad-leaved evergreen, deciduous, and coniferous species, including cycads, pines, sequoias, sumacs, poison ivy, birches, sycamores, laurels, plums, blackberries, and wild grapes (Wolfe 1971). According to Wolfe (1978), these plants grew in a climate with a mean annual temperature of 15°C (59°F), and the seasonal range of temperatures had increased to 7°C (11°F). The same trend can be seen in the Puget Group of Washington, and in the late middle Eocene floras of the Gulf of Alaska (Wolfe 1971, 1978). The soils, pollen, and floras from the San Diego area show that tropical palms, elms, and walnuts were replaced by sycamores, hickories, and abundant shrubs, herbs, and grasses, growing in a seasonally semi-arid climate with flash floods and thick caliche horizons in the soils, indicating long periods of dry climate (Peterson and Abbott 1979; Frederiksen 1991).

What kinds of mammals inhabited these changing forests of the middle Eocene? In North America, three land mammal ages (fig. 4.17) span the 12 million years of the middle Eocene: the Bridgerian (49–47 Ma), based on faunas from the Bridger Basin of Wyoming; the Uintan (47–40 Ma), based on faunas from the Uinta Basin of Utah; and the Duchesnean (40–37 Ma), based on faunas of the Duchesne River Formation, which overlies the Uinta Formation in the Uinta Basin (fig. 4.21). Mammals from these land mammal ages are also well represented in many other regions of North America, from Oregon to San Diego, and even in the Texas Gulf Coast (Robinson et al. 2004). These beds, which have been collected for more than a century, yield some of the most unusual and spectacular mammal fossils in North America.

First to catch one's eye in the Bridgerian and Uintan landscape would have been the huge uintatheres (fig. 4.22A), elephant-sized beasts that sported three pairs of bony knobs on the top of their skulls, and large canine tusks, but had remarkably undersized, simple molars with V-shaped crests. In chapter 3, we saw that the earliest uintatheres were known from the Paleocene of Asia, and by the late Paleocene they had reached North America too. What uintatheres were related to is still a mystery. McKenna and Manning (1977) suggested that they were related to the ungulates, or hoofed mammals (they did have large bodies and hooves, but not many other features that define the ungulates). Tong and Lucas (1982) have suggested that their teeth look like those of the rabbit-like anagalids of Asia, so they are essentially giant bunnies. Whatever they were related to, they were at their peak in the Bridgerian and Uintan, the largest land mammals in North America at the time. They remained common in Asia too, with

A

B

C

Figure 4.21. Typical outcrops of middle Eocene terrestrial beds. (A) The finely bedded lake shales of the Green River Formation, famous for its fossils of fish and many other organisms, near Hell's Hole Canyon, Utah. Photo by the author. (B) Exposures of the early middle Eocene Bridger Formation in the Bridger Basin, Wyoming, type area of the Bridgerian land mammal age. Photo courtesy of R. K. Stucky. (C) The sandstones and shales of the Uinta and Duchesne River formations in the Uinta Basin in northeast Utah, type area for the Uintan and Duchesnean land mammal ages. Photo by the author.

A

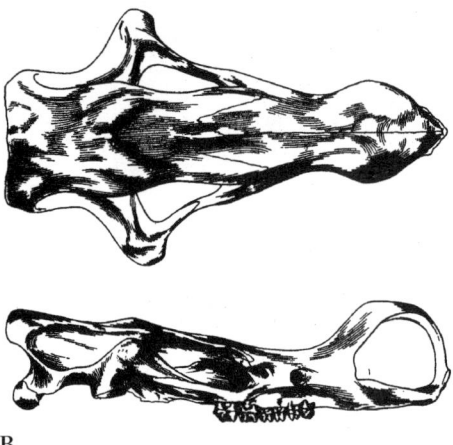

Figure 4.22. (A) Restoration of the horned Uintatherium, *the largest land mammal in the Bridgerian and Uintan of North America. It had six knobby horns on its skull, and huge tusks. From Scott 1913. (B) The middle Eocene Mongolian uintathere* Gobiatherium, *which had a bulbous snout instead of horns. From Osborn and Granger 1932.*

B

bizarre beasts such as *Gobiatherium,* which apparently had a huge bulbous inflated snout supported by bone (fig. 4.22B).

The next largest animals in the middle Eocene were a group of perissodactyls known as brontotheres, or titanotheres (fig. 4.23). In the middle Eocene, they resembled large tapirs and reached rhino size. By the late middle Eocene and late Eocene, they had grown to the size of elephants and had blunt bony paired battering-ram horns on their noses (fig. 1.1). In Asia, the brontotheres were common also and culminated in the embolotheres, which had fused the paired horns into a single broad battering

Brontotherium platyceras

Brontotherium leidyi

Manteoceras manteoceras

Dolichorhinus hyognathus

Mesatirhinus petersoni

Palaeosyops leidyi

Lambdotherium popoagicum *Eotitanops princeps* *Eotitanops gregoryi*

Figure 4.23. Evolution of the brontotheres from dog-sized beasts without horns to elephant-sized beasts with huge paired bony battering rams on their noses. This old diagram is now outdated: the last brontotheres are known to be late Eocene, and none survived into the Oligocene. From Osborn 1929.

ram. These animals apparently used the horns for sparring with other brontotheres (one specimen actually has a cracked rib that healed, probably broken by the butting of another brontothere) and for self-defense. They had low-crowned, simple teeth, even after they got to be huge, so they probably ate leaves and soft vegetation.

In addition to the brontotheres, other perissodactyls were diversifying in the middle Eocene as well. Early horses were represented by the beagle-sized *Orohippus* and *Epihippus*. These horses were still small, primitive, and three-toed like their ancestors. Tapirs were also diverse, with many forms, some of which began to show the signs of the proboscis or snout that defines their modern descendants. The earliest members of the rhino lineage also are common, although to the untrained eye, they look nothing like the living rhino. They had no horns and were about the size of a Great Dane, with slender, long legs and simple teeth for browsing on leaves.

Artiodactyls, too, diversified rapidly after their early Eocene spread around the world. From the tiny deer-like dichobunids (fig. 4.4), they evolved into more than a dozen families, most of which look like small deer to the untrained eye. However, some were more pig-like, including the early members of the entelodonts and achaenodonts (fig. 5.11). By the late Uintan, the first camels had appeared, although they too looked deer-like, with no sign of a hump yet (in fact, most fossil camels lacked humps).

All of these advanced large herbivores (uintatheres, perissodactyls, and artiodactyls) rapidly crowded out earlier occupants of the large herbivore niche. The archaic hoofed mammals that had dominated the Paleocene and early Eocene mostly disappeared. Only the dachshund-like hyopsodonts and the horse-like phenacodonts (fig. 4.24A, B) survived into the Bridgerian, and both were gone by the Duchesnean. The last of the pantodonts, *Coryphodon* (pl. 2), was already gone by the Bridgerian, and the rare taeniodonts and tillodonts both vanished by the end of the Uintan. Like the large herbivores, the large carnivorous vertebrate fauna was changing too. The bear-like hoofed mesonychids, source of the whales in the early Eocene, were already scarce by the middle Eocene. The last of the giant predatory *Diatryma* birds (fig. 3.18) vanished at the end of the Bridgerian. The archaic oxyaenid creodonts (fig. 3.16) were also declining and gradually being replaced by the true carnivorans, including the first members of the dog family.

Small mammals also showed dramatic changes. At the beginning of the Eocene, the trees were inhabited by the egg-laying rodent-like multituberculates and the archaic rodent-like plesiadapid primates. But the immigrant rodents quickly crowded out the multituberculates in the middle Eocene, so none is known from the Bridgerian, only two genera are known in the Uintan, and the last ones straggled on to the end of the Eocene. The plesiadapid primates were also displaced by the radiation of rodents, as well as by the two new immigrant groups of higher primates, the adapids and omomyids, closely related to modern lemurs and tarsiers. By the Uintan, rabbits had immigrated from Asia as well. Insectivorous mammals continued to change rapidly, with some groups evolving into anteater-like forms, presumably to feed on the abundant ant and termite nests in the middle Eocene forests.

Although mammals dominated the landscape, reptiles were still common in the Bridgerian as well, including many species of crocodilians (fig. 4.25A), and soft-shelled turtles. However, both groups began to disappear by the Uintan, suggesting that the climate was becoming colder and drier (Hutchison 1982). The famous Green River lake shales (fig. 4.21A) produce an incredible variety of freshwater fishes that lived in the middle Eocene, in-

A

B

Figure 4.24. (A) Reconstruction of the dachshund-shaped Hyopsodus, *one of the last of the archaic hoofed mammals, yet still common in the early middle Eocene of North America. After Gazin 1968. (B) The archaic hoofed mammal* Phenacodus, *common in the early Eocene and possibly related to horses and other perissodactyls. Painting by B. Horsfall, in Scott 1913.*

cluding skates, sturgeons, garfish, bowfins, mooneyes, herrings, suckers, catfish, trout, and perch (Grande 1980). These beautifully preserved fossil fishes are so common that they are commercially mined and sold in rock shops around the world. The Green River shales also preserve many other forms of life found in a freshwater lake setting, including frogs, lizards, boas, turtles, and crocodiles (fig. 4.25B), plus victims that drowned in the lake, such as a perfectly preserved bat and numerous kinds of birds.

In summary, the North American land vertebrate fauna shows several striking trends in the middle Eocene. Diversity in the Bridgerian was just as

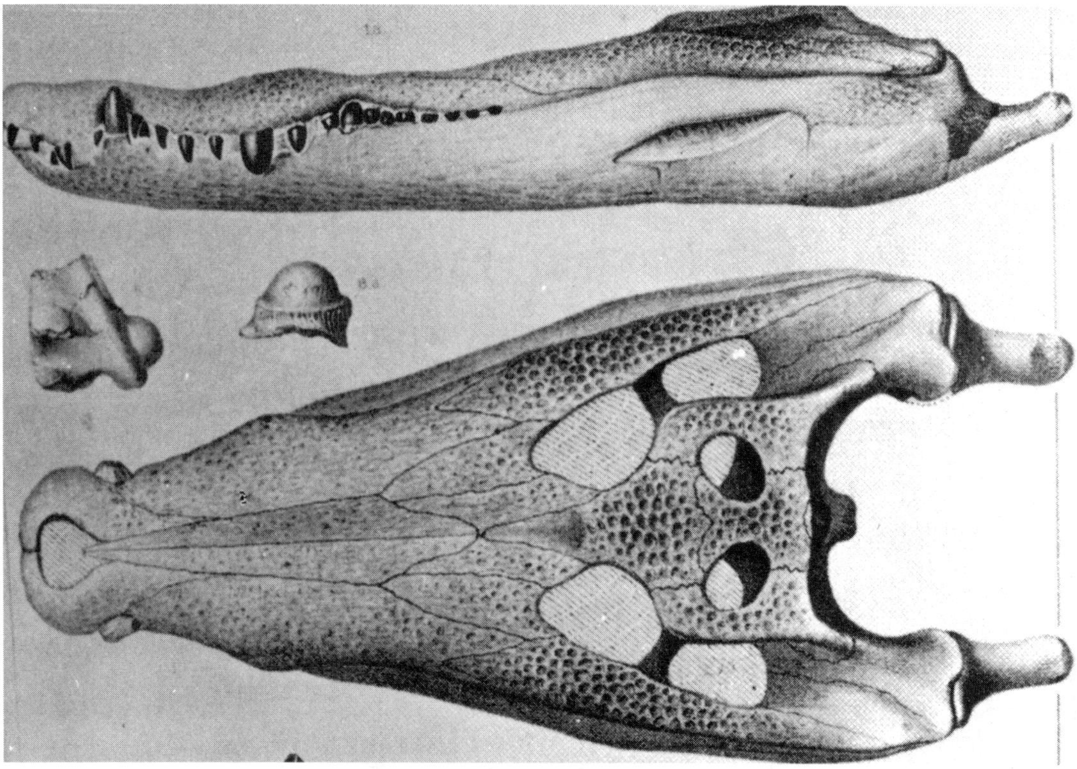

A

B

Figure 4.25. (A) Crocodile fossils are common in the swampy floodplain and river deposits of the middle Eocene Bridger Formation. This illustration is from "Cope's Bible." Cope 1884. (B) A crocodile skeleton from the middle Eocene Green River Formation of Wyoming. Photo courtesy of L. Grande.

high as in the early Eocene but declined rapidly in the Uintan and Duchesnean, so that 80% of the genera that were present in the Uintan were extinct by the end of the Duchesnean (Stucky 1990). Primitive groups, such as the pantodonts, tillodonts, taeniodonts, archaic hoofed mammals, multituberculates, and plesiadapid primates, were almost completely replaced

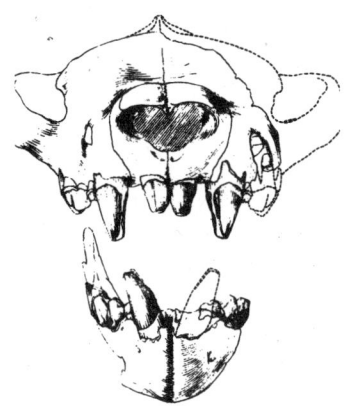

Figure 4.26. The bear-sized, bone-crushing creodont Sarkastodon, *from the middle Eocene of Mongolia. It was one of the largest carnivorous animals of its time. From Granger 1938.*

by more-advanced groups, such as perissodactyls, artiodactyls, rodents, rabbits, and advanced lemur-like primates. The replacement of arboreal small mammals with more ground-dwelling small mammals, and primitive herbivores with more-advanced leaf eaters, shows that the jungles of the early Eocene had dried up and been replaced by mixed forests with limited room for arboreal species like primates. The thick undergrowth of soft jungle plants that were eaten by the early large herbivores was replaced by a less dense scrubby brush that required more specialized teeth. The gradual disappearance of crocodilians and soft-shelled turtles confirms the cooling and drying trend of the middle Eocene.

Asian land mammals show many similarities to those of North America, but also some significant differences. Clearly, the free exchange of mammals of the early Eocene was over, and only limited interchange could take place via the Bering land bridge. In China, the anagalids, which had already given rise to the rabbits and rodents, persisted through much of the middle Eocene. Many other groups that had died out elsewhere lasted well into the middle Eocene of Asia, including the pantodonts (the huge *Hypercoryphodon* was a rhino-sized beast, the last of its kind), didymoconids, and the bear-sized scavenging oxyaenid creodont, *Sarkastodon* (fig. 4.26). The last of the uintatheres, the bulbous-nosed *Gobiatherium,* also occurs in the middle Eocene of Mongolia (fig. 4.22B). The predatory hoofed mesonychids survived in Asia long after they had died out elsewhere, culminating in the huge *Andrewsarchus* (fig. 4.27), which had an immense skull almost a meter long. This is the largest predatory land mammal the world had ever known, more than twice the size of the largest bear that has ever lived. The skeleton of this beast is unknown, but if it is reconstructed with wolf-like or bear-like proportions, it was about 4 meters long and more than 2 meters high at the shoulder!

In addition to all these early Eocene relict groups, Asia during the middle Eocene also saw more-advanced groups, including some of the earliest relatives of monkeys and apes, or anthropoid primates (Beard 2004). Horses were rare, but the perissodactyls included a great diversity of

Figure 4.27. The huge carnivorous-scavenging mesonychid Andrewsarchus *from the middle Eocene of Mongolia. Much larger than the largest modern carnivore (the Kodiak bear), it is known only from a fossil skull, so its body may have been more like that of its whale relatives than like the body of a bear or wolf, as shown here. Painting by Z. Burian.*

brontotheres and tapiroids, as well as advanced rhino groups such as the hippo-like amynodonts and the long-legged running hyracodont rhinos. There were a variety of early artiodactyls, including the pig-like entelodonts, the deer-like leptomerycids, and the aquatic anthracotheres (all discussed in the next chapter).

Cuvier's "Ancient Beast" and Plaster of Paris

The hilly region north of Paris is known as Montmartre ("martyr's mountain"). It is supposedly the site where St. Denis, in the third century A.D., was beheaded and then picked up his head and climbed to the summit helped by an angel. Montmartre has long been famous for its sidewalk cafes, artists, bohemian atmosphere, and nightclubs such as the legendary Moulin Rouge (home of the cancan and Toulouse-Lautrec). At the crest of the hill is the shiny white cathedral of Notre Dame du Sacré-Coeur, which was built to counteract the influence of all the nightclubs and sin in the streets. In the 1790s, Montmartre was an isolated hill well north of the city, covered with vineyards and pockmarked with mineshafts. Since Roman times, workers had known that underlying Montmartre were extensive deposits of gypsum (hydrous calcium sulfate), deposited in ancient lagoons that dried up during the later part of the Eocene. These workers had discovered that gypsum, when crushed and heated to 300°C, became a powder that set up quickly when water was added, forming an excellent building material. We now call this material plaster of Paris because it was first mined extensively in the northern region of Paris.

After the chaos of the French Revolution, construction of new buildings was proceeding at a torrid pace during the beginning of Napoleon's reign the late 1790s, and the gypsum mines were expanded to produce more plaster of Paris. Deep in one of the mines, workers found a nearly complete skeleton of a strange beast, which was promptly sent to the Baron Georges Cuvier to study. Cuvier already had the reputation of being one of the most brilliant men in French science, for he had founded the sciences of comparative anatomy and vertebrate paleontology. His prestige was such that he survived the French Revolution and Reign of Terror, Napoleon's wars, and the reigns of subsequent French kings without loss of power or position. He was particularly famous for his descriptions of the giant mammoths and mastodonts that had been sent to Paris, and for his description of the great lizard from the Meuse River of Belgium (now known as the mosasaur), captured by Napoleon's troops when they invaded the region. When Cuvier saw the skeleton from Montmartre, he noticed its similarities to the skeletons of the modern tapir and rhinoceros, but also the great differences. In 1804, he described it as *Palaeotherium* ("ancient beast") and noted that there was nothing alive today that closely resembled it (fig. 4.28). Soon Cuvier was describing fossils of ancient opossums from the same deposits, and strange even-toed hoofed mammals he called *Anoplotherium*.

All of these strange skeletons, none of which closely resembled living animals, led Cuvier to an inescapable conclusion: they were the remains of animals that were now extinct. At that time, the idea of extinction was an anathema, even to well-educated and less dogmatic gentlemen such as Thomas Jefferson. People still believed in the idea of God's Providence, and that God watched over even the tiny sparrow. Surely he would not allow any of his creations to die out? Jefferson had described some large fossil claws as those of a giant lion, and he directed Lewis and Clark to look for the beast during their western expeditions (we now know that they are claws of the extinct ground sloth *Megalonyx jeffersoni*). But Cuvier could not ignore the evidence of the *Palaeotherium,* and of the mastodont and mammoth fossils. These were somewhat like modern elephants, but also different. Clearly, they were so large that if they still lived today, they would have been discovered by the many expeditions that had occurred before 1800. In spite of the influence of the church in France, Cuvier boldly proclaimed that extinction was real, implying that God had allowed some of his creations to vanish.

Moreover, Cuvier noticed that the fossils from the Paris Basin revealed another pattern: each layer produced different kinds of fossils. Working

Figure 4.28. The tapir-like lophiodonts (left foreground) and horse-like palaeotheres (right foreground). The European archipelago in the middle Eocene was inhabited by many strange and endemic mammals, such as these, as well as by more familiar crocodiles and turtles. From Parley 1837.

with the fossil mollusk expert Alexandre Brongniart, Cuvier found that one could recognize the different layers of the Paris Basin by their distinctive fossils (fig. 4.29). The *gypsiferes de Montmartre* produced *Palaeotherium, Anoplotherium,* and other extinct mammals, whereas the layers of marine rocks below (today known as the type area of the middle Eocene Lutetian Stage, after Lutetia, the old Roman name for Paris) produced entirely different mollusks. By 1805, Cuvier and Brongniart were publishing the idea of faunal succession, although the concept had originated in England a few years earlier with the work of William Smith. To Cuvier, the differences between the ice age mammals in the upper layers and *Palaeotherium* from the Montmartre gypsum showed that "the older the beds in which these bones are found, the more they differ from those of the animals we know today" (Cuvier 1804). Cuvier, however, was not ready to suggest that animals had evolved through time, as his colleague Lamarck was saying. Instead, he interpreted these strange extinct animals as victims of a series of floods from the dark antediluvian period before Noah's flood in the Bible—thus avoiding conflict with the powerful theologians of his time.

Cuvier's *Palaeotherium* and *Anoplotherium* and other Montmartre fossils were among the first pre–ice age mammals described anywhere. As other fossil mammals were discovered in the 1800s, the peculiarity of the

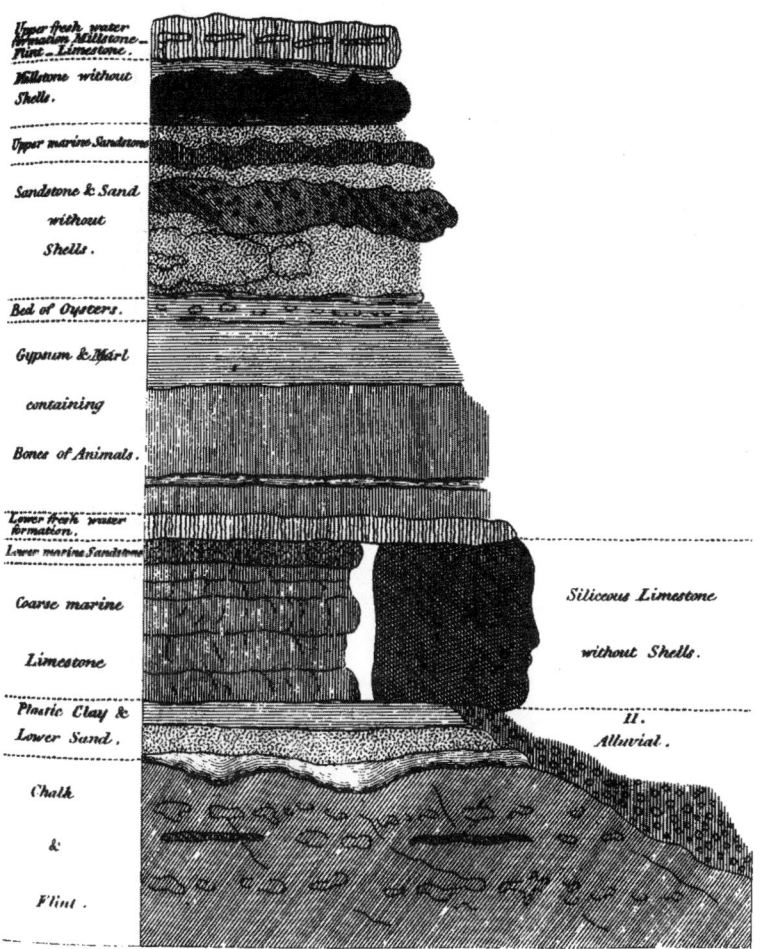

Upper fresh water formation Millstone Flint — Limestone.

Millstone without Shells.

Upper marine Sandstone

Sandstone & Sand without Shells.

Bed of Oysters.

Gypsum & Marl

containing

Bones of Animals.

Lower fresh water formation.

Lower marine Sandstone

Coarse marine

Limestone

Plastic Clay & Lower Sand.

Chalk

&

Flint.

Siliceous Limestone

without Shells.

II.

Alluvial.

Figure 4.29. Cuvier and Brongniart's stratigraphic profile of the Paris Basin section. The "coarse marine limestone" is the type section for the middle Eocene Lutetian Stage (named after the old Roman name for Paris, Lutetia). The gypsum and marl above it is upper middle Eocene and produced Cuvier's Palaeotherium. *The peculiar profile on the lower right is clearly that of Napoleon Bonaparte. From Cuvier 1818.*

European late Eocene mammals became more apparent. Although there is considerably faunal similarity between Asia and North America during the middle and late Eocene, Europe became isolated and its faunas more endemic as the early Eocene land bridges disappeared. High sea levels meant that much of the region was under water, forming an archipelago of subtropical islands (fig. 4.8). Floral analysis shows that the warm (23°C, or 73°F) average annual temperatures of the early Eocene cooled to only 16°C (61°F) at the beginning of the middle Eocene, and down to only 13°C (55°F) at the end of the middle Eocene (Boulter 1984; Collinson et al. 1981). This trend is similar to that observed by Wolfe

(1978) in North America, only the cooling and drying were not so severe, because of the effects of the warm seas bathing the isolated islands. Living on these forested islands was a mammal fauna that became highly endemic in the middle Eocene, after having moved so freely around to other continents in the early Eocene. Instead of horses, there were horse-like animals that Cuvier called palaeotheres (figs. 4.28, 4.30B). Today we know that they also include the British *Hyracotherium*, once incorrectly thought to be the first horse (Hooker 1989). There were also tapir-like perissodactyls known as lophiodonts—but no brontotheres or rhinos, which were found on the other continents. The artiodactyls (fig. 4.30C) were even more endemic, with nearly every family found only in Europe. There were some similarities with other regions, such as the endemic lemur-like primates, and the hyaenodont creodonts, which were found on the other northern continents in the late middle Eocene, and a huge diversification of rodents (all of which are endemic European families, however).

The most famous locality of this age in Europe is the legendary Messel site in western Germany, about 30 kilometers (20 miles) southwest of Frankfurt (fig. 4.30A). It was once a large open-pit oil-shale mine that was then abandoned and was supposed become a sanitary landfill site. Fortunately, an international scientific effort saved it, and the study of its amazing fossils intensified. Messel is an example of a *lagerstatt*, or "motherlode," of exceptional fossils that have not only the hard parts but even the soft tissues preserved. Most specimens are found as complete articulated skeletons in death poses, complete with the outlines of the soft tissues preserved as a black film on the surface of the shale (figs. 4.30B, C). In one fossil frog, the eyes, liver, and veins can be distinguished. Fossil birds show the details of their plumage; fossil bats (fig. 4.30D) show their wing membranes; and even the iridescent color patterns of beetle wing covers are still preserved! In some cases, the stomach contents of some animals are preserved as well. One kind of bat, for example, dined exclusively on butterflies. The horse-like palaeothere *Propalaeotherium* (fig. 4.30B) preserves a mass of clearly identifiable leaves and fruit seeds in its stomach. Another *Propalaeotherium* specimen shows tiny embryonic bones with baby teeth, so it was a pregnant mare with an unborn foal.

How did these extraordinary preservation conditions occur? The Messel sediments are muds (turned into shales) formed in a small (a few square kilometers) but deep lake trapped in a fault valley. The lake bottom had no circulation or overturning of the water, so it was stagnant and without oxygen, preventing scavenging fish and invertebrates from living there. In this warm tropical climate, frequent algal blooms further depleted the oxygen at the bottom. When animal carcasses sank to the bottom, they were buried before they could decay or be disturbed by currents or scavengers. The only decomposers were the bacteria adapted to oxygen-poor environments. In fact, the mysterious dark body outlines of these fossils are made entirely of the mineralized remains of these bacteria.

If European land faunas were endemic, those of the other continents

A

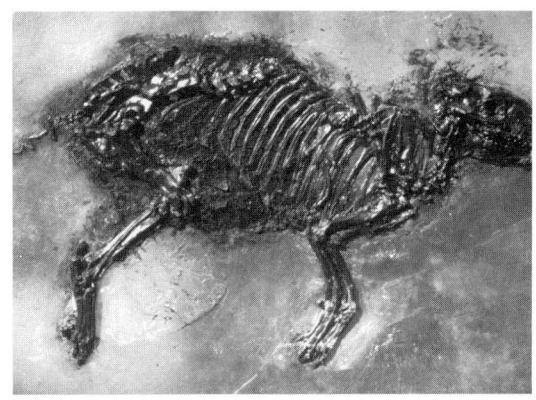

B

C

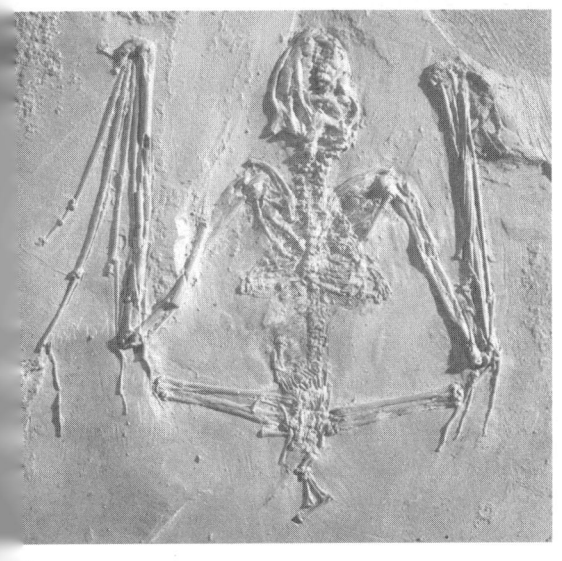

D

Figure 4.30. (A) An abandoned oil shale quarry that exposes the ancient lake beds of Messel, Germany, which preserve an extraordinary middle Eocene fauna. (B) A horse-like form, Propalaeotherium, *that preserves even the stomach contents of its last meal, as well as flesh and fur impressions. (C) The primitive artiodactyl* Messelobunodon *also showing the soft tissue preservation so typical of Messel. (D) The bat* Archaeonycteris trigodon, *which is a primitive form but already had the fully developed wings of a modern bat. Photos courtesy of the Senckenberg Institut Frankfurt.*

were even more distinctive because of even longer isolation. Africa started out with a Paleocene fauna that included endemic groups like early mastodonts (fig. 4.16) and insectivores, plus the ubiquitous hyaenodont creodonts (Gheerbrant et al. 1996); but by the middle and late Eocene, it had a highly endemic fauna with elements found nowhere else: mastodonts of many kinds, huge elephant-sized beasts known as arsinoitheres (see chapter 5), with giant paired spiked horns on their noses, plus giant hyraxes, elephant shrews, and the first relatives of Old World monkeys and apes. These are all known from North Africa (especially the Fayûm area of Egypt).

The Eocene of Australia is still poorly known, with only the ?early Eocene Tingamarra fauna (mostly primitive marsupials and one possible placental tooth) known so far (Godthelp et al. 1992).

Doubthouse World

The final 3 million years of the Eocene are known as the Priabonian Age on the global timescale, and they represent an interesting, poorly understood combination of conditions. The oxygen isotope records (figs. 3.22, 4.17) show a slight warming event of about 2°C by the middle of the late Eocene, temporarily reversing the long-term cooling trend of the previous 12 million years of the middle Eocene (Miller et al. 1987; Zachos et al. 1993, 1994, 2001). The globe had not yet dropped into full icehouse refrigeration, yet it was clearly not a greenhouse world either. Miller (1992) called this the "doubthouse world," because we have no simple greenhouse or icehouse models to explain the conditions of the globe. This slight warming event after the crash of the tropical forms at the end of the middle Eocene had only limited effects. Temperature-sensitive coccolithophorids show a slight recovery in the late Eocene (Aubry 1992), but most other marine organisms were unresponsive. There was no significant change in the planktonic or benthic foraminiferal faunas (Boersma et al. 1987; E. Thomas 1992), or in the mollusks of the Gulf Coast (Hansen 1988, 1992) or Pacific Coast (Squires 2003; Nesbitt 2003; Hickman 2003). The record of echinoids in Florida and the Gulf of Mexico region shows a slight increase in diversity in the late Priabonian after the crash at the end of the middle Eocene (Carter 2003).

However, the change in land floras was more dramatic. According to Wolfe (1978), mean annual temperature in the late Eocene of the Gulf Coast increased (fig. 4.20B) from 20°C (68°F) to 28°C (82°F), and in the Gulf of Alaska it increased from 12°C (54°F) to 18°C (64°F). This dramatic 6–8°C (10–14°F) warming was manifested in the floras. For example, in the Clarno Formation of central Oregon, the broad-leaved evergreen flora of the middle Clarno Formation (representing the cool late middle Eocene) is overlain by the paratropical late Eocene upper Clarno flora (Wolfe 1971). The same trend can be seen in the floras of the Puget Group in western Washington, and in Alaska as well. The floras of the

Rocky Mountains, however, do not show this dramatic fluctuation, probably because of the effects of altitude (Wing 1987). In the Big Badlands of South Dakota, by contrast, ancient soil horizons of the latest middle Eocene suggest dense woodland with over 100 centimeters of annual precipitation, whereas those of the upper Eocene Chadron Formation were produced by dry woodlands with open patches of brush (Retallack 1983, 1992).

What sort of animals lived in these final vestiges of forests of the doubthouse world? The best record comes from the upper Eocene Chadron Formation in the Big Badlands of South Dakota and adjacent Wyoming, Nebraska, and Colorado. The dry scrubby woodlands were dominated by the last of the huge brontotheres (fig. 1.1), which reached the culmination of their size and horn development. The rest of the mammals were early members of what has been called the White River Chronofauna, that is, the typical mammals of the White River Formation badlands. Most of the White River Chronofauna had immigrated to North America from Asia in the Duchesnean, but by the Chadronian they had become dominant. Among the perissodactyls, they included the huge brontotheres, plus the three-toed horse *Mesohippus* (fig. 5.12A) and three kinds of rhinos: the hippo-like amynodonts, the long-legged running hyracodonts, and the first members of the living family Rhinocerotidae (fig. 5.12B–D). Artiodactyls were also diverse, with abundant oreodonts (fig. 1.2), pig-like entelodonts and anthracotheres, the first peccaries (pig-like creatures known as javelinas in Latin America today), camels without humps, and deer-like leptomerycids. The small mammals included abundant rabbits, plus many kinds of rodents, including the first true squirrels, beavers, pocket gophers, and pocket mice. The predators included some of the last of the hyaenodont creodonts, along with true dogs, weasels, and a cat-like group of carnivorans known as nimravids. Some of these nimravids resembled sabertoothed cats, except that they are not true cats (despite decades of misconceptions), but may be related to dogs. Only a few rare relicts, like the last of the North American primates and multituberculates, remained from the early Eocene tropical fauna.

Late Eocene (Ulangochuian-Ergilian) mammals in Asia had some striking similarities to those in North America, and some important differences as well. As already mentioned, there were huge embolotheres, brontotheres with a single broad battering-ram horn. Other perissodactyls were at their peak in Asia in the late Eocene, including many archaic tapiroids, hippo-like amynodont rhinos (including the huge *Gigantamynodon*), a variety of long-legged hyracodont rhinos (some of which reached elephantine size), and primitive members of the living family Rhinocerotidae. Pig-like entelodonts and anthracotheres were also common, along with a diversity of deer-like artiodactyls. Rodents and rabbits (mostly families and genera endemic to Asia) were rapidly taking over as the archaic anagalids disappeared. The predators included a mix of the archaic hyaenodont creodonts (including the bear-sized *Pterodon*), the last of the

Figure 4.31. Exposures of upper Eocene beds in the Gran Barranca, on the flanks of the Andes in northwestern Argentina. The lowest beds in the photograph are Casamayoran (once thought to be lower Eocene, but now upper Eocene); the middle levels, labeled "Mustersan bench," are middle upper Eocene; and the uppermost levels are Divisaderan (latest Eocene). Courtesy of R. Madden.

hoofed predatory mesonychids, and more-advanced true carnivorans, including early dogs and civets, plus nimravids.

Europe continued to be an archipelago of endemic mammals in the late Eocene, dominated by unique groups that I have already mentioned: the horse-like palaeotheres (but no other perissodactyls such as the horses, rhinos, tapirs, and brontotheres found on other continents), twelve families of deer-like and pig-like artiodactyls (of which only the anthracotheres are found outside Europe), and dozens of endemic species of rodents, plus the last diverse primate fauna of adapids and omomyids, which had vanished everywhere else except Africa. The predators included abundant hyaenodont creodonts, along with primitive "miacid" carnivorans (extinct everywhere else by this time) and advanced carnivorans that belonged to the lineage that gave rise to bears and raccoons. Insectivorous mammals included a number of archaic lineages (leptictids, pantolestids, apatemyids) that were extinct or rare elsewhere in the world. As we shall see in the next chapter, the late Eocene was the last gasp for this endemic fauna, which would undergo radical rearrangement in the Oligocene.

South America had the most endemic fauna of all, because it was now truly isolated. On this continent, the Eocene had traditionally been divided into the Casamayoran (early Eocene), Mustersan (middle Eocene), and Divisaderan (late Eocene), although the records for all three of these intervals are poor compared to the records for other continents (Flynn and Swisher 1995). However, recent work by Kay et al. (1999) showed that there is huge, previously unsuspected gap in the record, with few lower or middle Eocene deposits in South America (fig. 4.31). The Casamayoran, Mustersan, and Divisaderan are mostly late Eocene in age (fig. 4.17). These South American faunas consist of the same elements that we saw established in the Paleocene: dog-like and hyena-like carnivorous marsupials, native hoofed mammals (notoungulates and litopterns), and abundant edentates (mostly primitive armadillos). None of these groups was found anywhere else in the world at the time except for the anteater *Eurotamandua*, which is known from the middle Eocene of Europe. Kay et al. (1999) pointed out, however, that these native herbivores were already developing high-crowned teeth for tough gritty vegetation long before the mammals of other continents developed them, which suggests that the climate and vegetation were much drier and harsher in South America in the Eocene.

Impacts without Impact

When the K/T impact theory struck in 1980, it quickly spurred scientists to look at other extinction events to see whether they too could have been caused by some impact. Sure enough, there was an iridium anomaly near the Eocene-Oligocene boundary (Alvarez et al. 1982; Asaro et al. 1982; Ganapathy 1982; Glass et al. 1982). The impact advocates quickly jumped on the bandwagon and declared the Eocene extinctions to be caused by the impact of an asteroid or a comet. End of story. Case closed. Scientists talked about impacts causing extinctions on a 26-million-year or 32-million-year cycle (Raup and Sepkoski 1984), blaming this cycle on periodic comet showers, the oscillation of the solar system through the galactic plane, an unknown Planet X that caused comet showers, or even an undetected companion star to the sun, dubbed Nemesis. Some scientists (e.g., Raup 1991) suggested that *all* extinctions were due to impact, regardless of whether there was direct evidence of impact or not. In the early days of the impact hypothesis, no idea was too outrageous. As Keith Thomson (1988, p. 59) put it, "Inevitably, with most subjects there is also a silly season, usually of unpredictable duration and of an intensity correlated with the state of the acceptance of the new idea . . . [including] proposal of ideas even more far-out than the original one."

But in the nearly quarter century since the original impact hypothesis and extinction cycle were proposed, the cold hard reality of fact has ruined the beautiful fantasies of the impact crowd. Only the K/T impact is well established, and the mass extinctions at the end of the Devonian, Permian, and Triassic are looking less and less like products of impact (judging from the presentations at the 2003, 2004, and 2005 meetings of the Geological Society of America). No one is looking for Planet X or Nemesis anymore, and most no longer look to astronomical cycles at all. There is no longer any mention of Raup and Sepkoski's (1984) periodic extinctions, as the numerous flaws in the analysis and the data have become apparent. Stanley (1990) provided the best explanation for the apparent cycle of extinction events. Mass extinctions can occur only when there is a high diversity of specialized organisms on earth to wipe out; and after a mass extinction, 10–15 million years are apparently required for the full recovery of life and the evolution of new organisms that would be vulnerable to another mass extinction. A catastrophe that occurred shortly after a mass extinction would have no obvious effect, because there would be few vulnerable taxa to be affected. Hence, mass extinctions cannot occur more frequently than every 15–30 million years, because that's how long it takes the world's biota to rebound from a previous event.

The excitement over the impacts of the late Eocene is a particularly good example of how scientists can get ahead of their data. When the iridium anomaly was first discovered *near* the Eocene-Oligocene boundary, the impact advocates said there must be a connection and bragged that they had found the cause. But those with better knowledge of the Eocene and Oligocene rocks and fossils dug in and did the hard detective work that the impact advocates had neglected (papers in Prothero and Berggren 1992; Prothero et al. 2003). The impact layers were dated at 35.5–36.0 Ma, in the middle of the late Eocene, when no extinctions had ever been proposed (fig. 4.17). The impacts occurred at least 1–2 million years before the Eocene-Oligocene boundary and (as we shall see in the next chapter) even earlier than the early Oligocene extinctions and at least 1–2 million years after the terminal middle Eocene extinctions at 37–38 Ma. In a few Caribbean sections that did contain the impact layer, there were five radiolarian species (out of dozens) that disappeared at the impact (Maurrasse and Glass 1976; Glass and Zwart 1977), but nothing else suffered—neither the rest of the microfossils (especially the sensitive planktonic foraminifera [Hut et al. 1987]), nor the marine invertebrates (mollusks, echinoids, crustaceans), nor the land plants or land mammals (Prothero 1994).

Then the sources of the impact debris and the iridium were identified. Deep drilling in the Chesapeake Bay region (Poag et al. 1992; Poag 1999) uncovered evidence of a crater buried deep beneath Maryland and Virginia, which was dated at 35.5 Ma (Poag et al. 2003). Eventually, scientists found evidence of a buried crater about 100 kilometers in diameter (about two-thirds the size of the K/T impact crater) and a second crater on the continental shelf 100 kilometers east of New Jersey in Toms Canyon (fig. 4.32), which was the same age and about 15 kilometers in diameter. Presumably, both craters were from the same late Eocene shower of rocks, and they left a debris field that was detected as far as the Indian Ocean and the South Atlantic (fig. 4.32). About the same time, a third crater was identified at Popigai in eastern Siberia (Masaitis et al. 1975; Bottomley et al. 1997), which was dated at 35.7 Ma and which left a crater about 90 kilometers in diameter. Thus, there were now three well-dated large impacts in the late Eocene, yet the evidence was still clear—these huge impacts had no effect on life.

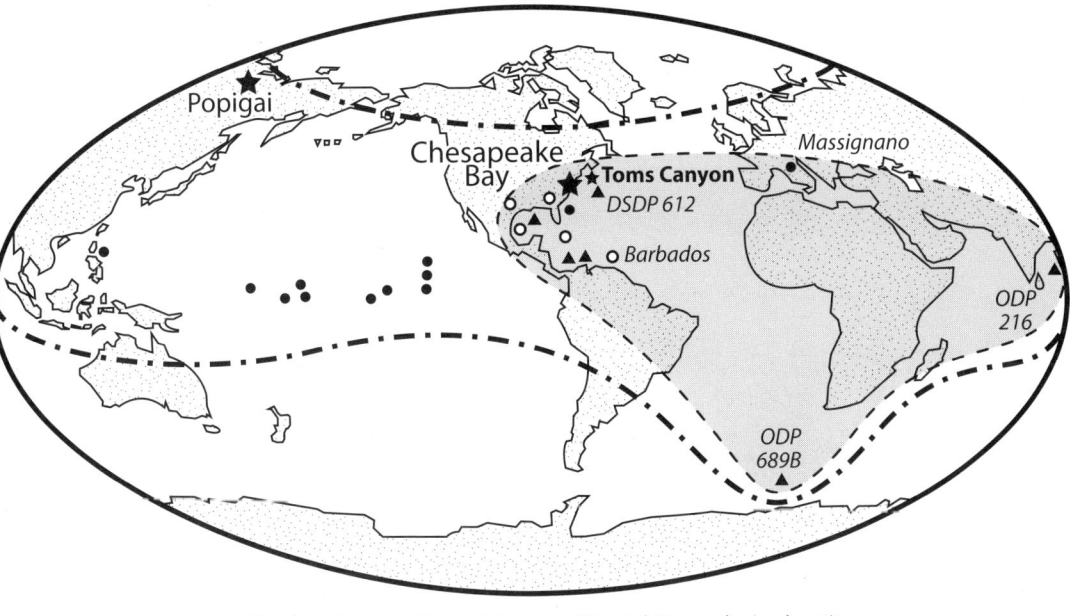

● Microkrystites   ○ Microtektites   ▲ Microtektites and microkrystites

Figure 4.32. Stars show the locations of the major late Eocene impact sites (Chesapeake Bay, Toms Canyon, and Popigai), and the shaded pattern shows the distribution of the tektites field scattered from the Chesapeake Bay and Toms Canyon impacts. After Poag 1999.

Even more embarrassing for the impact crowd are the implications of these discoveries. If the 180-kilometer K/T impact at Chicxulub supposedly had such a great effect (but see chapter 2), but the 100- and 90-kilometer craters at Chesapeake Bay, Toms Canyon, and Popigai had no effect, what does this imply about the idea that impacts cause extinctions? Raup (1991) had originally fit a "kill curve" of size of impact versus number of species wiped out, using the K/T crater as a model (fig. 4.33). Poag (1997) had to replot the kill curve with the Eocene crater data, and its shape changed dramatically. If craters over 100 kilometers in diameter have no effect, then only the biggest impacts have any chance of causing extinction, and the kill curve must shoot up rapidly between diameters of 100 and 180 kilometers. This was further confirmed when other Cenozoic impact sites were re-examined. The Montagnais impact structure, off the coast of Nova Scotia, was formed by an impact at 50 Ma that had no effect on life (Bottomley and York 1988; Aubry et al. 1990; Jansa et al. 1990). Likewise, the Miocene Ries crater in Germany had no effect on life either (Heissig 1985). Prothero (2005a) surveyed of all the Cenozoic impacts in the impact database on the Web and found that none corresponded with peaks of extinction on any diversity curve; Alroy (2002) came to a similar conclusion.

Naturally, the impact advocates have not taken this bad news lightly.

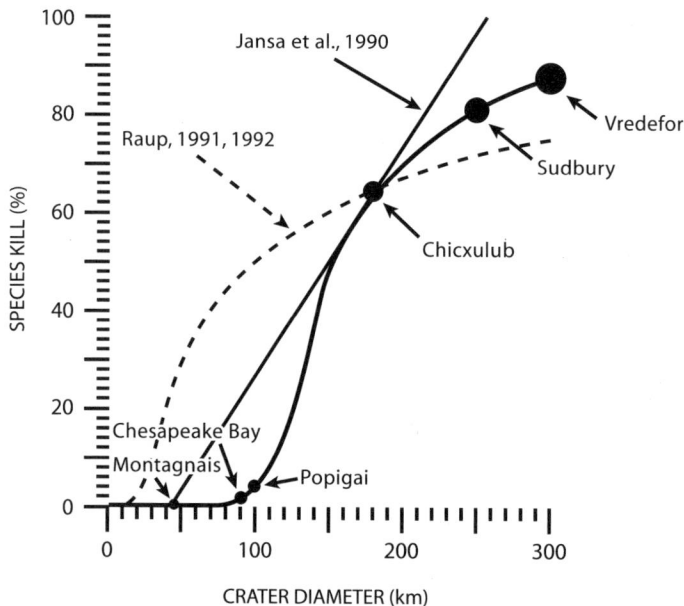

Figure 4.33. Raup's "kill curve." The "kill curve" (dashed line) of Raup (1991) was originally fit to the K/T impact data (60% of species wiped out with a crater of about 180 kilometers in diameter), and it predicted that much smaller impacts should cause significant extinctions. However, when the late Eocene impacts (which caused almost no extinctions) are plotted, the "kill curve" takes a different, S-like shape, and suggests that only the largest impacts have the potential to cause mass extinction. After Poag 1997.

Because of the overwhelming evidence that the late Eocene impacts had little or no biotic effect, impact advocates have instead argued that impacts might have caused some of the late Eocene climatic perturbations that preceded the extinctions in the earliest Oligocene (Poag 1999; Vonhof et al. 2000; Coccioni et al. 2000; Poag et al 2003; Fawcett and Boslough 2002). What is peculiar about these explanations is that they predict opposite effects. The direct effect of an impact should produce a debris ring and global cooling (Vonhof et al. 2000; Fawcett and Boslough 2002), yet the isotopic and paleoclimatic records of the late Eocene show that the exact opposite, a short-term warming event, actually occurred (Poag 1999; Poag et al. 2003). Clearly, impacts cannot cause global warming and cooling simultaneously. Even if they could, there is no clear explanation for how either climatic change might have caused extinctions in the early Oligocene, more than 2 million years later, given that the effects of such impact events diminish over a period years or decades, not millions of years.

For more than twenty years now, scientists have been intensively studying the record of the Eocene-Oligocene transition, and the verdict is clear: the huge impacts in the late Eocene had no significant effect. More importantly, the theme of the recent presentations at several scientific meetings is that most mass extinctions were not caused by impacts. Only the K/T extinction presents a good case for the impact hypothesis, and its effects may not have been as severe as some have suggested (see chapter 2). The impact-extinction theory of the 1980s may have generated a lot of heat and publicity, but all scientific hypotheses must eventually be subjected to the crucible of careful testing and analysis of the data—and now, twenty-five years later, this theory has been found wanting.

For Further Reading

Aubry, M.-P., S. G. Lucas, and W. A. Berggren, eds. 1998. *Late Paleocene–Early Eocene Climatic and Biotic Events in the Marine and Terrestrial Records.* New York: Columbia University Press.

Beard, K. C. 2004. *The Hunt for the Dawn Monkey: Unearthing the Origin of Monkeys, Apes, and Humans.* Berkeley: University of California Press.

Gunnell, G. F. 2001. *Eocene Biodiversity: Unusual Occurrences and Rarely Sampled Habitats.* New York: Kluwer Academic/Plenum Publishers.

Poag, C. W. 1999. *Chesapeake Invader: Discovering America's Giant Meteorite Crater.* Princeton, N.J.: Princeton University Press.

Prothero, D. R. 1994. *The Eocene-Oligocene Transition: Paradise Lost.* New York: Columbia University Press.

Prothero, D. R., and W. A. Berggren, eds. 1992. *Eocene-Oligocene Climatic and Biotic Evolution.* Princeton, N.J.: Princeton University Press.

Prothero, D. R., L. C. Ivany, and E. A. Nesbitt, eds. 2003. *From Greenhouse to Icehouse: The Marine Eocene-Oligocene Transition.* New York: Columbia University Press.

Schaal, S., and W. Ziegler, eds. 1992. *Messel: An Insight into the History of Life and of the Earth.* Oxford: Clarendon Press.

Wing, S. L., P. D. Gingerich, B. Schmitz, and E. Thomas, eds. 2003. Causes and consequences of globally warm climates in the early Paleogene. *Geological Society of America Special Paper 369.*

Figure 5.1. Typical outcrops of lower Oligocene beds (Scenic Member of the Brule Formation) in the Big Badlands of South Dakota. Photo by the author.

5

The Icehouse Cometh:
The Oligocene

Blow, blow, thou winter wind;
Thou art not so unkind
As man's ingratitude;
Thy tooth is not so keen
Because thou art not seen,
Although thy breath be rude. . . .
Freeze, freeze, thou bitter sky,
That dost not bite so nigh
As benefits forgot;
Though thou the waters warp
Thy sting is not so sharp
As friend remember'd not.

William Shakespeare, *As You Like It*, 1599

The Big Chill

Almost a century ago, the differences between the Eocene and Oligocene
were apparent to paleontologists. Stehlin (1909) called the differences between Eocene and Oligocene European mammal faunas *la grand coupure*
("the great break"). Osborn (1910) noted the transformation of North
American land mammal faunas (pl. 3) of the Big Badlands of South
Dakota (fig. 5.1) from those dominated by leaf eaters with primitive feet
and limbs to those with more-specialized teeth for harsher vegetation, and
with longer limbs for more running. Of course, the original reason that
von Beyrich (1854) proposed that the Oligocene be designated as an
epoch in the first place was that the molluscan faunas of units like the
Boom Clay in Belgium (fig. 5.2A) were distinctive from those of the tropical Eocene. Similar tendencies were noted in early research on mollusks
of the Gulf Coast and Pacific Coast of the United States.

A

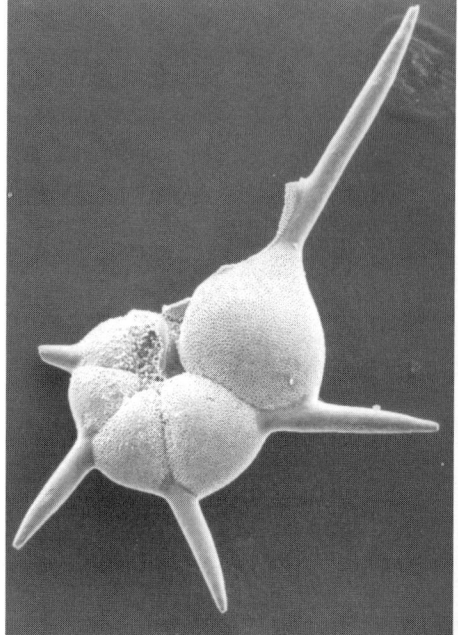

B

C

Figure 5.2. (A) Exposures of the type Rupelian (lower Oligocene) Boom Clay in Belgium. The alternating dark and light color bands are due to climatic cycles on 100,000-year scales (Milankovitch cycles). Photo courtesy of S. Van Simaeys. (B) The spiny planktonic foraminiferans known as hantkeninids. These foraminiferans were typical of the late Eocene, and their final extinction officially marks the Eocene-Oligocene boundary. Photo courtesy of G. Keller. (C) The officially designated type section for the Eocene-Oligocene boundary in an abandoned quarry in Massignano, near Ancona, on the Adriatic coast of Italy. The boundary (as recognized by the last appearance of hantkeninids) occurs at the uppermost left side of the central saddle. Photo courtesy of A. Montanari.

But the details of how and why the world transformed during the Eocene-Oligocene transition could not be studied until the 1970s, when the Deep Sea Drilling Project retrieved hundreds of sediment cores from the deep oceans around the world. The record of geologic time preserved in the deep sea is the most complete of any in the world, so the details of each million-year period of the transition can be precisely studied. Chock full of microfossils, these cores were dated by magnetic stratigraphy, and not only did their microfossils tell of changes in abundance, diversity, and dominance as water temperature changed, but their shells were a good source of oxygen and carbon isotope data to analyze the temperature and chemistry of the world's oceans. And in cores from around the Antarctic, there was another clue: large grains of sand and gravel (ice-rafted detritus) that could not have reached the deepest oceans unless they had been dropped there by melting icebergs.

By the mid-1980s, geochemists had good evidence from the oxygen isotope curve (figs. 3.22, 5.3) that there had been a dramatic drop in temperatures in the deep sea near the Eocene-Oligocene boundary. Current estimates (Miller et al. 1987; Zachos et al. 1993, 1994, 2001) are that deep-sea mean temperatures dropped 5–6°C (9–11°F), so that the global mean deep-sea temperature in the early Oligocene was only 5°C (41°F). This temperature change was by far the most dramatic in the entire Cenozoic until a similar cooling event in the middle Miocene signaled the onset of the modern East Antarctic ice sheet (see chapter 6). Until the mid-1980s, conventional wisdom had been that there were no glaciers on Antarctica before the middle Miocene, but the oxygen isotopes suggested otherwise.

But oxygen isotopes are not the only evidence of cooling. Ice-rafted detritus has been recovered from both lower and upper Oligocene sediments in numerous deep-sea cores recovered in the Southern Ocean (Kennett and Barker 1990; Barron et al. 1989; Zachos et al. 1992). Drill holes on the Antarctic margin at the Ross Sea and at Prydz Bay contain thick lower and middle Oligocene glacial deposits, suggesting that by the early Oligocene, a major ice sheet covered much of Antarctica and that it lasted for several hundred thousand years. That ice sheet retreated, but had returned again by the middle Oligocene, as indicated by the oxygen isotope record (fig. 5.3) and by the pulse of middle Oligocene glacial deposition.

As we saw in the last chapter, scientists in the 1980s (e.g., Pomerol and Premoli-Silva 1986) talked about a Terminal Eocene Event (a term coined by Wolfe [1978] on the basis of North American floral change) and treated the Eocene-Oligocene boundary as if it were an abrupt impact horizon like the K/T boundary. In 1989, international committees voted to define the end of the Eocene on the disappearance of a characteristic group of Eocene planktonic foraminiferans, the spiny hantkeninids (fig. 5.2B). The committees picked a particular section in the Massignano area of the Adriatic coast of central Italy (fig. 5.2C) as the type section for the boundary and formally defined it that way by a vote in 1989. However,

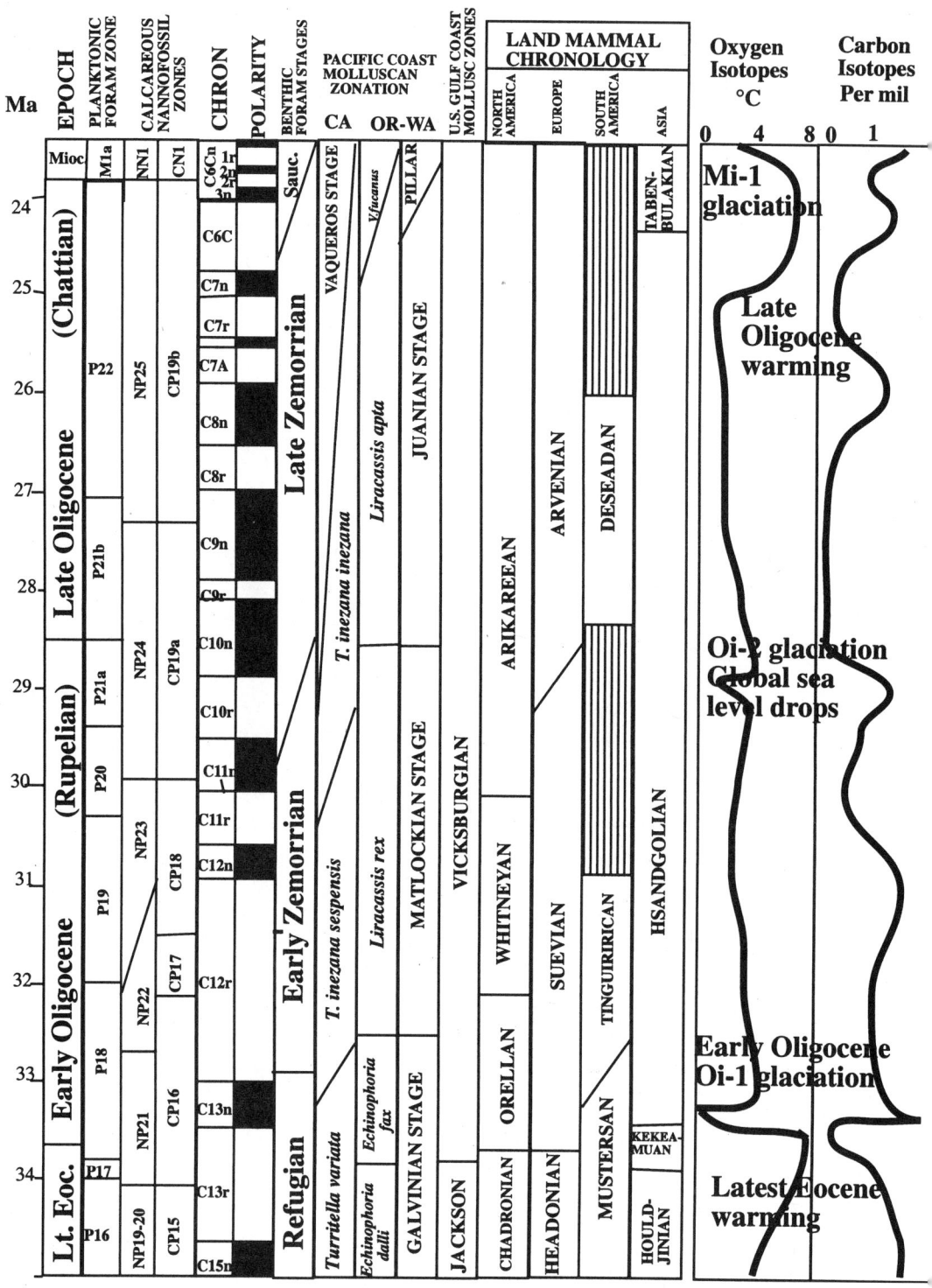

Figure 5.3. The Oligocene timescale (left columns), as recently calibrated by Berggren et al. (1995). The North American land mammal record is after Prothero (1995) and Woodburne and Swisher (1995); the South American land mammal ages are after Flynn and Swisher (1995) and Kay et al. (1999); the European record is after Schmidt-Kittler (1997); and the Asian record is after Wang et al. (1998) and Meng and McKenna (1998). Note that the boundaries between North American land mammal ages are precisely calibrated, so they are drawn as synchronous horizontal lines; whereas those of South America, Asia, and Europe are less well dated, so they are drawn as time-transgressive diagonal boundaries. The majority of the South American land mammal record is unknown, as indicated by the vertically ruled pattern. The isotopic records are after Zachos et al. (2001), but they are very generalized and smoothed out. Abbreviations: CA = California; Lt. Eoc. = Late Eocene; Mioc. = Miocene; OR-WA = Oregon-Washington; Sauc. = Saucesian.

almost as soon as the boundary had been agreed upon, problems arose. Examination of other microfossils from the type section of the late Eocene Priabonian Stage showed that part of the type late Eocene (as defined by the hantkeninid datum) was early Oligocene (Brinkhuis 1992). As the records of oxygen isotopes from deep-sea cores improved, it became clear that the transition had been an abrupt event (developing over a few thousand to tens of thousands of years), comparable to the glaciation events of the Pleistocene ice ages (Zachos et al. 1993, 1994). More importantly, microfossils showed clearly that the event was not at the Eocene-Oligocene boundary, as had long been assumed, but in the earliest Oligocene (33 Ma), about a million years later. For scientists who wanted the Eocene-Oligocene boundary to coincide with the global oxygen isotope event and glaciation, the event was too late. The boundary had been set in beds that are now by definition about a million years older than the climatic change, and the decision was final. However, where boundaries are placed is not crucial, as long as scientists agree on what they are calling Eocene and Oligocene, and are not arguing over problems caused by miscorrelation. Ironically, the K/T boundary is defined now on the iridium anomaly, and the Paleocene-Eocene boundary has just been defined on the carbon isotope anomaly, so those two important boundaries correspond to global events. However, the Eocene-Oligocene boundary definition is now set so that it is about a million years too old for its global oxygen isotope event and climatic signal.

What could have caused such a dramatic shift in deep-sea temperatures? I have already pointed out (chapter 4) that there was a long-term cooling trend throughout the middle and late Eocene, with small glaciers appearing on the Antarctic Peninsula as early as 49 Ma (Birkenmajer 1987; Birkenmajer et al. 2005). Nevertheless, floras (Kemp 1978; Case 1988) show that dense forests still covered most of the continent well into the late Eocene. The long-term Eocene cooling may be explained by gradual removal of greenhouse gases, possibly due to reduced seafloor-

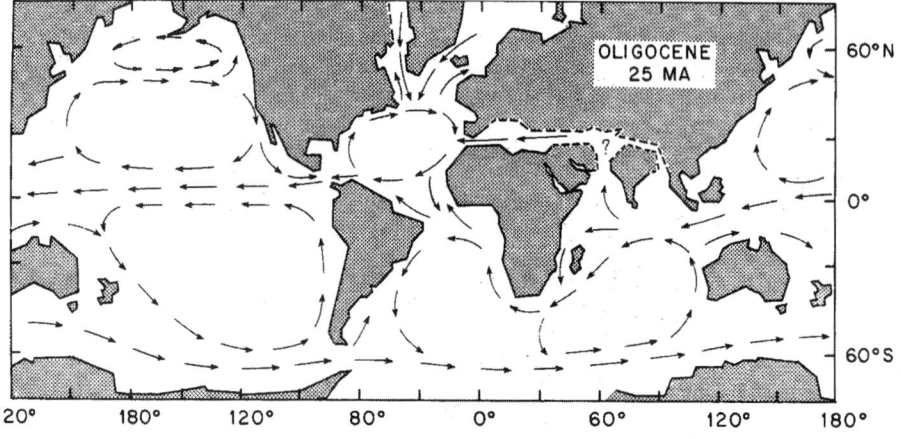

Figure 5.4. Oligocene pattern of oceanic circulation, showing the development of the Antarctic Circumpolar Current, which dominates oceanic circulation patterns today. From Frakes 1979.

spreading rates (Berner et al. 1983). However, such processes cannot explain the abrupt onset on glaciation in the early Oligocene.

Instead, we must turn to the role of ocean currents and their effects on climate. Today the Antarctic Circumpolar Current (ACC) circulates around Antarctica in an easterly, or clockwise, direction (fig. 5.4). It is one of the fastest and most voluminous of all currents in the ocean, traveling about 25 centimeters per second, and with a volume of 233 million cubic meters per second, or more than 1,000 times the flow of the Amazon River (Callahan 1971)! As this shallow surface current moves around Antarctica, it acts like a refrigerator door, locking in the cold over the South Pole and preventing the mixing of polar cold water with warmer subpolar and temperate waters. The ACC also causes circumpolar deep water to rise to the surface, producing huge blooms of diatoms that take advantage of the silica and other nutrients brought up from the deep ocean. This diatom bloom, in turn, means that the Southern Ocean is incredibly rich for plankton feeders, which is why most of the world's great whales (especially the blue whale) feed there.

Finally, the water brought to the surface by the ACC is chilled as it reaches the frozen surface, and it descends into the deep ocean again to form the Antarctic Bottom Water (AABW), the deepest water mass in the ocean. This cold oxygenated water mass flows along the bottom of the ocean all the way to the North Atlantic and North Pacific, forming the modern global thermohaline circulation system, known as the psychrosphere. The AABW makes up 59% of the water in the oceans, so it is extremely influential in global climate (Warren 1971). As we saw in chapter 4, during the Paleocene and Eocene there was no such circulation, and instead tropical and temperate waters mixed with polar waters, substantially reducing the difference between polar and tropical temperatures.

When did the ACC and AABW originate? We have already seen that Antarctica and Australia were connected in the Early Cretaceous but had begun to break apart by the Late Cretaceous and continued to separate, with a shallow embayment between them, during the early Cenozoic. By the middle Eocene, there was a shallow marine gulf between the two continents, but no deep-water connection. Deep-sea cores drilled around Antarctica show no evidence of deep-water circulation prior to the early Oligocene (Kennett et al. 1972, 1975). Prior to 40 Ma, the deep waters of the ocean came from the North Pacific, but neodymium isotopes taken from deep-sea cores show that the deep waters began to originate in the Southern Ocean at around 40 Ma (Thomas 2004). Cores taken just south of New Zealand show particularly well the blast of deep cold water coming through between Tasmania and East Antarctica during the earliest Oligocene (Murphy and Kennett 1986; Kamp et al. 1990; Exon et al. 2002). This certainly had an important role in the development of the ACC. What about the other half, the connection between South America and the Antarctic Peninsula now severed by the Drake Passage? Originally, the seafloor-spreading evidence appeared to indicate that the Drake Passage did not open until the latest Oligocene (Barker and Burrell 1977, 1982; Sclater et al. 1986). However, there is now evidence (Diester-Haass and Zahn 1996) that the Drake Passage did open in the early Oligocene. If so, then the complete ACC could have developed, and this development would explain the sudden worldwide drop in temperatures, as well as all the evidence for dramatic changes in oceanic circulation.

The story does not end there. In addition to the deep cold bottom waters that originate in the Antarctic and flow north along the bottom of the ocean, there is a second source of deep cold bottom water, the North Atlantic Deep Water (NADW). This water chills in the Arctic Ocean and then flows out through the Norwegian (Greenland) Sea into the deep North Atlantic. When did this current originate? As we saw in previous chapters, the Arctic Ocean had been isolated by land masses through most of the Cenozoic, with perhaps a brief episode of breaching in the early Eocene. A number of deep-sea cores show that the NADW must have originated in the early Oligocene (Miller and Tucholke 1983; Miller and Fairbanks 1983; Miller and Thomas 1985; Miller 1992), and this has been further confirmed by recent oil-company exploration in the region (Davies et al. 2001). Thus, both the AABW and NADW abruptly began in the early Oligocene, dramatically changing ocean currents, transferring heat between the Equator and the poles, and triggering the early Oligocene icehouse conditions worldwide.

How did marine life respond to the sudden onset of bottom water currents and the rapid change in temperature? Contradicting earlier notions that the Terminal Eocene Event was some global catastrophic extinction

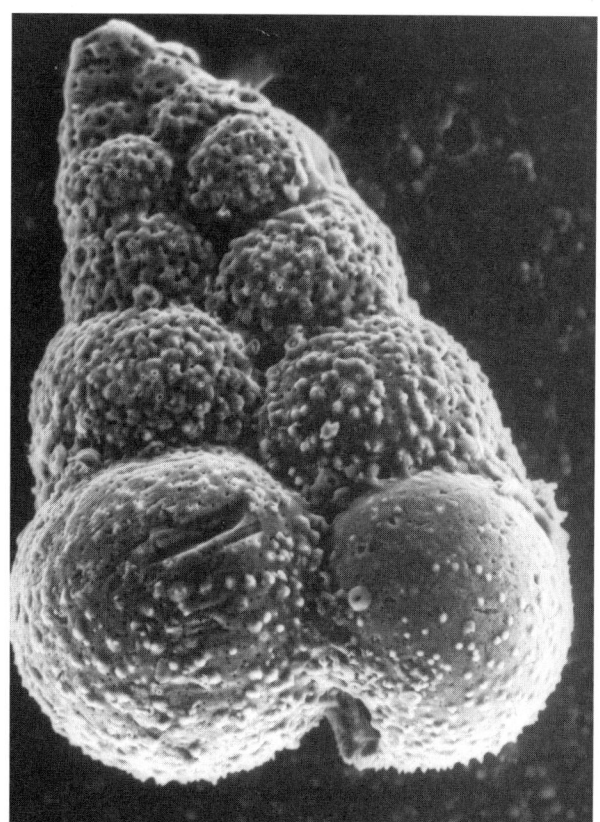

Figure 5.5.
Chiloguembelina cubensis.
Biserial heterohelicid
foraminifera like this are
particularly abundant in
the early Oligocene. Photo
courtesy of G. Keller.

like that of the K/T event, recent work (papers in Prothero and Berggren 1992; Prothero et al. 2003) has shown that the most severe mass extinction took place at the end of the middle Eocene (37 Ma), as discussed in chapter 4. Tropical taxa were largely gone by the late Eocene, and those that survived into the Oligocene were already adapted to the colder, harsher icehouse world, so the extinction did not hit as severely as once supposed. According to Boersma et al. (1987), planktonic foraminiferans of the early Oligocene were a homogeneous fauna that was small in size, low in diversity, and found almost worldwide. Only a handful of species, mainly the hantkeninids (fig. 5.2A), went extinct at the Eocene-Oligocene boundary, or in the earliest Oligocene. These foraminifera suggest that the cold oceans were well circulated, not stratified by depth or temperature. The abundance of deep-water, cold-tolerant biserial heterohelicid foraminiferans (fig. 5.5) is particularly diagnostic of the upwelling and mixing of cold bottom waters with surface waters. In the tropics, however, some depth stratification remained, as the carbon isotope differences between shallow- and deep-water foraminiferans remained high.

The cooling definitely affected temperature-sensitive marine algae. Among the coccolithophorids (Aubry 1992), there was extinction among temperate-zone taxa, and the cooling was the final blow for the tropical

Eocene forms. All but one of the long-ranging species that had evolved in the early Eocene, several of the long-ranging species that had evolved in the middle Eocene in response to the climatic deterioration, and all the short-lived taxa that had evolved in the late Eocene became extinct, reducing diversity to 30% of the original number of species since the extinctions began in the middle Eocene. Cold-tolerant coccolithophorids bloomed in huge numbers, owing to the release of nutrients by the vigorous deep-water circulation. By the end of the early Oligocene, the division of the oceans into latitudinal water masses led to increased coccolith provincialism as species became segregated into different temperature water masses. The diatoms also responded dramatically. About 45% of the Eocene diatom species became extinct by the end of the early Oligocene, and were replaced by almost the same number of new species (Baldauf 1992). This turnover occurred in all latitudes, but especially in the Antarctic, where there was a great increase in siliceous productivity due to the cold deep-water currents that brought silica to the surface waters (Baldauf and Barron 1990). There was also significant extinction in the dinoflagellates at the beginning of the Oligocene (Brinkhuis 1992).

The cold bottom waters naturally greatly affected the benthic foraminifera that experienced these currents directly. Earliest Oligocene benthic foraminifera had a much lower diversity than they did during the Eocene, with deep-water *Nuttalides umbonifera* making up 70% of some assemblages (E. Thomas 1992). Most species retreated to living within the sediment, which suggests that the deep cold waters were corrosive to their calcite shells. The United States Gulf Coast saw a high rate of extinction of benthic foraminiferans as well (Gaskell 1991; Fluegeman 2003). In the tropical Tethys, the last of the warm-water taxa, such as the disk-shaped nummulitids (fig. 4.10C), disappeared (Adams et al. 1990). Agglutinated foraminiferans, which build their shells from cemented sand grains, were also severely affected, again because of the cold corrosive bottom water (Kaminski 1987).

Larger marine life was not spared either. On the Gulf and Atlantic coasts of the United States, there was another major extinction of mollusks, with about 95% of the snail species and 89% of the clam species that survived the middle Eocene extinction dying out in the early Oligocene (Hansen 1988, 1992; Dockery and Lozouet 2003; Campbell and Campbell 2003; Hansen and Kelley 2004). Similar trends occurred on the Pacific Coast, where there was mass extinction in the snails (Squires 2003), with most of the new species coming down from the cold North Pacific and Arctic regions (Oleinik and Marincovich 2003), and clams were strongly affected as well (Nesbitt 2003; Hickman 2003). There was also a dramatic drop from forty-three late Eocene echinoid species to only fifteen species found in lower Oligocene deposits of the Gulf Coastal Plain (Carter 2003), and about 50% of echinoid species became extinct worldwide (McKinney et al. 1992). Tropical echinoids were the most severely affected, and the groups that flourished were the cold-adapted marsupiate echinoids, which protect their eggs in a pouch; these managed to survive and spread during the chilly Oligocene.

Another indicator of conditions in shallow marine habitats is the oxygen isotopes recorded not only in mollusk and foraminiferan shells but even in the tiny ear bones of fish, known as otoliths. Ivany et al. (2000, 2003, 2004) found the warmest temperatures and the least seasonality in the early Eocene, followed by cooling in the middle Eocene. However, the otolith isotopes do not show a dramatic temperature drop in the Gulf Coast of Alabama and Mississippi at the end of the Eocene, or at the Eocene-Oligocene boundary. Instead, there was an increase in seasonality, from a range of only 2°C (25–27°C) between summer and winter in the early Eocene to 5°C (15–20°C) in the late Eocene. In the early Oligocene, winter temperatures dropped another 4°C (to 11°C), but summer temperatures were not affected. Thus, the increased seasonality was enhanced by dramatically cooler winters in the Gulf Coast in the Oligocene, even though average temperatures had not changed that much.

At the top of the marine food pyramid are the whales. As we saw in chapter 4, they had evolved into huge archaeocetes by the middle and late Eocene, which apparently flourished in low-latitude areas such as the warm Tethys Seaway and the Gulf Coast of the United States (Fordyce 2003). By the latest Eocene, however, archaeocetes had vanished, and they were rapidly replaced by two new groups that dominate the oceans today: the odontocetes, or the toothed whales (including sperm whales, dolphins, orcas, and porpoises), and the mysticetes, or the baleen whales (blue whales, humpbacks, fin whales, gray whales, and many others), which use a filter made of a tough horny protein known as baleen in their mouth to screen out plankton from the seawater. Early members of both groups are found mostly at high latitudes in the Oligocene (especially New Zealand and the North Pacific). According to Fordyce (1980, 1989, 1992, 2003), their evolution and diversification in the early Oligocene were triggered by the explosive blooms of plankton when rising deep bottom waters (especially in the Southern Ocean) brought up nutrients. This population explosion in diatoms and coccolithophorids fed the crustaceans, which are the main source of nourishment for baleen whales.

In summary, the early Oligocene extinctions of marine life were not as severe as those of the middle Eocene, but still there was tremendous turnover, especially with tropical and warm-adapted taxa being replaced by cold-adapted taxa. By the late early Oligocene, diversity in the oceans had stabilized at a new, much lower level than that during the Eocene, and held on at these lower numbers of species throughout the cold harsh conditions of the rest of the Oligocene.

Bad Lands, Good Fossils

A visit to the Big Badlands of South Dakota today reveals a harsh, forbidding landscape (fig. 5.1), with numerous exposures of bare rock. In the blazing summer sun, the temperatures often top 40°C (104°F). These conditions can be lethal, especially when one is out all day collecting fossils. The region was so harsh and forbidding that the French fur trappers avoided these *mauvaises terres,* or "bad lands," as did the Native Americans. The early white settlers disliked any area of bare rock and little grass, and the Big Badlands were famous as "a bad place to lose a cow." Chief Bigfoot and his band fled to the Big Badlands in the dead of winter in 1890 to escape the U.S. Cavalry, and were eventually trapped and massacred just to the south at Wounded Knee. When land was divided up in the west, much of the Big Badlands region was given to the Native Americans for their reservations, because the white Americans viewed it as worthless.

But these lands are not worthless to scientists. For geologists and paleontologists studying the rocks and collecting the fossils, plants get in the way, and the more rock exposure, the better. Geologists have learned that regions with names like Devil's Punchbowl and Hell's Half Acre are heavenly, because the bare rock that drove away the early settlers is perfect for earth scientists. As early as 1846, the first government surveys passed through the Big Badlands, bringing back fossils; and by the 1850s, the fossil riches were being described by the founders of vertebrate paleontology in the United States: Joseph Leidy, Edward D. Cope, and Othniel C. Marsh. Pioneering geologist Ferdinand V. Hayden even led expeditions into the region, where Native Americans called him "man who picks up stones running" and left him alone because they thought he was crazy to be picking up old bones in the desert heat. The best and most abundant vertebrate fossils around the world are found in badlands regions, such as those in Mongolia and China and Kazakhstan in Asia, or in southern Argentina, or in many places in North America. However, the most famous are the Big Badlands of South Dakota, which are the richest fossil beds of all. The term "badlands" is a general one referring to a topographic feature, but "Badlands" with a capital "B" refers to the original Big Badlands of South Dakota.

Today, these rocks still yield fossils at an astounding rate because every winter snow and every spring rain weathers out new specimens entombed in the soft silts deposited on Oligocene floodplains. The quality and quantity of fossils is truly amazing, with most species known from hundreds of specimens and often complete skeletons. By contrast, most of the Paleocene and Eocene mammals discussed in previous chapters are known only from a few teeth and jaws. Badlands fossils are so common that they form a huge commercial market and are sold in most rock shops. Unfortunately, this ready availability also means that many specimens are obtained illegally and poached from either Badlands National Park or the reservation land to the south.

The fossils from the White River Group of the Big Badlands of South Dakota, and similar deposits in North Dakota, Nebraska, Wyoming, and Colorado, are important for another reason as well. The White River

Group preserves the entire late Eocene through middle Oligocene, so it provides a detailed record of how climate changed, and what happened to the creatures that lived there. At first, however, detailed studies comparable to those that have been discussed elsewhere in the book were impossible. The older collections of fossils stored in most museums had been made without record of the precise level from which the fossils had come, so paleontologists could not study the record in detail. (Poor record keeping is the major problem with rock shop specimens, because poachers have little regard for recording data like the level or geographic location from which the fossil was obtained; consequently, most such specimens are scientifically useless.) But in the 1940s and 1950s, Morris Skinner and his crews from the Frick Laboratory built the largest White River collections ever assembled, and, more important, they recorded the exact level from which every specimen came. The Frick Laboratory was a private organization built by Childs Frick, son of steel and railroad magnate Henry Clay Frick (Andrew Carnegie's partner). From the 1930s until he died in 1965, Childs Frick used his immense wealth to build the largest collection of fossil mammals ever assembled. His crews worked in the northern Plains every spring through fall, and in southern California, Arizona, New Mexico, and Texas each winter; and most of the workers carefully recorded the exact levels from which their fossils were obtained. They collected almost every important locality until they had hundreds of specimens, and today this collection occupies seven floors of storage in the American Museum of Natural History in New York. There is a whole floor of mastodonts, a whole floor of horses, a whole floor of rhinos, a whole floor of camels, and so on. Where paleontologists once had only a few isolated teeth, now they have complete skeletons and often hundreds of jaws and skulls, so everything that was once believed about North American later Cenozoic mammals must be completely rethought as these huge collections are gradually studied and published.

The other missing link in the study of the White River fossil collections was dating. When I first came to the American Museum in 1976, the new technique of magnetic stratigraphy was just beginning, so I studied the fossils and paleomagnetism of the White River Group as my dissertation project. Over the past twenty-five years, I have visited nearly every important middle Eocene through Oligocene fossil locality in western North America to take paleomagnetic samples, and from this research we now have a high-resolution correlation of most of the key Eocene-Oligocene fossils in this part of the world. (This work has since been published in Prothero et al. 1983; Prothero 1985; Prothero and Berggren 1992; Prothero and Emry 1996; and Prothero and Whittlesey 1998.) The final piece of the puzzle was the radiometric dates to calibrate these magnetic correlations. When I first started, we had only the old potassium-argon dates from Evernden et al. (1964), but by 1989, we had the new method of argon-argon dating, which gave much more precise and reliable results (Swisher and Prothero 1990; Prothero and Swisher 1992). When these new dates came out, we had to revise all our long-held notions, and I had to recalibrate all the work I had done in the 1980s. For example, the Chadronian land mammal age had been considered early Oligocene for more than a century, but the argon-argon dates indicated that it was actually late Eocene. Once these changes were made, however, we had a timescale for North America that was detailed and reliable and, better still, could be correlated with the magnetic stratigraphy of the marine record, so that events could be matched up precisely for the first time (further discussion of these changes can be found in Prothero 1994).

With revised correlations, what can we say about the terrestrial record at the Eocene-Oligocene transition? The White River sections preserve details of the history of many kinds of organisms. For example, Greg Retallack (1983) studied the color bands visible in the Badlands sections and found that they were paleosols, or ancient soil horizons. Those from the upper Eocene Chadron Formation were formed under forests (fig. 5.6A) with closed canopies of large trees (the huge root casts are particularly conspicuous) with 50–90 centimeters (20–35 inches) of rainfall per year. In the overlying lower Oligocene (Orellan) Brule Formation, the paleosols indicate more open, dry woodland with only 50 centimeters (20 inches) of rainfall per year (fig. 5.6B). In eastern Wyoming, Emmett Evanoff studied the sediments (Evanoff et al. 1992) and found that the moist Chadronian floodplain deposits had abruptly shifted to drier, wind-blown deposits by the Orellan. In the same beds are climate-sensitive land snails. According to Evanoff et al. (1992), Chadronian land snails are large-shelled taxa similar to those found in wet subtropical regions, like modern Central America. On the basis of modern analogues, these snail fossils indicate a mean annual temperature of 16.5°C (63°F) and a mean annual precipitation of about 45 centimeters (18 inches), and these values are similar to those obtained by Retallack (1983) for neighboring South Dakota. By contrast, Orellan land snails are drought-tolerant, small-shelled taxa indicative of warm-temperate open woodlands with a pronounced dry season. Their living analogues are found today in Baja California.

The amphibians and reptiles suggest similar trends of cooling and drying in the early Oligocene (Hutchison 1982, 1992). The Eocene is dominated by aquatic species (especially salamanders, pond turtles, and crocodilians) that (as noted in chapter 4) declined steadily in the middle and late Eocene. Crocodiles were gone from North America by the Chadronian, but a few fossil alligators have been recovered from the Chadron Formation. By the Oligocene, only land tortoises (fig. 5.7) were common, indicating a pronounced drying trend. In fact, these tortoises (*Stylemys nebraskensis*) are so common in the Orellan that these beds were originally called the "turtle-oreodon beds" after their two most common vertebrate fossils.

Land plants are not well preserved in the highly oxidized beds of the Big Badlands (except for the durable hackberry seeds, which are calcified while they are alive), so we must look to other regions to understand the

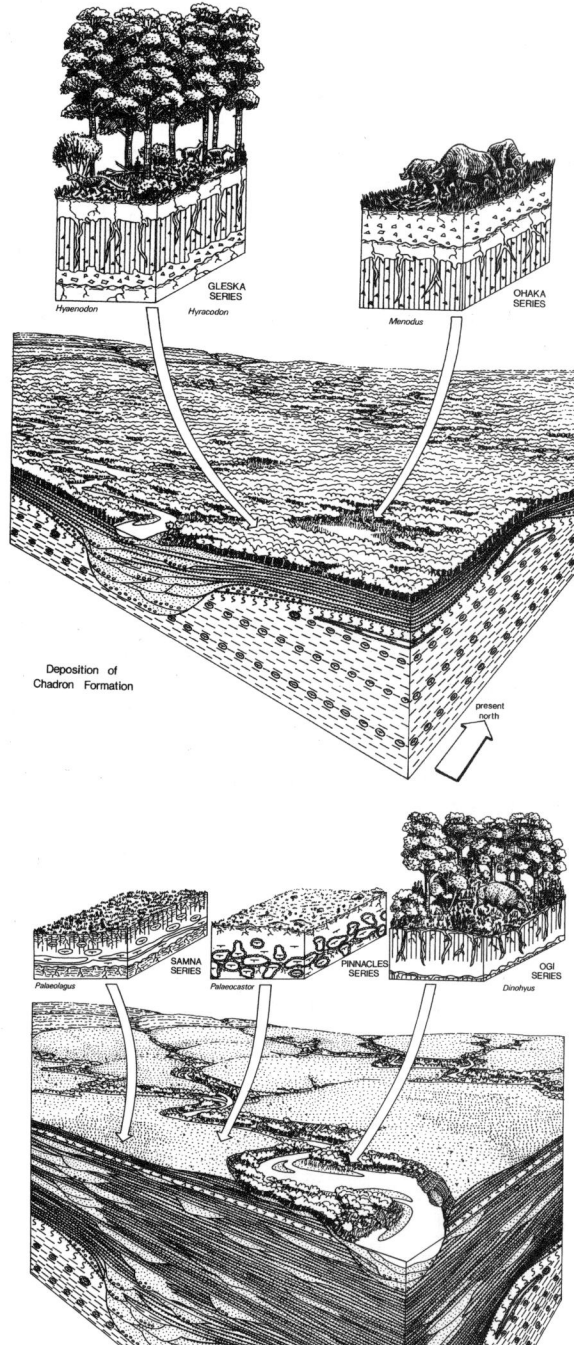

Figure 5.6. Reconstructions of the Big Badlands of South Dakota in the late Eocene and the late Oligocene, based on ancient soil horizons preserved in the rocks. (A) In the late Eocene, forests still dominated in this wet landscape, with limited open brushy areas. (B) By the late Oligocene, the paleosols show that forests had completely vanished, and open scrublands and early grasslands had developed on a much drier, more open landscape. From Retallack 1983.

154 • After the Dinosaurs

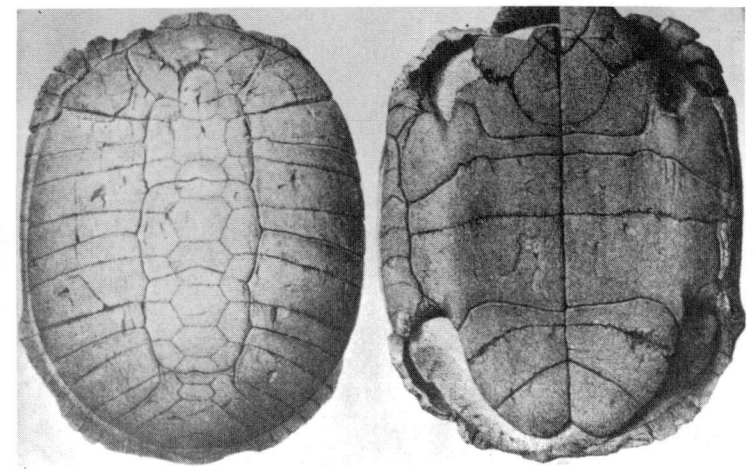

Figure 5.7. The common early Oligocene dry-land tortoise of the Big Badlands, Stylemys nebrascensis, *whose remains are so abundant that these rocks used to be known as the "turtle-oreodon" beds. From Leidy 1854.*

floral change. But the rest of North American floras show a clear trend (fig. 4.20B). On the basis of leaf-margin analysis, Wolfe (1971, 1978, 1985, 1992) suggested that mean annual temperatures in North America cooled by about 8–12°C (13–23°F) in less than a million years. This cooling event is by far the most dramatic of the entire North American floral record and, as noted above, was the original basis for the so-called Terminal Eocene Event (even though revised dating now places the event in the early Oligocene). Perhaps Wolfe's (1971) earlier phrase, "Oligocene deterioration," would be a better term. This dramatic change can be seen in floras from the Gulf Coast to Alaska (figs. 4.20B, 5.8). For example, the pre-deterioration Rex Creek flora in Alaska was a broad-leaved evergreen forest composed of conifers, holly, and Labrador tea, living at a mean annual temperature of 15°C (59°F) with a range of no more than 10°C (36°F). Post-deterioration Alaskan floras were mixed northern hardwood forests composed of alders, beeches, hickories, and *Metasequoia*, indicating a mean annual temperature of 7°C (45°F). Even more striking, these floras could tolerate temperatures as much as 28°C (100°F) above and below the mean, conditions that are comparable to the severe freezing conditions and hot summers that now occur in northern hardwood forests of eastern Canada and New England.

The same dramatic floral changes occurred in the Pacific Northwest. In western Oregon, the pre-deterioration floras formed in paratropical rain forests that experienced mean annual temperatures of 20–22°C (68–72°F) and about 180 centimeters (70 inches) of annual rainfall and that tolerated a temperature range of only 7°C (25°F). The plants included oaks, figs, magnolias, pawpaws, holly, soapberry, laurels, myrtles, ebony, roses, legumes, and herbs such as borages and heliotropes, as well as a number of less familiar families typical of the paratropical rain forests of Central America and Asia. The post-deterioration floras include oaks, laurels,

A

Figure 5.8. Typical late Eocene and early Oligocene floras. (A) Typical floras of the late Eocene are broad-leaved and tropical, including the tree fern Anemia *(center) and* Allantodiopsis. *(B) By contrast, early Oligocene floras consist of plants whose leaves were much smaller, and many had jagged margins. They included poplar, alder, willow, walnut, winter hazel, holly, and plum trees. From Wolfe 1977, courtesy of the U.S. Geological Survey.*

B

alders, *Metasequoia*, ash, hawthorns, sycamores, walnuts, hackberries, elms, lindens, beeches, bracken ferns, and horsetails, similar to the floras in modern cool-temperate redwood forests. According to Wolfe (1978), the mean annual temperature indicated by these plants is 12°C (54°F), a decline of at least 10°C (14°F) since the earliest Oligocene; and the seasonal range of temperature approached 25°C (almost 80°F) above and below the mean.

Although because of their high altitude, some Rocky Mountain floras did not reflect the trend as much (Wing 1987), others show it dramatically (Wolfe 1992). For example, the famous late Eocene Florissant lake beds in central Colorado preserve a high-altitude streamside flora of firs, pines, spruce, cypress, elms, hackberries, cottonwoods, mountain mahogany, hawthorn, apples, plums, and huge giant sequoias (fig. 5.9) with abundant cattails, horsetails, and mosses as well (Myers 2003). These plants lived at a mean annual temperature of 12.5°C (55°F)—much like the plants of the Pacific Northwest. But the early Oligocene Antero flora, just half a million years younger than Florissant and only a few miles away, is a northern hardwood forest (pines, oaks, blueberries, mountain mahogany, and some subalpine plants) tolerant of a mean annual temperature of 4.5°C (40°F). Even the high Rockies experienced about 8°C (15°F) of cooling. Finally, Gulf Coast floras and pollen show dramatic cooling as well, from subtropical conditions in the upper Eocene Jackson Group with a mean annual temperature of 28°C (82°F) to a much cooler oak-dominated forest in the early Oligocene (Vicksburg Group), according to Frederiksen (1988).

Despite these dramatic changes in the soils, land plants, land snails, and reptiles and amphibians, the change in the mammalian fauna is not that impressive (Prothero 1994; Prothero and Heaton 1996; Prothero 1999). Most of the archaic Eocene taxa (especially the forest dwellers and arboreal forms) were already gone by the last Eocene, with only a few multituberculates (fig. 3.7) straggling on to the middle Chadronian. A few groups, such as the brontotheres (fig. 1.1), the camel-like oromerycids, the mole-like epoicotheres, and two groups of rodents, did die out near the end of the Chadronian, but none was around to witness the big early Oligocene climatic deterioration. Most of the sixty-two lineages that were present before the climatic crash showed no change whatsoever, except for a dwarfing event in one lineage of the oreodont *Miniochoerus*. Apparently, the groups that were present in the late Chadronian were already adapted to the drier, more open woodlands habitats, so the vegetational change did not make that much difference—or else the responsiveness of mammals to short-term changes in climate has been oversold, and they are not as sensitive as we have long believed (Prothero and Heaton 1996; Prothero 1999).

The mammals that lived in what is now the Big Badlands during the Orellan (pl. 3) were much like those of the Chadronian, minus the huge brontotheres. Oreodonts (fig. 1.2) were by far the most common crea-

Figure 5.9. Giant sequoia stumps. Late Eocene pre-deterioration floras from Florissant Fossil Beds National Monument, Colorado, include this huge set of giant sequoia stumps, still on display in the park. Photo by the author.

tures, with hundreds of their skulls known from almost every meter of the section. Early gazelle-like camels (*Poebrotherium* and *Paratylopus*) without humps (fig. 5.10) were also common, as were the huge pig-like entelodont *Archaeotherium* (fig. 5.11) and the tiny deer-like artiodactyls without antlers, *Leptomeryx* and *Hypisodus*. Among perissodactyls, the primitive three-toed horses *Mesohippus* (fig. 5.12A) and *Miohippus* were

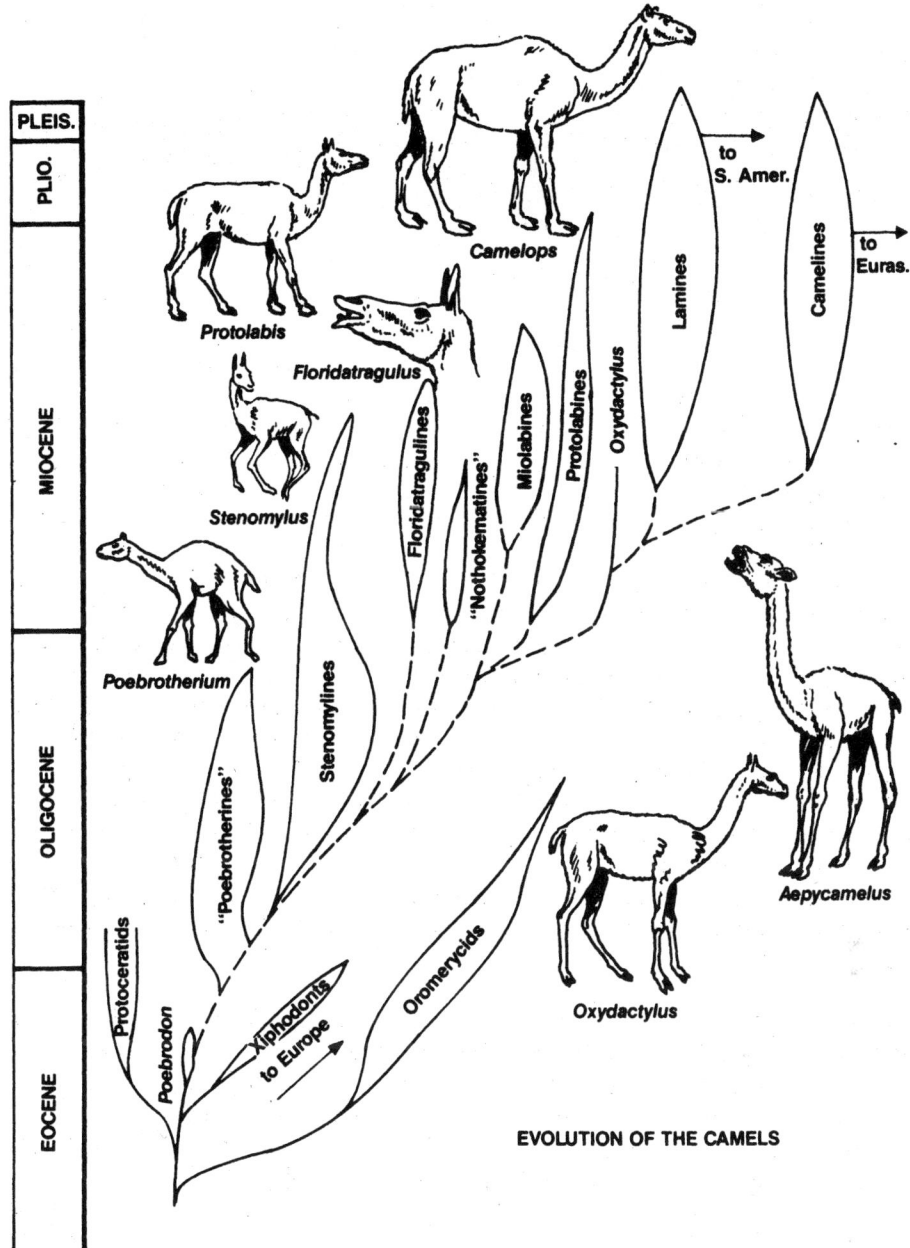

PLEIS.

PLIO.

MIOCENE

OLIGOCENE

EOCENE

Camelops

to
S. Amer.

to
Euras.

Protolabis

Floridatragulus

Lamines

Camelines

Stenomylus

Floridatragulines

"Nothokematines"

Miolabines

Protolabines

Oxydactylus

Poebrotherium

"Poebrotherines"

Stenomylines

Protoceratids

Poebrodon

Xiphodonts
to Europe

Oromerycids

Oxydactylus

Aepycamelus

EVOLUTION OF THE CAMELS

Figure 5.10. Evolution of camels. Camels first evolved in North America during the late middle Eocene and were gazelle-like in shape through most of the Oligocene. However, by the late Oligocene and Miocene, they had diversified into a wide variety of shapes, including gazelle-like camels, giraffe-like camels, and short-legged camels (protolabines and miolabines). Drawn by C. R. Prothero.

Figure 5.11. Archaeotherium mortoni, *the common early Oligocene entelodont from the Big Badlands. The huge pig-like beasts known as entelodonts were common both in Eurasia and North America during the Oligocene. They had bony warts on their faces and lower jaws, huge canines, and blunt teeth for crushing bones and vegetation. Painting by B. Horsfall, in Scott 1913.*

common, as were the hippo-like *Metamynodon* rhinos (fig. 5.12B) wading in the rivers; the long-legged, Great Dane–sized rhino *Hyracodon* (fig. 5.12C) running across the landscape; and the true rhinoceros *Subhyracodon* (fig. 5.12D). Tapirs, on the other hand, were rare after their great diversity in the middle Eocene. Small mammal faunas were dominated by rodents (especially modern groups like hamsters, pocket gophers, beavers, and squirrels), plus abundant rabbits and a whole suite of insectivorous mammals. Preying upon the herbivores was the last of the archaic creodonts, *Hyaenodon* (fig. 5.13A), plus a much greater diversity of true carnivorans: the early dog *Hesperocyon* (fig. 5.13B), which looked more like a weasel; the first members of the weasel family; primitive amphicyonids (known as "beardogs," they are a unique extinct family); and abundant nimravids (fig. 5.13C), which converged on cats and sabertooths, even though they are not closely related to cats.

Most of the herbivorous mammals had fairly primitive, low-crowned (brachydont) dentitions for eating leaves and shrubs, and there is no evidence of grasslands, or of animals with high-crowned (hypsodont) teeth for eating grasses yet. However, animals with extremely primitive, low-

A   B

C   D

Figure 5.12. Perissodactyls (odd-toed hoofed mammals) of the early Oligocene. (A) The common early Oligocene three-toed horse Mesohippus, *which was about the size of a German shepherd dog. (B) The hippo-like rhinoceros* Metamynodon, *found in many of the lower Oligocene river-channel deposits of the Big Badlands. (C) The running rhinoceros* Hyracodon, *which was about the size and proportions of a Great Dane. (D) The true rhinoceros (family Rhinocerotidae)* Subhyracodon, *also common in the early Oligocene of the Big Badlands. Painting by B. Horsfall, in Scott 1913.*

crowned teeth (such as brontotheres and pantodonts) were gone by this time, so that whatever plants they fed on, the softest vegetation must have disappeared. The absence of tree-dwelling mammals like primates and multituberculates also shows that the dense forests must have vanished: almost all the small mammals (even the squirrels) appear to have been adapted to life on the ground. This White River Chronofauna (described in chapter 4) was stable and well established, and would remain relatively unchanged from the late Eocene (Chadronian) until well into the Miocene.

La Grande Coupure

In Europe, there are no great desert badlands exposures rich with fossils like those in North America or Mongolia. Instead, the best fossil record of

A

B

C

Figure 5.13. Carnivorous mammals of the early Oligocene. (A) The primitive creodont Hyaenodon, *one of the last of its kind in North America, with the tiny deer-like* Leptomeryx *in the background. (B) Early true dogs, like this* Hesperocyon, *looked more like weasels than their descendants did. (C) The false sabertooth* Hoplophoneus primaevus, *with the horse* Mesohippus *in the background. Despite their appearance, the cat-like predators known as nimravids were not true cats and may have been more closely related to dogs than cats. Painting by B. Horsfall, in Scott 1913.*

terrestrial life in the Eocene and Oligocene comes from pockets and caverns in the limestone bedrock that have trapped dense concentrations of mammal bones that fell in or were washed in. Known as "fissure fills," these locations can yield thousands of extraordinarily preserved bones, but no complete skeletons, and because each fissure could have opened at any time in the Eocene or Oligocene, there is no stratigraphic superposition. Nevertheless, European paleontologists have made the best of this record and have sorted out the complex history of fissure formation and filling in the Cenozoic. The overall picture was apparent to Hans Stehlin in 1909, when he described the differences between the Eocene and Oligocene faunas as *la grande coupure*. Late Eocene faunas (discussed in chapter 4) were still dominated by the endemic groups of the European archipelago, including the horse-like palaeotheres (fig. 4.30B), dozens of endemic families of deer-like (fig. 4.30C) and pig-like artiodactyls, many endemic rodents, the last of the lemur-like adapid and omomyid primates, and both carnivorans and hyaenodont creodonts. By contrast, the Oligocene faunas were strikingly different. Some 60% of these Eocene endemic groups had vanished. The new replacement groups were dominated by immigrants from Asia. The carnivorous mammals included leftover hyaenodonts, plus some of the earliest members of the bear, raccoon, weasel, and mongoose families, as well as beardogs and nimravids. Among perissodactyls, the last of the palaeotheres survived alongside true rhinoceroses, fleet hyracodont rhinos, and aquatic amynodont rhinos. The few surviving endemic artiodactyls had to compete with immigrant pig-like entelodonts, the earliest peccaries, hippo-like anthracotheres, and a variety of advanced ruminants. The archaic Eocene rodents almost vanished, and were replaced by immigrant groups, including the aplodontids (represented today by the sewellel), beavers, squirrels, dormice, and hamsters, plus the rabbits.

As we saw on other continents, the most striking difference is the loss of arboreal mammals (especially primates) and the replacement of obligatory leaf-eaters with mixed feeders, adapted to tougher vegetation and mixtures of leaves and scrub brush. One way to detect this difference is to examine the distribution of body sizes. If you plot all the herbivorous mammals in a fauna in order of increasing size, they form a distinctive plot called a cenogram (fig. 5.14). In a tropical forest, the plot has a gentle slope with no break, because there is a high diversity of mammals of almost every body size. But in drier, scrubbier habitats, the shape of the cenogram is remarkably different. The slope is steeper (because large hoofed mammals are bigger), the line is shorter (because there are fewer total species), and there is a distinct break in the middle size range, reflecting the absence of medium-sized arboreal species. Thus, the trends we have seen in qualitative form can be translated into graphical form as well.

Even though the faunal transformation was dramatic during *la grande coupure*, the transformation was largely caused by wholesale immigration of new groups replacing the old ones, not by a dramatic climatic change.

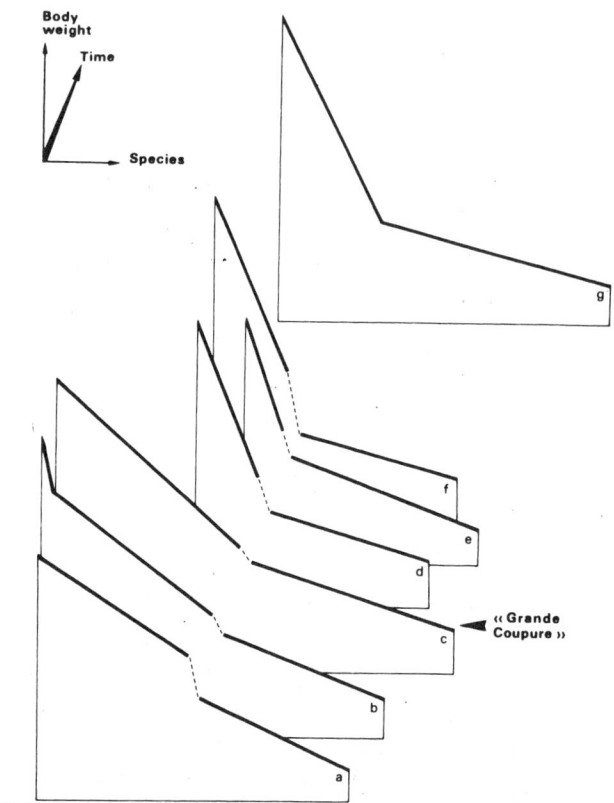

Figure 5.14. Cenograms of European mammalian faunas from the Phosphorites de Quercy, France, showing the sharp change in the size distributions after la grande coupure. *Before that event (cenograms a and b), the slope is gradual with no inflection and only a slight break in the midsize range. After the event (cenograms d–f), there is a sharp inflection, with a steeper left side (reflecting a few, much larger mammals at the high end of the distribution), a large break in the middle (where the tree-dwellers, such as primates, would fall), and an overall shorter cenogram (reflecting a much smaller total diversity of species). After Legendre and Hartenberger 1992.*

Floral analysis bears this out (Collinson and Hooker 1987; Collinson 1992). The floras of the late Eocene were dominated by evergreens, bald cypresses, and reed marshes, and those of the Oligocene were mixed deciduous and evergreen floras indicative of a warm-temperate climate. The pollen shows that the last of the tropical and subtropical vegetation disappeared, replaced by temperate plants and conifers (Boulter and Hubbard 1982, 1983; Boulter 1984; Collinson 1992). Thus, there is clear evidence of cooling in Europe, but not of the drying trend seen in North America, probably because these small humid islands were not likely to undergo the drying seen in the center of the North American continent.

For a long time, European paleontologists correlated *la grande coupure* with the Eocene-Oligocene boundary (Cavelier 1979; Savage and Russell 1983), and some European paleontologists insist on that correlation even today. But Hooker (1992) showed that *la grande coupure* occurred in the early Oligocene, which is consistent with what we now know about the early Oligocene global cooling event. When the Oligocene glaciers advanced, they caused a global drop in sea level as they locked up a huge amount of water. This sea level drop probably opened land-bridge corridors between the European archipelago and Asia, allowing mammals to immigrate in huge numbers in the early Oligocene.

Figure 5.15. Typical outcrops of the Oligocene in the Kumbulak Cliffs of Kazakhstan. The upper Eocene beds (Chegan) are shown in the right distance at the base of the section. The lower Oligocene (Hsandgolian) Kutanbulak Formation makes up the badlands in the left distance, and they are capped by the upper Oligocene Chilikta Formation in the foreground. Photo courtesy of S. Lucas.

The Asian Oligocene fossil record is also excellent, not only in China and Mongolia but also in many regions of the former Soviet Union, such as Kazakhstan (fig. 5.15) (Russell and Zhai 1987; Lucas et al. 1998; Emry et al. 1998: Meng and McKenna 1998). The early Oligocene in eastern Asia is known as the Hsandgolian, after the famous locality of Hsanda Gol in Mongolia. In this unit, the faunas were very different from those of the late Eocene, and also from those of the rest of the world. The embolotheres were gone, as were the last of the mesonychids, and most of the tapiroids that dominated in the late Eocene. However, the antique didymoconids persisted, as did the entelodonts (fig. 5.11) and the hyaenodont creodonts (fig. 5.13A). Both hyracodont (fig. 5.12C) and true rhinos were common, as were deer and a variety of deer-like artiodactyls. Several new groups occur for the first time, including various ruminant artiodactyls and several advanced rodents, plus a whole suite of advanced carnivorans: beardogs, weasels, and the first true cats (sharing the predator niche with the hyaenodonts, the cat-like nimravids, and civets, which were already established).

Even though this list of characters seems similar to the early Oligocene faunas of other continents, there are important differences. The diversity of late Eocene large mammals is very reduced, with only rhinos and entelodonts occupying the large-body-size niche. And huge they were! Hsanda Gol is famous for producing the largest land mammals that ever lived, the indricotheriine hyracodont rhinos such as *Paraceratherium* (formerly known as *Baluchitherium*). This beast was 6 meters (18 feet) tall at the shoulder and probably weighed 20 tons (fig. 5.16). Its head was so high off the ground that it probably browsed the tops of trees, and it

Figure 5.16. The immense hyracodont rhinoceros Paraceratherium *(formerly known as* Baluchitherium *or* Indricotherium*).* Paraceratherium *was the largest land mammal that ever lived, and here it towers above the elephants in this display. Although it was large and heavy, it still had the long legs and toes of its hyracodont ancestors (fig. 5.12C). It was a common fossil in the Oligocene beds of Asia. Photo courtesy of the University of Nebraska State Museum.*

dwarfed modern elephants. Even its skull was over 1.5 meters (5 feet) long. Yet despite its gigantic size, it retained the hallmarks of its ancestry in the running rhino family Hyracodontidae. Its limbs were long and relatively slender, with long toe bones, very different from the short, compressed limbs and toes seen in other huge land animals, such as elephants. Its teeth were still like those of a hyracodont, only huge. Like all other primitive rhinos, it had no roughened spot on the forehead for attachment of the horn, so it was probably hornless.

These few large rhinos and entelodonts were the only big species, and there were almost no mid-sized species. The Hsanda Gol fauna is overwhelmingly dominated by rodents and rabbits, many with high-crowned teeth for eating gritty vegetation. According to Meng and McKenna (1998), the overall faunal composition suggests that the environment was arid, with few tall trees for arboreal species. Almost all the land mammals were adapted either for feeding on the tops of the trees (like the indricotheres) or for living close to the ground and within the limited brush cover (most of the deer-like artiodactyls, entelodonts, and their predators, plus the great variety of ground-dwelling and burrowing rodents and rabbits). Leopold et al. (1992) studied the pollen from the Oligocene of China and concluded that much of the region was covered by a woody scrubland, with many arid-adapted plants, such as saltbush.

In summary, although Hsandgolian mammals share many families and genera with North America (especially the rhinos, entelodonts, anthracotheres, and ruminants, plus many of the same weasels, beardogs, and nimravids), the fauna was much more arid-adapted than that of the Orellan in North America (pl. 3). In Asia, there were many more mammals bearing high-crowned teeth, and much floral evidence for grittier diets and scrubbier vegetation. And although many of these Asian groups migrated to Europe during *la grande coupure* to drive out the native fauna, Europe in the early Oligocene was still a much wetter and milder place than either Asia or North America.

Lost Worlds

Asia, North America, and Europe were partially connected during the early Oligocene and shared many mammals in common, but the same is not true of the rest of the continents. Africa remained isolated from the rest of the world, and continued to develop a peculiar endemic fauna as well. As we saw in chapter 4, the continent was already inhabited by a strange mixture of native endemic mammals. Foremost among these were the proboscideans, including the pig-like *Moeritherium* and the more mastodon-like *Numidotherium*. The most common mammals were the hyraxes, which today include several species of marmot-like mammals that are distant relatives to the proboscideans. However, the hyraxes of Africa in the Oligocene were very diverse, with some reaching rhino-sized proportions! Even more striking were the arsinoitheres, which reached elephantine size but had a pair of sharp bony horns on their noses (fig. 5.17A). These animals had once been placed in their own order (Embrithopoda) and were considered a zoological mystery, but recent work has shown that they are closely related to proboscideans and sirenians. At one time, they were known only from *Arsinoitherium* itself from the Fayûm beds of Africa (fig. 5.17B), but now they are known from the Eocene of Turkey (Maas et al. 1998) and the Paleocene of China (McKenna and Manning 1977) as well. The only artiodactyls in the continent were the ubiquitous aquatic anthracotheres, which managed to travel freely between Europe, Asia, and North America in the Oligocene.

In addition to these large mammals, the small mammal fauna of the Fayûm beds of Africa included many different insectivores (including one, *Ptolemaia,* that became a wolf-sized predator), but the rodents were all from an endemic family, the Phiomyidae. There were no rabbits such as were found everywhere in the Northern Hemisphere by the early Oligocene. Even more striking are the flesh eaters. No members of the order Carnivora were living in Africa in the Oligocene. Only the ubiquitous hyaenodont creodonts (fig. 5.13A) fed on early mastodonts, hyraxes, arsinoitheres, and anthracotheres. But the most distinctive (and important to us humans) element of the fauna was the abundance of primates, including the first true apes and many relatives of Old World monkeys (fig. 5.17A). Primates had vanished from Europe by the late Eocene and were gone from much of Asia and North America by the late Eocene, so their last refuge was Africa, where they flourished and diversified into apes and monkeys. Eventually, by the late Oligocene, they would spread to South America, where they founded the huge radiation of New World monkeys, the family Cebidae (spider monkeys, howler monkeys, marmosets, capuchins, and many others). They spread to Eurasia in the middle Miocene, but primates did not return to North America (where they were dominant in the Paleocene) until the arrival of humans.

South America, too, continued its isolation in the Oligocene. For a long time, there were no known faunas of the Eocene-Oligocene transition or earliest Oligocene known from South America. Supposedly early Oligocene Deseadan faunas are now known to be late Oligocene (Flynn and Swisher 1995). But recent discoveries in the Chilean Andes have produced a whole new set of faunas that are the basis for the latest Eocene–early Oligocene Tinguiririan (fig. 5.3) South American land mammal age (Wyss et al. 1990, 1993, 1994). These faunas, which are just now being studied and described, contain the usual suspects: native ungulate groups (notoungulates and litopterns), plus edentates and marsupial carnivores. As we noted for the late Eocene in South America, the herbivores already had very high-crowned teeth (long before this phenomenon occurred in Asia, and especially early compared to Europe or North America). And these herbivores are important for another reason: the first South American rodents now known date to the late Eocene in age (Frailey and Camp-

A

B

Figure 5.17. The Fayûm beds of Egypt preserve some of the best late Eocene and Oligocene fossils in Africa. (A) Reconstruction of the forested Fayûm area in the early Oligocene. The elephant-sized two-horned Arsinoitherium wanders around, while the early monkey Apidium dozes on the branches. From Turner and Anton 2004. (B) Today, the Fayûm region is a hostile, forbidding desert just west of the Nile Valley. Photo courtesy of D. T. Rasmussen.

bell 2004). Today, this group (the Caviomorpha) is tremendously diverse, with everything from the giant capybara (the largest rodent alive, as big as a pig) to agoutis and pacas to the tiny Guinea pig. They are closely related to the African porcupines and are now thought by most paleontologists to have rafted over from Africa or Europe. Once established in South America, they quickly radiated and came to dominate the small mammal fauna.

Cold, Dry World

The late Oligocene continues the tendencies that marked the beginning of the Oligocene. On some oxygen isotope curves (figs. 3.22, 5.3) near the Antarctic (Miller et al. 1987; Zachos et al. 1992), there is another big drop in temperatures, suggesting that the glaciers had returned to Antarctica after having melted back in the late early Oligocene. Early Oligocene glaciers may have lasted only a few hundred thousand years, and they were found mostly on West Antarctica and in the Indian Ocean sector of the Southern Ocean (Kennett and Barker 1990). But about 30 million years ago, the temperatures dropped again, and major glaciers formed on Antarctica that lasted almost 4 million years (Miller et al. 1991; Miller 1992). Glacial sediments of middle Oligocene and late Oligocene age can be found from many drill cores taken from the edge of the Antarctic continent. Deep seismic exploration has even found a huge middle Oligocene erosional surface buried beneath hundreds of meters of upper Oligocene glacial sediments, a surface that can be traced from the deep sea up onto the Antarctic continent, and across at least 100,000 square kilometers beneath the Ross ice shelf. This surface suggests that an enormous amount of seawater was locked up in ice sheets, causing sea level to drop dramatically. The surface was then buried as those same ice sheets grew and dumped a huge load of their sediment upon it.

The effect of this huge ice sheet can be seen worldwide. Wherever we have a good sedimentary record of the Oligocene, there is usually a big erosional surface at about 30 Ma. In most cases, it is the largest and most deeply eroded unconformity in the entire sequence (Haq et al. 1987). Even in the deep sea, with its nearly continuous rain of fine clays and microfossils, there are deep-sea erosional and dissolution events (Keller et al. 1987). Seismic profiles off the continental shelf of New Jersey and Virginia reveal deeply incised submarine canyons, cut by rivers when the retreating oceans left most of the continental shelf exposed in the middle Oligocene (Miller et al. 1985, 1987).

Despite the occurrence of this major new glaciation, global cooling, and a huge drop in sea level, the response by most organisms was not impressive. There was a diversification of cold-adapted planktonic foraminiferans (Keller 1983a, 1983b) and coccoliths (Haq and Lohmann 1976; Aubry 1992), but there were relatively few extinctions in the plankton. There were no extinctions in the diatoms (Baldauf 1992). Although data are sparse, there were few changes in the marine mollusks, except that cold-water forms became dominant on the Pacific Coast (Hansen 1987, 1992; Dockery 1986, 2003; Squires 2003; Oleinik and Marincovich 2003). Nor do the echinoid faunas show much change (McKinney et al. 1992).

If the late Oligocene climatic change was so severe, why didn't marine life respond? Most paleontologists who work on these fossils suggest that they were already cold-adapted forms that could survive another episode of cooling. Most of the vulnerable temperature-sensitive tropical Eocene forms were long gone, and all that remained were hardy survivors, tolerant of a wide range of temperatures as well as cooling. For example, planktonic foraminiferans are represented during the middle Oligocene by only a handful of species that tolerated cold water (Keller 1983a, 1983b). As we saw in chapter 3, this pattern is typical of most mass extinctions, where the recovery world was inhabited by hardy generalists and opportunists (weedy species) that could tolerate further bad times without going extinct (Stanley 1990).

Land plants tell the same story. Wolfe (1978) detected no significant drop in temperature or seasonality in late Oligocene plants after the early Oligocene deterioration, probably because most late Oligocene plants were already cold-tolerant. In the Big Badlands paleosols, Retallack (1983, 1992) detected a transition from Orellan (earliest Oligocene) wooded grasslands and gallery woodland to late Oligocene (Whitneyan and Arikareean) open grasslands with trees only along watercourses (fig. 5.6B). There is also a marked drying trend, from average annual rainfall in the early Orellan of 50–90 centimeters (20–35 inches) to only 35–45 centimeters (14–18 inches, just wetter than a desert) in the late Oligocene. Whitneyan and early Arikareean deposits of the Great Plains were predominantly wind-blown silts and volcanic dust, indicating very dry conditions.

What sort of animals lived in these cool, dry conditions? For North America, in the upper part of the Badlands section in South Dakota and especially Nebraska we have an excellent record of the Whitneyan and Arikareean land mammal ages (fig. 5.18A). The best record of the Arikareean occurs in the volcaniclastic deposits of the John Day Formation in central Oregon (fig. 5.18B). The diversity of land mammal faunas hit an all-time low during this time (Prothero 1985; Stucky 1990, 1992). In addition to there being fewer total species (lower species richness), the faunas were composed of just a handful of common mammals (higher dominance). By far the most abundant in the Whitneyan and Arikareean was the little oreodont *Leptauchenia* (fig. 5.19A), which had high-crowned teeth for eating tough, gritty vegetation. It is found in large numbers in the wind-blown deposits of the Whitneyan and Arikareean of the High Plains, which suggests that it inhabited these dune deposits. However, its ears, eyes, and nostrils were at the top of its head (fig. 5.19B), leading some paleontologists to suggest that it was partially aquatic like a hippo or tapir, both of which have similar features. The absence of *Leptauchenia*

Figure 5.18. (A) The classic upper Oligocene (Whitneyan) Whitney Member of the Brule Formation and Arikareean deposits of the Gering and Monroe Creek formations, Arikaree Group, at Helvas Canyon on Wildcat Ridge in western Nebraska. (B) The most fossiliferous late Oligocene record in North America: the John Day beds of central Oregon, which entombed hundreds of Arikareean fossils along with volcanic ashes that allow precise dating. Photos by the author.

A

B

A

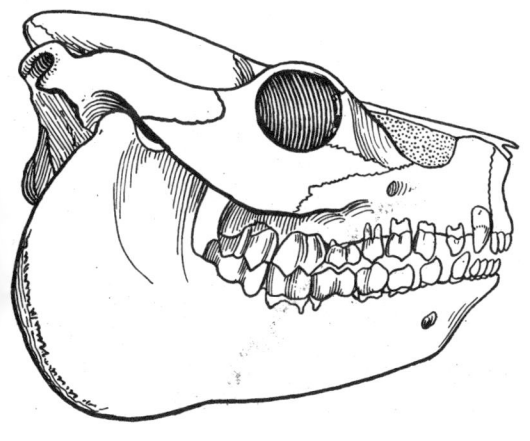

B

C

Figure 5.19. (A) The small oreodont Leptauchenia, *by far the most common mammal in the late Oligocene of North America. This restoration of* Leptauchenia, *shows the old notion that the eyes and ears high on the skull were hippo-like features for swimming. (B) The skull of* Leptauchenia, *showing the high placement of the eyes and ears, the large bony ear cavities, the opening on the face, and the very high-crowned teeth. (C) The Whitneyan horned artiodactyl* Protoceras, *distantly related to camels. Painting by B. Horsfall, in Scott 1913.*

in floodplain and river-channel deposits, and their abundance in the dune deposits, seem to rule this out (Prothero and Sanchez 2006). Moreover, there are living desert animals with high-placed nostrils and eyes for burrowing, and with openings on the snout like the facial vacuities of leptauchenines, which may have served as an area for a filtering system to keep dust out of the lungs. A dwarfed leptauchenine known as *Sespia* was particularly common in the early Arikareean both in the Great Plains and in California.

In addition to the highly specialized leptauchenine oreodonts, there were abundant oreodonts (*Eporeodon* and *Mesoreodon*) of the main oreodont lineage as well. Some got to be almost cow-sized, and their huge skulls are among the largest fossils found in the Whitneyan. There were abundant camels (fig. 5.10), which began to diversify in the Arikareean to take advantage of their long necks and legs. These included the gazelle-like stenomyline camels (fig. 6.6E), which by the Arikareean had developed extremely high-crowned teeth as well. Closely related to the camels was a group of artiodactyls known as protoceratids, which had first appeared in the great middle Eocene artiodactyl radiation. But Whitneyan *Protoceras* (figs. 6.8, 5.19C) was a distinctive beast. Male skulls have two pairs of knob-like horns over the eyes and over the nose, whereas the females were apparently hornless. Small hornless deer-like ruminants such as *Hypertragulus* were also common in the Arikareean.

Among perissodactyls, there were a handful of species of the three-toed horse *Miohippus* during the Whitneyan, and by the later Arikareean we can see the beginning of the great Miocene horse radiation with the appearance of *Kalobatippus*, *Archaeohippus*, and *Parahippus*. Rhinos, however, remained low in diversity, with the hippo-like amynodont (fig. 5.12B) and the running hyracodonts (fig. 5.12C) dying out. Only a few species dominated by one genus, the true rhinoceros *Diceratherium*, the first horned rhino in North America, lived for the 10 million years of the Whitneyan and Arikareean. *Diceratherium* had paired flange-like horns on its nose and reached the size of modern rhinos.

Like the oreodonts, camels, ruminants, horses, and rhinos, the carnivores were also low in diversity during the Whitneyan and Arikareean. The last hyaenodont creodonts (fig. 5.13A) died out by the early Arikareean, so for the rest of the Cenozoic, the predator niche in North America would be occupied only by members of the order Carnivora. The cat-like nimravids (fig. 5.13C) straggled on through the Whitneyan and Arikareean with sabertooths like *Nimravus* and *Pogonodon*, then vanished at the end of the Arikareean at 23 Ma, leaving a "cat gap" of about 6 million years when there were no cat-like forms on this continent until the arrival of the true cats in the early Miocene (at around 17 Ma). Weasels and beardogs also persisted, along with a radiation of many types of dogs (at least a dozen genera in the Arikareean), which quickly filled the niches of bone-crushing hyenas as well as small- and medium-sized carnivorous mammals.

A

B

Figure 5.20. (A) Mylagaulids were horned rodents distantly related to the living sewellel, or Aplodontia. *Painting by B. Horsfall, in Scott 1913.* (B) They left corkscrew-shaped burrows, known as devil's corkscrews (Daemonelix), *in the Arikareean dune deposits.*

Finally, the dune deposits of the Whitneyan and Arikareean housed a great radiation of rodents that adapted to the harsher, drier climates with higher-crowned teeth and a greater emphasis on burrowing. Nearly all the archaic rodent groups of the Eocene and early Oligocene were gone, replaced by the aplodontids, cricetids (now represented by hamsters and New World mice), pocket mice, jumping mice, and pocket gophers. One group of rodents, the horned mylagaulids (fig. 5.20A) became highly specialized for burrowing and dug distinctive helical-shaped burrows. Known as "devil's corkscrews," or *Daemonelix,* they are among the most distinctive burrows found in the Arikaree Group (fig. 5.20B).

In Asia, mammals were already adapted to the dry climates of the early Oligocene, so those of the late Oligocene (Tabenbulukian) were not markedly different from those of the Hsandgolian (Russell and Zhai 1987; Meng and McKenna 1998). Giant indricotheres (fig. 5.16) still towered above the trees, along with a number of smaller true rhinoceroses and several kinds of small hornless ruminants, including the first deer, *Eumeryx*. However, many characteristic Eocene and early Oligocene groups were finally gone, including the hyaenodont creodonts (their disappearance in Asia was simultaneous with their disappearance in North America), the cat-like nimravids (they became extinct in Asia slightly earlier than in North America), and huge pig-like entelodonts (fig. 5.11), along with several archaic groups of rodents. In the absence of hyaenodonts and nimravids, the beardogs, weasels, civets, and the first Asian dogs performed the role of flesh eaters. Finally, the rodents diversified just as they had in North America, with many species of cricetids, beavers, jumping mice, and squirrels complementing the many endemic Asian rodent groups.

Europe, too, saw a lower diversity of mammals than at any other time in the Cenozoic (Vianey-Liaud 1991; Legendre and Hartenberger 1992). The last of the archaic palaeotheres, anthracotheres, entelodonts, and several other archaic artiodactyl groups vanished, as did several of the endemic European rodent groups. This time period marked the end of Europe's provincialism as more and more of the mammal fauna was shared with Asia, and most of Europe's Eocene endemics disappeared. True rhinoceroses continued to diversify (some with paired nasal horns, as in North America), although the hornless hyracodont and amynodont rhinos still straggled on. The last of the European amynodonts, *Cadurcotherium*, probably had a short trunk like that of a tapir. However, several modern groups (including the first true tapirs, pigs, peccaries, deer, and advanced ruminants) replaced these European endemics (Brunet 1977; McKenna 1983; Agusti and Anton 2002), accompanied by an evolutionary radiation of advanced rodents, including glirids (dormice) and cricetids. As in North America, the late Oligocene was the last gasp for hyaenodont creodonts and cat-like nimravids. Their role as flesh eaters was taken over by a radiation of beardogs and the first true bears (a small raccoon-like omnivore known as *Cephalogale*) and raccoons, along with civets and weasels (but no dogs or cats yet).

What about the island continents that were separated from the Northern Hemisphere migration routes between Asia, Europe, and North America? Africa's isolation continued into the late Oligocene, although few fossil mammals are known after the early Oligocene faunas of the Fayûm in Egypt (Rasmussen et al. 1992). As far as the spotty late Oligocene fossil record (and the early Miocene faunas) shows, Africa was still dominated by its peculiar fauna of endemic rodents, mastodonts, hyraxes, and early monkeys and apes. Australia, too, lacks any early or

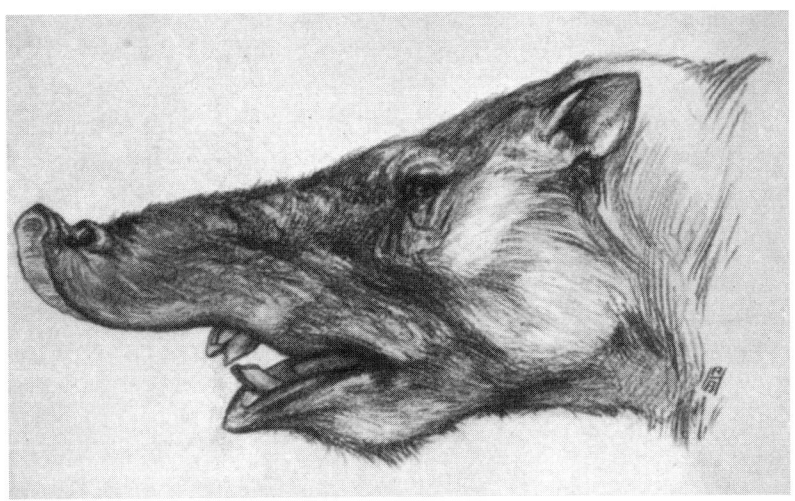

Figure 5.21. The elephant-like extinct hoofed mammal known as Pyrotherium *was one of the most characteristic mammals of the late Oligocene (Deseadan) of South America. After Scott 1913.*

late Oligocene faunas, except for possibly one locality of latest Oligocene age, which has primitive members of several living native marsupial families (Tedford et al. 1975).

Only one land mammal age, the Deseadan, represents the late Oligocene in South America (Flynn and Swisher 1995; Kay et al. 1999). According to Marshall and Cifelli (1989), the Deseadan fauna had even more mammals with high-crowned teeth than we see in earlier times, and climatic clues indicate that there were abundant savanna woodlands in the Deseadan. Many of the archaic groups of leaf-eating native hoofed mammals, rodent-like marsupials, and primitive edentates were on their last legs. In some cases, the Deseadan yields huge end-members of long-established lineages, such as gigantic carnivorous marsupials. Another Deseadan giant is the peculiar *Pyrotherium,* which looked remarkably like a small mastodont, yet had no relation to the proboscideans (fig. 5.21). By contrast, lineages of smaller species—and the smaller, less specialized marsupial carnivores—began to diversify to cope with harsher, drier conditions. Finally, the Deseadan saw two new groups take off. The caviomorph rodents had arrived in the late Eocene, but by the Deseadan, they had diversified into at least seven families, including the chinchillas and New World porcupines. More importantly, the Deseadan marks the first appearance of New World monkeys, the Cebidae, which presumably rafted over from Africa, the only place that had primates by the late Oligocene. As discussed above, they quickly became established, and today they are represented by a huge diversity of spider monkeys, howler monkeys, marmosets, capuchins, and their kin. Thus, most of South

America's living native fauna (edentates, marsupials, caviomorph rodents, and New World monkeys) had arrived there by the late Oligocene.

As the Oligocene concluded, there was a slight warming event in the oxygen isotope curve (Zachos et al. 2001), then an abrupt glacial advance and cooling at the Oligocene-Miocene boundary. When this ended, the world warmed into one of the longest of the epochs, the Miocene, which lasted from 23 to 5 Ma (more than 18 million years). Only the Eocene (55–34 Ma, 21 million years) was longer.

For Further Reading

Agusti, J., and M. Anton. 2002. *Mammoths, Sabertooths, and Hominids: 65 Million Years of Mammalian Evolution in Europe.* New York: Columbia University Press.

Prothero, D. R. 1994. *The Eocene-Oligocene Transition: Paradise Lost.* New York: Columbia University Press.

Prothero, D. R., and W. A. Berggren, eds. 1992. *Eocene-Oligocene Climatic and Biotic Evolution.* Princeton, N.J.: Princeton University Press.

Prothero, D. R., and R. J. Emry, eds. 1996. *The Terrestrial Eocene-Oligocene Transition in North America.* Cambridge: Cambridge University Press.

Prothero, D. R., L. C. Ivany, and E. A. Nesbitt, eds. 2003. *From Greenhouse to Icehouse: The Marine Eocene-Oligocene Transition.* New York: Columbia University Press.

Figure 6.1. The modern Antarctic ice sheet first developed in the middle Miocene and has covered the South Pole for the last 15 million years. Photo courtesy of L. C. Ivany.

6

The Savanna Story: The Miocene

And now there came both mist and snow,
And it grew wondrous cold:
And ice, mast-high, came floating by,
As green as emerald.

Samuel Taylor Coleridge, *Rime of the Ancient Mariner,* 1798

Warm Again

The Miocene (from 23 to 5 Ma), the second longest epoch of the Cenozoic, represented a time of tremendous change. Global climate went through numerous fluctuations, from a severe icehouse at the end of the Oligocene to a warming interval in the early Miocene, then the full-fledged onset of the modern Antarctic ice cap (fig. 6.1) by the middle Miocene, culminating in the cooling and drying of climate in temperate latitudes to form grassy savannas (pls. 4, 5). The surface of the globe changed too, as continents collided, oceans dried up, and land bridges formed and broke. Finally, the marine and terrestrial faunas began to take on their modern appearance, as most of the living families of animals and plants were on the scene by the end of the epoch.

Let us start with the global tectonic picture (fig. 6.2). During the Paleocene, Eocene, and Oligocene, the dominant theme was the continuing separation of the major continents from the fragments of the Pangaea supercontinent, a process that had begun back in the Mesozoic. The Atlantic had been opening since the Triassic, and India, Africa, and South America had separated from Antarctica-Australia by the Late Cretaceous. High spreading rates and global plate reorganizations characterized the Eocene and Oligocene and also may have been responsible for the cooling trend (Berner et al. 1983). Australia and Antarctica had begun to split apart by the Late Cretaceous, and by the early Oligocene they were fully separated, allowing deep-water circum-Antarctic circulation (Exon et al.

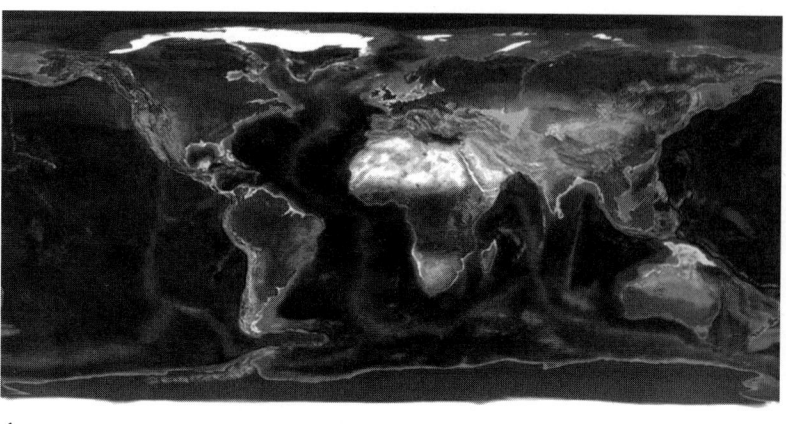

Figure 6.2. (A) Paleogeographic map of the world during the Miocene. Courtesy of R. Blakey. (B) Collision of Africa (AF) and India with Eurasia during the later Cenozoic closed up the ancient Tethys Seaway and caused a crumpling of continental crust that formed a mountain belt running from the Alps (A) through the Himalayas (H) to Indonesia and New Guinea. From Prothero and Dott 2003.

A

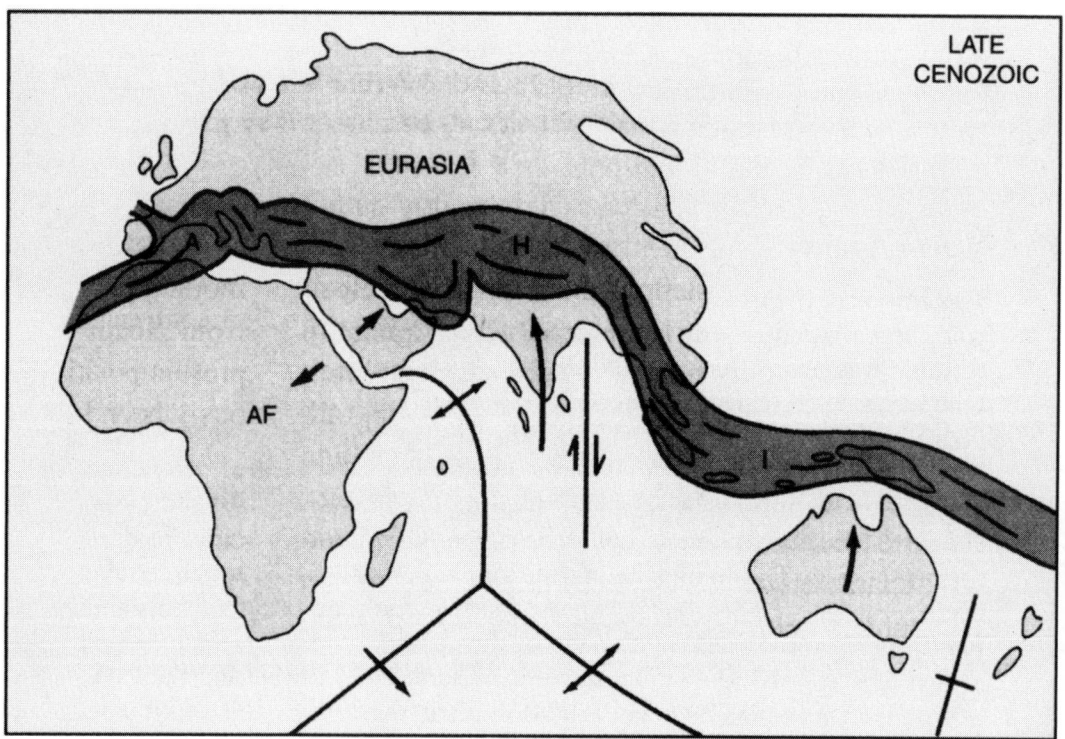

B

2002). South America may have finally pulled away from Antarctica by the early Oligocene as well (Diester-Haass and Zahn 1996). Meanwhile, India completed its mad dash across the Indian Ocean in the Late Cretaceous and Paleocene, colliding with Asia by the early Eocene to begin the uplift of the Himalayas.

In contrast to all this continental separation and seafloor spreading in the early Cenozoic, the effects of tectonic collisions and mountain uplift characterize the Miocene. The rate of influx of sediments from the Himalayas into the deep Indian Ocean (Rea 1992) indicates that the Himalayan uplift accelerated by the middle Miocene as the docking of India concluded. Increased weathering triggered by this uplift may have taken excess carbon dioxide out of the atmosphere and further enhanced the global cooling trend (Ruddiman and Kutzbach 1991; Raymo and Ruddiman 1992; Ruddiman 1997). By the early Miocene, Africa had nearly closed the Tethyan gap and was beginning its collision with Europe to form the Alps. First, though, was the impact of the Arabian Peninsula on the Middle East, forming the Zagros crush zone in Iran and closing the Tethys in the middle, so the eastern end of the Mediterranean was closed as well. We will soon see what happened when Africa had a land corridor to the rest of the Eurasian continent. Finally, by the middle Miocene, the long-quiescent western United States (which hadn't seen mountain building since the middle Eocene, when the Laramide Orogeny ended) became tectonically active, with the ripping open of the Basin and Range Province, the eruption of the Cascade Mountains and the Columbia River basalts, the formation of the San Andreas fault, and the renewed uplift of the Rocky Mountains (Prothero and Dott 2003). In short, the Miocene was a time of mountain building and uplift, with several continents feeling the effects not only of global cooling and drying but also of local topographic uplift and its climatic effects.

Next, let us look at the global climatic signal (figs. 3.22, 6.3). After the brief pulse of Antarctic glaciation that marked the Oligocene-Miocene boundary, the oxygen isotope record between 24 and 14 Ma is marked by a steady increase in global temperatures, peaking at about 14–16 Ma, in what Zachos et al. (2001) called the "mid-Miocene climatic optimum." Deep-sea bottom-water temperatures rose from about 3°C (37°F) to as high as 7°C (45°F), an increase that implies a much more dramatic increase in shallow marine settings and on the land. According to Woodruff (1985), the early Miocene ocean was relatively warm and not stratified by depth, with a diminished surface-to-bottom thermal gradient, so the effects of the cold bottom waters that formed in the Oligocene diminished. The bottom was also less oxygenated, consistent with a decrease in the production of cold Antarctic Bottom Waters.

The terrestrial record corroborates these trends in the deep sea. Floras such as the early Miocene Brandon lignite in Vermont (Tiffney 1994) suggest that even this far north, the climate was frost-free and warm-temperate to subtropical, with a mean annual temperature as high as 17°C (63°F); today the mean annual temperature in this part of Vermont is only 7.6°C (46°F), with severe winters. In addition, it was much wetter than today, with about 166 centimeters of annual precipitation, contrasted with only 95 centimeters today. Even up in the Canadian Arctic, the lake sediments that filled the Haughton impact crater on Devon Island contain a cool-temperate hardwood flora, which lived at a mean annual

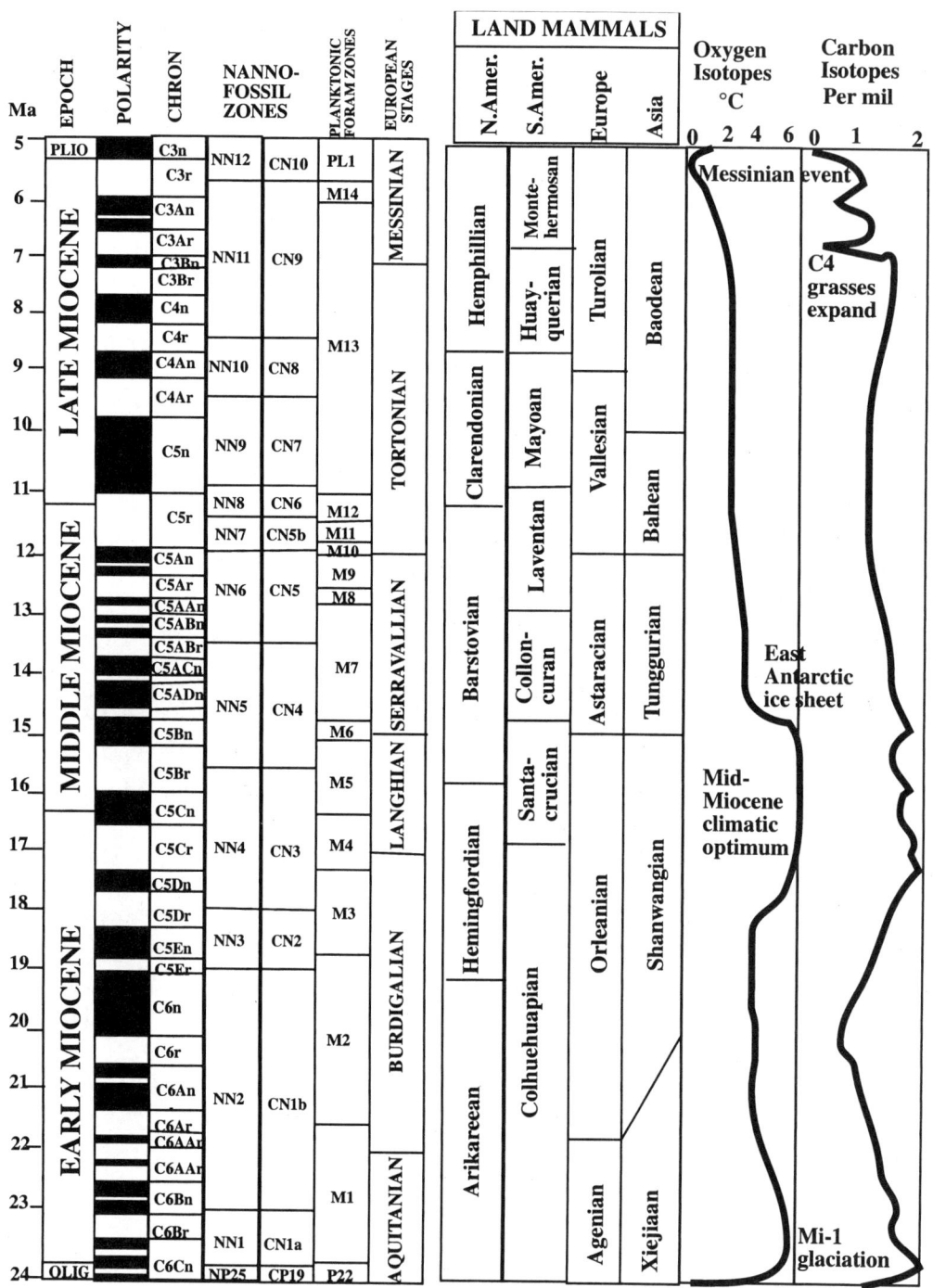

Figure 6.3. The Miocene timescale (left columns), as recently calibrated by Berggren et al. (1995). The North American land mammal record is after Tedford et al. (2004) and Woodburne and Swisher (1995); the South American land mammal ages are after Flynn and Swisher (1995); and the European and Asian record is after Qiu and Qiu (1995). The isotopic records are after Zachos et al. (2001), but they are very generalized and smoothed out. Abbreviations: OLIG = Oligocene; PLIO = Pliocene.

temperature of 11–15°C (52–59°F). Today, this region at 75° north latitude is Arctic tundra and permafrost, but then it supported a New England–style boreal forest with a peculiar mixture of turtles and mammals (Whitlock and Dawson 1990). However, the southeastern Gulf Coast was still warm-temperate to subtropical.

On the High Plains, there was a mixture of scrublands and woodlands in the early Miocene, and the same was true in the Rockies and the Great Basin of Nevada (Graham 1999). For example, the Buffalo Canyon flora of western Nevada (Wolfe et al. 1997) consisted of abundant pines, willows, poplar, and members of the rose family and experienced a mean annual temperature of 10°C (50°F) and an annual temperature range of 14°C (from below freezing to 75°F), and about 90–100 centimeters (35–40 inches) of annual precipitation. In the Pacific Northwest, famous floras from Alvord Creek in Oregon and Seldovia Point in Alaska show that these regions were covered by a mixed deciduous hardwood and coniferous forest. The Seldovia Point flora comes from the Cook Inlet region near Anchorage, Alaska (60° north latitude), and contains hackberry, oaks, walnuts, magnolia, liquidambar, and other trees not associated with the freezing modern Alaska climate. Its mean annual temperature was 6–7°C (43–45°F), but with an annual temperature range of 26°C (below freezing to over 90°F). Thus, climates were still cool-temperate even above the Arctic Circle in North America until the end of the late Miocene. There is no indication yet of the cooling that would come in the late Miocene, or the growth of the Arctic ice cap in the Pliocene.

The climate in rest of the world was also warmer and milder in the Miocene than in the Oligocene (Agusti and Anton 2002). Warm climates melted the Oligocene Antarctic glaciers, so the Antarctic was again mostly ice-free. The water released by melted glaciers raised sea levels, and many parts of Europe became islands again. Shallow seas covered the Rhône and Aquitaine basins in France and the Tagus Basin in Spain and Portugal, and there was extensive connection between the Mediterranean and Parathethys seas and the Indian Ocean. These European islands were cloaked by rich broad-leaved evergreen woodlands, dominated by oaks and laurels, cinnamon, magnolia, podocarps, pines, hemlocks, figs, and rattan palms. Such plants indicate a mixture of temperate and subtropical conditions, with many species that did not tolerate freezing. The coastal regions supported extensive mangrove swamps and palms, again indicating subtropical to warm-temperate conditions. As we saw in the Alaskan and Canadian Arctic, subtropical forests also extended into eastern Siberia and Kamchatka (Volkova et al. 1986). Subtropical mollusks occurred in shallow marine waters of these regions as well (Gladenkov 1992).

Warm-water tropical fishes and nautiloids were common in the epicontinental seas of Europe, and water temperature has been estimated at 25–27°C (77–81°F). Large reefs again grew around the Mediterranean after their near disappearance in the cold Oligocene. Reef-building corals even became established around North Island, New Zealand, suggesting that the water off this now cold-temperate part of the world had to have been in the 19–28°C (66–82°F) range for reefs to survive (Hornibrook 1992). The early to middle Miocene was the beginning of an explosive diversification of reef corals after the late Oligocene extinctions, especially in regions like the Caribbean (Coates and Jackson 1985; Budd et al. 1994).

Warm-water larger benthic foraminiferans and mollusks reached their southernmost Neogene distribution during the early Miocene. Planktonic foraminiferans also underwent a major morphological diversification after their low point in diversity in the early Oligocene (Kennett and Srinivasan 1983; Tappan and Loeblich 1988; MacLeod et al. 2000), with many morphological forms that had vanished during the Oligocene reappearing by convergent evolution during the Miocene (Cifelli 1969). For example, Oligocene extinctions had reduced the morphological diversity of planktonic foraminiferans to the simplest, most primitive forms, the globigerines and hastigerines (fig. 6.4). From these primitive stocks, the early Miocene planktonic foraminiferans re-evolved almost all the more complex shapes that had previously occurred in the Eocene, including the rapidly expanding chambers of the globigerinoids, the ball-shaped orbulines, and the keeled turboratalids and globorotalids.

Mollusks were also on the rebound after the crash in diversity and the influx of cold-water taxa in the Oligocene. Durham (1950) and Addicott (1969, 1970, 1976) studied the mollusks of the San Joaquin Basin of California, and also those of coastal Oregon and Washington. Late Oligocene molluscan faunas in both regions were at their lowest diversity, as represented by cold-tolerant faunas of formations such as the Pysht and Blakeley in Washington (*Liracassis apta* Zone). But the latest Oligocene to early Miocene faunas of the Vaqueros Formation show a renewed diversification of mollusks of the "Vaquerosian" stage, with abundant scallops and *Turritella inezana* in many outcrops of the Vaqueros Sandstone throughout California (Loel and Corey 1932). In addition to diversity trends, the effects of early Miocene warming are even more obvious in the northward advance of warm-water clams such as the venus clam *Dosinia*, the ark shell *Anadara*, the oyster *Crassostrea*, and the lucinid clam *Miltha* and snails such as the fig shell *Ficus*. There were also parallel trends in the retreat of cold-water species from Alaska that had once lived much further south in the cold waters of the Oligocene (Addicott 1969; Squires 2003; Oleinik and Marincovich 2003). The same trend has also been observed for benthic foraminifera from California (Kleinpell 1938, p. 102). Bryozoans also recovered after Oligocene lows in diversity (Horowitz and Pachut 1996; Taylor 2000). Echinoids (Kier 1975; Poddubiuk and Rose 1984; Smith and Jeffrey 2000) show a slight increase in diversity in the early Miocene after their low diversity in the middle Oligocene.

In summary, the early Miocene saw a marked warming trend from the end of the final Oligocene Antarctic glaciation at 23 Ma until the peak of

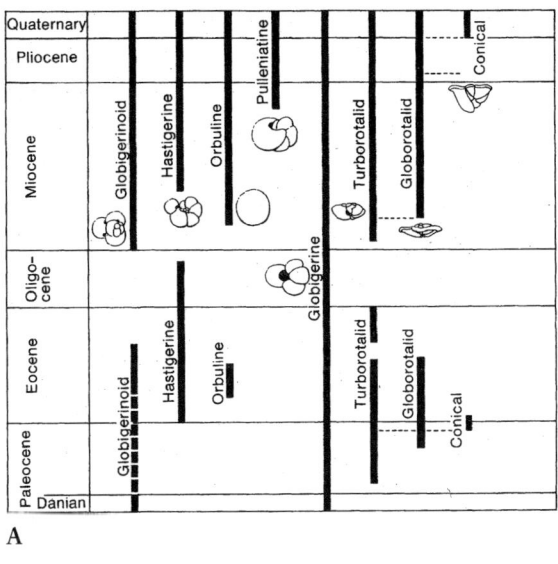

A

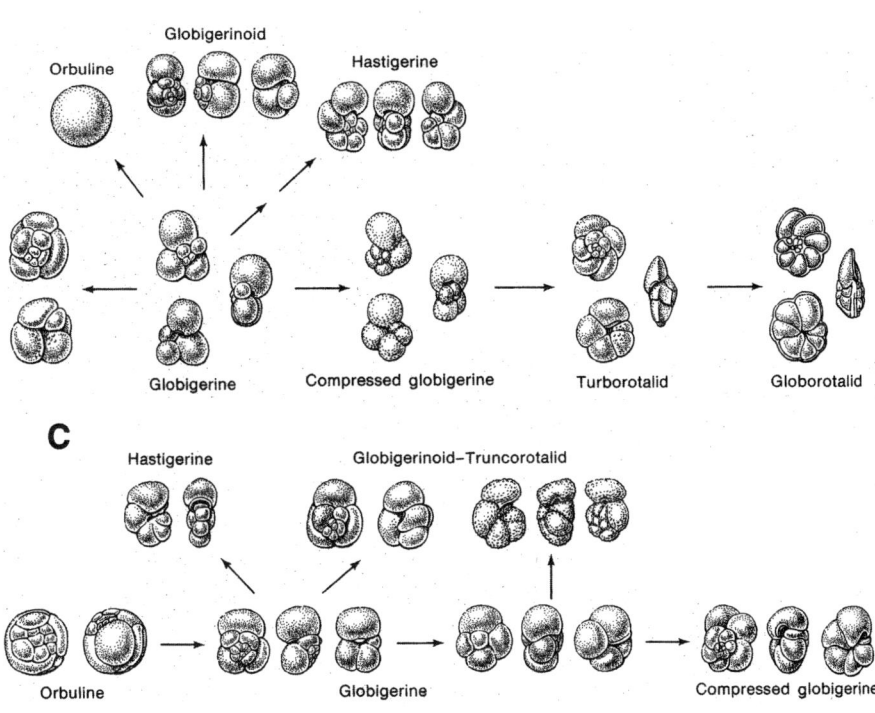

C

B

Figure 6.4. (A) Time ranges of major shell types of planktonic foraminifera. Note that many of the shell types vanished during the Oligocene extinctions, only to re-evolve during the Miocene warming. (B) Radiation of Miocene planktonic foraminifera from simple common ancestors in the Oligocene. After Cifelli 1969.

the mid-Miocene climatic optimum at 14–16 Ma. Nearly every paleoclimatic indicator shows that the world was again becoming warmer and milder, with temperate and subtropical vegetation at higher latitudes, and vegetation instead of ice caps on the poles.

Mass Migrations

The early Miocene was marked by another distinctive phenomenon—the greatest migration of mammals between continents since the early Eocene (Webb 1985; Woodburne and Swisher 1995). Perhaps the warm, mild conditions and abundant vegetation all the way to the Bering Strait made passage between Eurasia and North America easier than ever before. Perhaps tectonic changes made the land bridge more accessible. Whatever the reason, the Bering land bridge was a veritable freeway that allowed many groups of Eurasian mammals to reach North America, and a few from North America to reach Eurasia. By the end of the early Miocene (at about 18 Ma), another land bridge opened: the Arabian Peninsula had collided with Asia, closing off the Tethys to form the Mediterranean. This allowed the migration of animals back and forth from the former island continent of Africa.

The first wave of immigration to North America can be seen in latest Arikareean faunas, which are earliest Miocene in age (the rest of the Arikareean is Oligocene). The most famous locality of this age is the legendary Agate Springs Quarry (now Agate Fossil Beds National Monument) in western Nebraska (fig. 6.5). Discovered in 1906, Agate Springs has produced hundreds of skeletons of early Arikareean fossil mammals over the years, mostly collected by the University of Nebraska, the Carnegie Museum in Pittsburgh, and the American Museum of Natural History in New York. Apparently, the fossil beds were deposited in an ancient river channel and sand bars, where bones washed in by the thousands. One typical slab of bone covering 44 square feet contained 4,300 skulls and bones, suggesting that the Agate bone-bearing layer may have once contained at least 3 million bones from about 17,000 different animals! The most common mammal in the Agate fauna (pl. 4) was the immigrant rhinoceros *Menoceras arikarense* (once mistaken for the native paired-horned rhinoceros, *Diceratherium*). *Menoceras* was the size of a sheep but had a pair of small horns on the tip of the nose in males (fig. 6.6A). Its closest relatives evolved in Eurasia in the late Oligocene and early Miocene, so it was clearly an immigrant from the Old World. Other earliest Arikareean immigrants included two new kinds of weasel-like carnivorans, *Oligobunis* and *Zodiolestes*; the beardogs *Ysengrinia* and *Daphoenodon*; and the early bear *Cephalogale* (which was about the size and shape of a raccoon).

Another immigrant was the bizarre chalicothere *Moropus* (fig. 6.6B), a perissodactyl distantly related to tapirs (Schoch 1989). Chalicotheres first evolved in Asia in the late Eocene and were found in Europe in the

A

B

Figure 6.5. The classic earliest Miocene beds at Agate Fossil Beds National Monument in western Nebraska. (A) View of the quarries from the north, showing the square-topped Carnegie Hill (right) where the Carnegie Museum of Pittsburgh excavated, and the pointed University Hill (left), site of the University of Nebraska excavations. (B) Close up of the quarries at University Hill. Photos by the author.

Oligocene after *la grande coupure.* They did not cross into North America until the early Arikareean (27 Ma), and they were never abundant (except at Agate Springs Quarry). Typically, they existed in small numbers for the rest of the early and middle Miocene. *Moropus* was a peculiar beast. It was built like a large, muscular horse, except that it had long forelegs and short hind legs; and instead of hooves, it had long heavy claws! When chalicotheres were first found in Europe, scientists refused to believe the claws belonged to the same animal, since the rest of its skeleton is that of a hoofed mammal. There have been many ideas about how a herbivore with simple, low-crowned (brachydont) teeth used these claws. Some have suggested they were for digging up roots or ripping up bark, or even wilder notions. However, Coombs (1983) has shown that the *Moropus*

A

B

C

D

Figure 6.6. Typical mammals of the earliest Miocene.
(A) The immigrant rhinoceros Menoceras, with its
slender legs and pair of small horns. It is by far the
most common fossil at Agate Springs Quarry. From
Scott 1913. (B) The chalicothere Moropus, here
reconstructed dragging down branches with its long
clawed muscular forelimbs. Drawing by H. Galiano,
in Coombs 1983. (C) The Eurasian chalicothere
Chalicotherium, which knuckle-walked like a gorilla
with its long forelimbs and short hind limbs. From
Agusti and Anton 2002. (D) The rhino-sized
entelodont Daeodon (formerly Dinohyus), a huge
bone-crushing scavenger and predator found at
Agate Springs Quarry. Photo by the author of a
diorama in the Denver Museum of Nature and
Science. (E) The gazelle-like camel Stenomylus,
particularly common in the Arikareean. From Scott
1913.

E

probably used these claws to hook tree branches and bring them down to the level its mouth, just as the giant ground sloth used its claws to feed. This use is confirmed by the most extreme member of the family, *Chalicotherium* itself, which lived in Eurasia in the Miocene (fig. 6.6C). In addition to claws, it had extraordinarily long arms for grasping and hanging on tree branches, so that, like a gorilla, it knuckle-walked with its claws curved upward and inward.

Two other important latest Arikareean immigrants were ruminant artiodactyls, which had four-chambered stomachs and chewed their cud. One group, the blastomerycids, looked much like the tiny chevrotains and musk deer that still live in Asia today. Like musk deer, the blastomerycids lacked antlers, but the males had large canines instead; and they probably lived in the underbrush of deep forests, hiding from predators. The other group, the dromomerycids, is distantly related to deer. Unlike true deer, however, dromomerycids had permanent bony horn cores rather than antlers that were shed every winter and grown again in the spring. However, this group underwent a large evolutionary radiation in the early and middle Miocene of North America (since true deer had not yet arrived), with a great diversity of horns shapes (fig. 6.7).

These immigrants were found side by side with native groups that had been in North America since the early Arikareean, or earlier. Agate Springs Quarry yields not only the rhino *Menoceras* and the chalicothere *Moropus* but also the last of the archaic pig-like entelodonts, *Daeodon* (fig. 6.6D; formerly called *Dinohyus*), which was the size of a rhino and had a huge massive skull covered with warty bony knobs. The latest Arikareean faunas also included a variety of gazelle-like and antelope-like camels (still found only in North America). The most remarkable of these were the stenomyline camels, which had gazelle-like proportions and extraordinarily high-crowned (hypsodont) molars that allowed them to eat tough, gritty vegetation (fig. 6.6E). The distant relatives of camels, the protoceratids, included the distinctive *Syndyoceras*, which had a V-shaped pair of horns on its nose and long curved horns over each eye (fig. 6.8). Oreodonts were in their heyday, with many forms evolving from the primitive sheep-sized *Merycoidodon* of the Orellan (fig. 1.3). They include a variety of forms with high-crowned teeth; some had long legs for running, and others (e.g., *Promerycochoerus*) had short legs and pig-like bodies, and a proboscis or trunk, so they were apparently semi-aquatic like tapirs (fig. 6.9). These shared the rivers and forests with the last of the pig-like anthracotheres, and the peccaries (the New World javelinas). Three-toed horses were beginning to diversify, with the last of the small, archaic *Miohippus* living side by side with more-advanced horses such as the equine with higher-crowned teeth, *Parahippus*, and two anchitherines that retained low-crowned teeth, *Anchitherium* and *Archaeohippus*.

In addition to the immigrant predators mentioned already, there was a huge radiation of native dogs, including the borophagines, which had bone-crushing teeth like those of hyenas. Native beardogs and bears were

Douglass' deer,
from Nebraska

Lull's deerlet,
from Nebraska

Rak's deer,
from California

Sinclair's deer,
from Nebraska

Procranioceras,
from Nebraska

Gregory's deerlet,
from Nebraska

Cranioceras,
from Nebraska

Sinclairomeryx,
from Nebraska

A

Figure 6.7. The deer-like ruminants known as
dromomerycids were common in the North
American Miocene. (A) Dromomerycids sported
a variety of cranial appendages, which were
permanent bony horns, not deciduous antlers like
those found in true deer. From Scheele 1955. (B)
Restoration of Dromomeryx borealis. After Scott
1913.

B

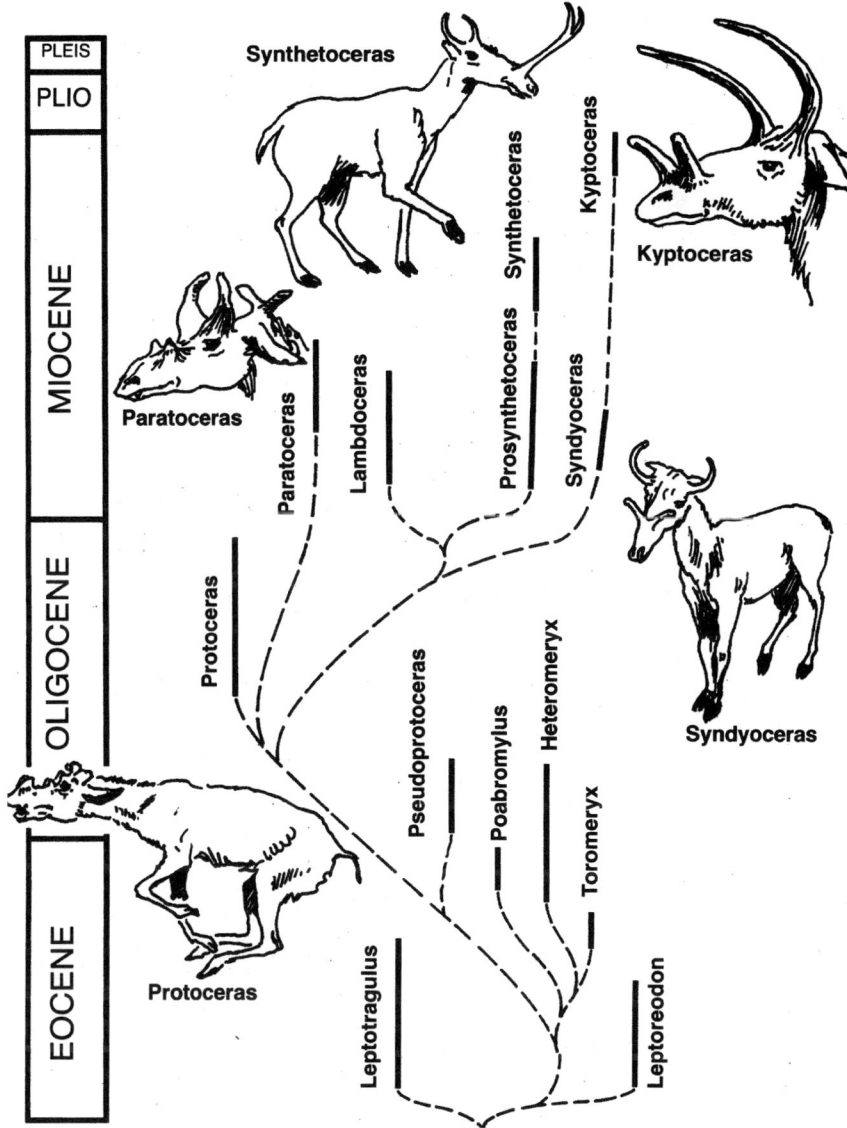

Figure 6.8. Evolution of the horned tylopods known as protoceratids, which originated in the middle Eocene. By the Miocene, there were several genera, with interesting combinations of horns on their noses and heads. Their broad snouts and low-crowned teeth and their occurrence in more-forested regions suggest that they were browsers of leafy vegetation, like moose, rather than specialized grazers. Drawing by C. R. Prothero.

also diverse, as were the mustelids (members of the weasel family). The last of the cat-like nimravids disappeared at the end of the Arikareean, leaving a "cat gap" of no cat-like forms in North America for several million years. Finally, rodents and rabbits continued to diversify. Late Arikareean beds all over western North America yield a diversity of aplodontids (sewellels, or "mountain beavers"), horned mylagaulids (fig. 5.20),

A

B

Figure 6.9. The oreodonts Brachycrus *(A) and* Promerycochoerus *(B). The oreodonts (see fig. 1.2) evolved in a variety of forms in the Miocene, including this pig-like or tapir-like form* Promerycochoerus, *which was common in the early Miocene. After Scott 1913.*

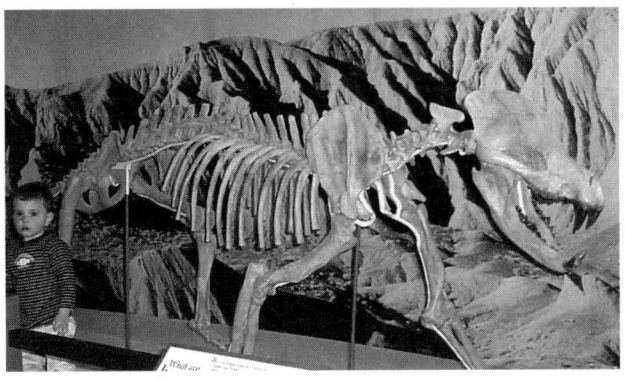

Figure 6.10. Mounted skeleton of the huge Miocene beardog Amphicyon ingens. *Photo courtesy L. Spoon.*

cricetids (New World mice), jumping mice, squirrels, beavers, pocket gophers, and pocket mice.

If the earliest Miocene (late Arikareean, 19–23 Ma) was characterized by a significant number of Asian immigrants to North America, the late early Miocene (Hemingfordian, 16–19 Ma) was marked by a veritable flood of new Eurasian arrivals. These included new kinds of shrews, rabbits, the huge beardog *Amphicyon* (fig. 6.10), the long-legged running bears *Hemicyon* and *Ursavus*, several kinds of mustelids (related to weasels and otters), primitive raccoons, and two types of rodents, the petauristine squirrels and the eomyid *Eomys*. The most striking immigrants were the two major groups of rhinoceroses to dominate the rest of the Miocene (Prothero 2005b). One group, the aceratherines, were primitive generalized hornless rhinoceroses that had a long prehensile upper lip (and maybe a short proboscis) for stripping off leaves from trees and bushes (fig. 6.11A). At least four genera of aceratherines (*Floridaceras, Galushaceras, Peraceras,* and *Aphelops*) occur in the Hemingfordian, and the latter two continued through the rest of the Miocene. The other immigrant is the one-horned teleoceratine rhino *Teleoceras,* which had a barrel-shaped body and stumpy legs, and high-crowned teeth for grinding grasses (fig. 6.11B). Some paleontologists believe that it was aquatic like a hippo, living in the streams and lakes in the daytime for protection and coming out at night to graze. By the end of the Hemingfordian, the previously established paired-horned rhinos *Menoceras* (fig. 6.6A) and *Diceratherium* were extinct, replaced by these immigrants, which would flourish during the rest of the Miocene.

The other important large mammals to appear in the Hemingfordian were the pronghorns (Janis and Manning 1998). Although called "antelopes," they are unrelated to the true antelopes of Africa, which are

A

B

Figure 6.11. Two lineages of rhinoceroses that replaced Menoceras *in the early Miocene and dominated North America for the rest of the Miocene. (A) The long-legged aceratherine rhinoceros* Aphelops *had a short proboscis and prehensile lip and probably roamed the underbrush browsing on leaves, as do many of the living rhinos. Drawing by B. Nafus. (B) The hippo-like* Teleoceras *had a barrel-like chest, short limbs, high-crowned teeth, and a small nasal horn and is thought to have grazed near watercourses, as the hippopotamus does today. Painting by Z. Burian.*

members of the cattle family, Bovidae. Instead, pronghorns became a primarily American group; they arrived in North America in the Hemingfordian from Eurasia and then quickly diversified in the early and middle Miocene of North America, forming a huge radiation with a great variety of horn types (fig. 6.12). Today, only one species, *Antilocapra americana*, is left of this great native radiation.

Although the Hemingfordian saw many new immigrants, the established groups were numerically dominant. Among the carnivorans, they

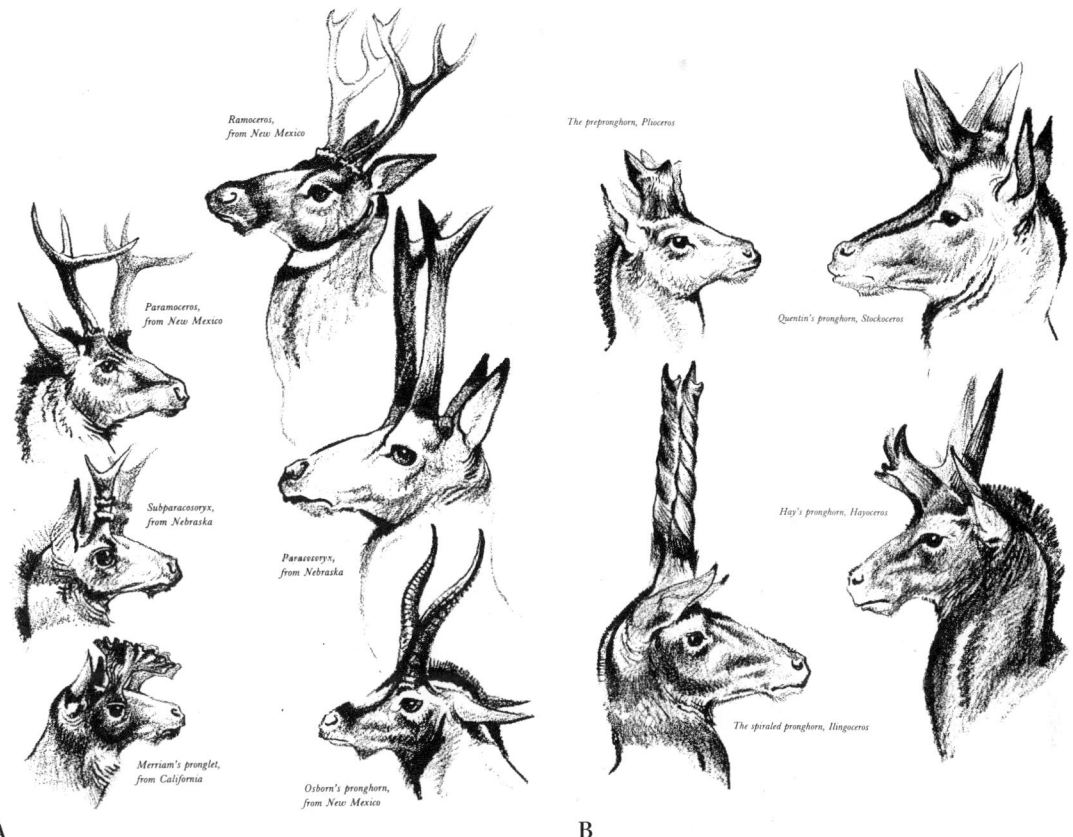

Figure 6.12. Pronghorns. Pronghorns were diverse during the Miocene, with dozens of genera sporting a variety of cranial appendages. From Scheele 1955.

included a great diversity of beardogs, bears, mustelids, and dogs. A greater variety of horses included the anchitherines (with low-crowned teeth for leaf eating) *Anchitherium*, *Hypohippus*, and *Archaeohippus*, plus more-advanced, grazing horses, such as *Parahippus* and *Merychippus*, with higher-crowned teeth. The chalicotheres were rare, but a peculiar form called *Tylocephalonyx* evolved, which had a big bony dome on its forehead. The highly specialized oreodonts continued to flourish, as did the wide variety of gazelle-like and giraffe-like camels, and especially the deer-like blastomerycids and dromomerycids, which diversified into dozens of species.

At the end of the Hemingfordian, two more new immigrant groups were added to the North American landscape from Eurasia. One was *Pseudaelurus*, the first true cat of the family Felidae. The other was the mastodonts, which had escaped Africa earlier in the Miocene (see next section), gradually showing up in the latest Hemingfordian and early Barstovian (but not until the late Barstovian in the High Plains).

What about the other side of the Bering land bridge? Some native North American groups, such as camels, pronghorns, oreodonts, and protoceratids, apparently never made the crossing. But Asia saw some immigrants from North America in the early Miocene. Most prominent of these was the three-toed browsing horse *Anchitherium*, which spread from North America to Asia to Europe by the early Miocene. Asia also had rhinos like the aceratherines and teleoceratines of North America, although they were different species of closely related genera. Many of the same beardogs and weasels found in North America were found in Asia during the early Miocene. In addition, the early relatives of the North America dromomerycid radiation (*Palaeomeryx*) were found in Eurasia as well. However, Asia had many unique endemic groups that were not found in North America and were also either rare or absent from Europe. These included archaic relict groups that had vanished from the rest of the world, such as the hippo-like amynodont rhinos and the giant indricotheres, which survived in places like Dera Bugti in Pakistan right up until about 17 Ma, before finally going extinct. Chalicotheres also thrived in Asia, with only occasional escapes to other continents. The archaic hyaenodont creodonts (including the huge bear-sized *Megistotherium*) also survived in Asia (fig. 6.13), even though they were long extinct elsewhere in the world (except Africa). Anthracotheres were also hugely diverse in Asia, even though they were rare on other continents. In addition, Asia saw the first appearance of many new groups that would soon come to dominate the Old World continents. These included the first true pigs (which had evolved in Eurasia in the late Oligocene), the first true deer (family Cervidae), and early members of the giraffe family (Giraffidae) and cattle family (Bovidae). Finally, the early mastodonts, which had escaped from Africa at about 18 Ma, were also abundant in Asia.

Because there was a connection by land corridors during the early Miocene, Europe had similarities with Asia, and its own unique mammals as well (Agusti and Anton 2002). Europe did not have the huge diversity of anthracotheres found in Asia, or the archaic amynodont and indricotherine rhinos. Instead, there were archaic groups that persisted from the late Oligocene, including the tiny cainotheres, which were chevrotain-like artiodactyls more primitive than the ruminants. The holdovers also included the last of the paired-horned rhinos (*Menoceras*), also seen in North America (fig. 6.6A). The cat-like sabertoothed nimravids (*Prosansanosmilus*) persisted in Europe long after they had vanished elsewhere. However, many more-advanced groups were also undergoing a spectacular evolutionary radiation, including true pigs, deer, early bovids and giraffids, and the advanced aceratherine and teleoceratine rhinos. There was also a wide variety of advanced carnivorans (figs. 6.10, 6.13): beardogs (including the huge 300-kilogram *Amphicyon*); raccoon-like bears; early raccoons, weasels, and civets; and the first true cats (*Pseudailurus*) seen also in North America. Of the immigrants from North America, only the three-toed browsing horse *Anchitherium* made it to Europe in the early Miocene.

Figure 6.13. Restoration of typical Miocene carnivorous mammals. In the middle is the last of the hyaenodont creodonts, Megistotherium, compared with a modern gray wolf (foreground). In the background is the huge beardog Amphicyon, which was the largest predator of the Miocene and is larger than most living bears and dogs. From Agusti and Anton 2002.

Finally, what about Africa? As I have suggested already, the key event is the connection of the Arabian Peninsula to Asia, forming the first land escape route from Africa to the rest of the world at around 18 Ma. Before this time, Africa was the home of a variety of endemic groups: a wide diversity of mastodonts (which had always been native to Africa), along with native hyraxes and anthracotheres performing many different ecological roles. Archaic groups like the insectivorous pantolestids (long extinct everywhere else) and the native tenrecs, golden moles, aardvarks, and elephant shrews were also present. There were abundant primates, which had become restricted to Africa in the Oligocene after a nearly worldwide Eocene distribution, including primitive lorises, apes, and Old World monkeys. Africa was also the home of the peculiar phiomyid rodents, and the archaic creodonts were the main predators. After the Arabian land bridge was established, there was a virtual flood of migration. Gomphothere mastodonts (fig. 6.14A, pl. 6) spread quickly out to Eurasia at 18 Ma, and by 16.5 Ma they finally crossed the Bering land bridge and reached North America. Another group of proboscideans, the deinotheres, which had no upper tusks but had peculiar lower tusks that curved down below the jaw, also escaped, but remained restricted to Eurasia (fig. 6.14B). Apes, monkeys, and lorises also migrated to Eurasia at around 19 Ma, although they never became common outside of Africa. The hyena-like creodont Hyainolouros escaped from Africa and became resident in Europe long after other creodonts had vanished from Europe and North America. A few of the peculiar large hyraxes also managed to become established in Eurasia by the middle Miocene. But although proboscideans, primates, creodonts, and hyraxes managed to escape from

A

Figure 6.14. Typical proboscideans of the Miocene. (A) Gomphotherium, *the main lineage of mastodonts in the Miocene. After Scott 1913. (B) The huge* Deinotherium, *with its distinctive down-turned lower tusks, was one of the commonest large mammals of the Miocene of Eurasia and Africa. After Scheele 1955.*

B

Africa, the wave of immigration of Eurasian mammals into Africa was much greater in the early Miocene. Both aceratherine and teleoceratine rhinos crossed the land bridge, establishing rhinos in Africa for the first time. Advanced Eurasian carnivorans, such as beardogs, dogs, civets, and true cats, came over to compete with the relict hyaenodont creodonts. A wide variety of artiodactyls also made the crossing, including true pigs, primitive deer, giraffes, and bovids, laying the foundation for the huge diversity of these families in the modern African savanna. Thus, many of the beasts that we think are typical of the modern African fauna—rhinos, giraffes, antelopes and Cape buffalo, musk deer, and pigs—arrived in the early Miocene (albeit with primitive members of those families, not the modern genera).

Although Africa had finally ended its long isolation, South America and Australia had not. As I mentioned in earlier chapters, both regions had been island continents since the Paleocene, isolated from the animals of the northern continents. South America had seen the immigration of caviomorph rodents in the late Eocene and the New World primates in the

Figure 6.15. Exposures of the lower Miocene Santa Cruz beds along the coast of Argentina. Photo courtesy of R. Kay.

late Oligocene. Otherwise, the continent was still the home of its native endemic groups: marsupial carnivores, edentates, and native hoofed mammals (notoungulates and litopterns). The earliest Miocene in South America is known as the Colhuehuapian (19 to 21 Ma), named after the faunas of Lake Colhue-huapi in Argentina; and the late early Miocene is the Santacrucian (15 to 17 Ma), after the Santa Cruz Formation (fig. 6.15) in Argentina (Flynn and Swisher 1995; Kay et al. 1999). Both faunas have similar components, which differ at the generic level from the Deseadan fauna. The New World monkeys, with a number of primitive genera, were well represented. The rodents, befitting their much longer residence, had diverged even more. Primitive members of at least six caviomorph families had appeared, including the New World porcupines, the chinchillas, the spiny rats, and the agoutis.

Much more impressive, however, is the tremendous diversity of native groups in the Santacrucian (fig. 6.16). These included seven families of armadillos, ground sloths, and anteaters representing the edentates (fig. 6.16A), plus the first members of the huge armadillo-like glyptodonts. Marsupials still performed the roles of the carnivorous mammals, from smaller weasel-like forms such as *Cladosictis* to much larger creatures such as the wolf-like *Prothylacynus* and the hyena-like *Borhyaena*. Most

A

B

C

Figure 6.16. Typical mammals of the lower Miocene
Santa Cruz beds in South America. (A) The ground sloth
Hapalops *and the armadillo-like glyptodont*
Propalaeohoplophorus. *(B) The toxodont* Nesodon, *built
like a hippopotamus but completely convergent. (C) The
litoptern* Thoatherium, *which was more horse-like and
single-toed than true horses of the early Miocene. (D) The
mastodon-like* Astrapotherium, *which had a trunk and
enlarged tusks, even though it was unrelated to elephants
or mastodonts. After Scott 1913.*

D

impressive of all, however, was the radiation of native hoofed mammals. The notoungulates included toxodonts like *Nesodon*, which looked like a pygmy hippo (fig. 6.16B), and the rabbit-like hegetotheres. The litopterns developed forms such as the llama-like *Thesodon* and the horse-like *Thoatherium*, which was already more one-toed than any horse would ever become (fig. 6.16C). There were also relict groups from the early Cenozoic, including the mastodont-like astrapotheres (fig. 6.16D). Like the mammal faunas of the rest of the world, those of South America represent the establishment of the major groups that would dominate the rest of the Miocene, and the disappearance of the final archaic groups that were important in the Eocene or Oligocene.

Finally, I have said little about Australian land mammals so far—for good reason. Until recently, we knew almost nothing of the Paleogene mammals in Australia, with the exception of the scrappy faunas from the early Eocene Tingamarra fauna (Godthelp et al. 1992) and the latest Oligocene Geilston travertine (Tedford et al. 1975). Thanks to spectacular specimens preserved mostly in caves and sinkholes in the Riversleigh area of Queensland, we now have a good window on life in Australia in the latest Oligocene and early and middle Miocene (Archer et al. 2001). Except for bats that flew over from Asia, all the mammals were marsupials, members of many of the families that are native to Australia today. These included a diversity of kangaroos (some of which were carnivorous, others rat-sized), huge wombats known as diprotodonts, and the palorchestids, which converged on mastodonts in having a short proboscis. There were wolf-like thylacines (related to the Tasmanian wolf, which went extinct in the twentieth century), marsupial moles, a great variety of bandicoots, dasyurids, and possums, as well as primitive bettongs, koalas, and cuscuses. The top predator was the marsupial lion, *Priscileo*, which had huge stabbing lower canines, short upper canines, and long blade-like cheek teeth. Riversleigh also yields well-preserved skulls of the ancient platypus *Obdurodon*, previously known only from a single isolated specimen about 25 Ma in age from the Tirari Desert (Woodburne and Tedford 1975).

Along with the mammals, there were many of the other vertebrates still found in Australia today, including several kinds of flightless birds, such as early cassowaries and emus, plus bowerbirds, cockatoos, and passerine birds. There were also the extinct dromornithids, which reached 3 meters in height and weighed up to 400 kilograms. The lower vertebrates consisted of crocodiles and goannas (monitor lizards), pythons, geckos, and side-necked turtles, plus abundant frogs, lungfish, and a variety of other fish. All of these Riversleigh animals lived in a tropical rain forest not too different from that found along the Queensland coast today.

In summary, the early Miocene was not only a warming interval but also a peak of migration between the northern continents and, for the first time, between those continents and Africa. During the early Miocene, most continents acquired the mammalian families that would dominate the rest of the Miocene. In many cases (such as Eurasia, Africa, and Australia), those families are still the dominant ones on the continent today.

Climatic Optimum

The warming trend that began in the early Miocene peaked at the beginning of the middle Miocene (14–16 Ma), the last gasp of warm conditions on earth before glaciation returned to Antarctica permanently at 14 Ma. This mid-Miocene climatic optimum (Zachos et al. 2001) shows up on all the deep-marine oxygen isotope curves (fig. 3.22) as bottom-water temperatures as high as 6°C (43°F) (Miller et al. 1987, 1991; Zachos et al. 1993, 2001). Ocean margin surface productivity, especially that of the siliceous plankton (diatoms and radiolarians) increased at 16 Ma, suggesting that atmospheric-oceanic circulation and upwelling intensified compared to the sluggish, warm-water conditions of the early Miocene (Woodruff 1985).

Responding to the increasing warmth, the middle Miocene was also the Neogene peak of diversity in marine life. Planktonic foraminiferal speciation was at a Neogene maximum as the diversification in the early Miocene continued (fig. 6.4), with more species than at any other time in the past 30 million years (Kennett and Srinivasan 1983; Tappan and Loeblich 1988; MacLeod et al. 2000). The warming is also shown by the spread of warm-water larger foraminiferans and calcareous nannoplankton to high latitudes. The high diversity of corals persisted from the early Miocene (Coates and Jackson 1985; Budd et al. 1994), as did the diversity of bryozoans (Horowitz and Pachut 1996; Taylor 2000) and echinoids (Kier 1975; Poddubiuk and Rose 1984; Smith and Jeffrey 2000).

But the changes were particularly apparent in the mollusks. In the shallow-marine basins of Europe, a uniform tropical and subtropical mollusk fauna developed over a large area. These subtropical conditions are apparent on the shorelines of the European seas, which were inhabited by crocodiles and palm trees as far north as Poland. In North America, there are numerous middle Miocene marine deposits that yield an incredible diversity of mollusks. One of the most famous of these is the dense shell beds of the Calvert and Choptank formations in the Chesapeake Bay region (figs. 6.17A, B), which are famous for their diverse subtropical to temperate mollusk fauna (Ward 1992). Over a hundred species of clams and snails have been recorded from these shell beds, including a great variety of scallops, venus clams, and *Turritella pilsbryi*. The heavily corrugated snail *Ecphora* is particularly distinctive of this time (fig. 6.18).

Similar trends can be seen in the west. On the Pacific Coast, many formations bear famous middle Miocene shell beds, including the Astoria Formation in Oregon and Washington (Newportian Stage of Addicott 1976) and the Topanga Canyon Formation in Los Angeles, the Olcese

A   B

Figure 6.17. (A) View of the middle Miocene beds of the Choptank Formation in the Calvert Cliffs, on the shores of Chesapeake Bay, showing the dense shell beds. (B) Close up of one of the Choptank Formation shell beds. Photos courtesy of S. Kidwell.

Sand (fig. 6.19) and Round Mountain Silt northeast of Bakersfield, and the Temblor Formation in the Coast Ranges of California (Temblor Stage of Addicott 1976). These yield an incredible abundance of mollusks, including the index fossil *Turritella ocoyana*; a great diversity of carnivorous snails; and a wide variety of clams as well. The early middle Miocene was the peak of the warming trend observed by Durham (1950) and Addicott (1969, 1970). Warm-water mollusks such as the clams *Dosinia* and *Anadara* and the snails *Ficus, Cancellaria,* and *Turritella* reached their northernmost distribution, and also were more speciose than at any other time in the Miocene (Addicott 1970). For example, in the early Miocene

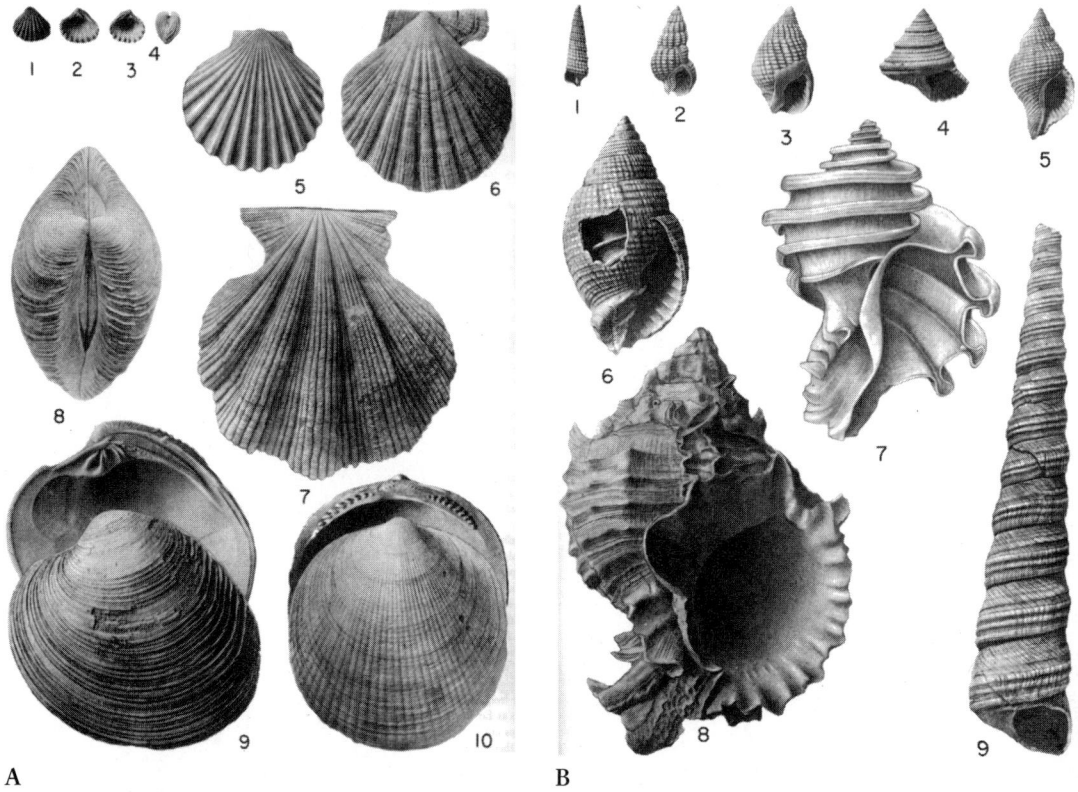

Figure 6.18. Typical Miocene mollusks. (A) 1–4. The tiny nut shell Glans decemcostata; *5. the scallop* Pecten poulsoni; *6. the scallop* Chlamys decemnarai; *7. the scallop* Lyropecten ernestsmithi; *8–9. the quahog* Venus berryi; *10. the clam* Glycimeris americana. *(B) 1.* Triphora bartschi; *2.* Uzita neogenensis; *3. the mud snail* Illyanassa grandifera; *4. the top shell* Calliostoma mitchelli; *5. the oyster drill* Urosalpinx trossula; *6.* Cancellaria rotunda; *7. the distinctive whelk* Ecphora quadricostatus; *8. the muricid* Murex pomum; *9. the high-spired* Turritella pilsbryi. *Courtesy of the U.S. Geological Survey.*

there were only six species of *Cancellaria*, but by the middle Miocene there were eighteen; early Miocene *Turritella* had only two species, but five are recognized in the middle Miocene (Addicott 1970).

Back to the Water Again

These warm, rich seas were also good hunting grounds for a variety of marine predators. Middle Miocene marine deposits, such as the Calvert Cliffs of Chesapeake Bay and Sharktooth Hill in California, are famous not only for their huge numbers of fossil mollusks but especially for their marine vertebrates. For example, Sharktooth Hill produces teeth of over two dozen species of sharks, including members of most of the living families. By far the most impressive of these, however, was the giant great

Figure 6.19. Close up of shell beds from the middle Miocene Olcese Sand, near Bakersfield, California. Photo by the author.

white shark, *Carcharocles megalodon* (fig. 6.20), known mostly from its huge teeth, which were 16 centimeters (6 inches) long. Since the rest of the shark's skeleton is cartilage and does not fossilize, we can only estimate its total size, but most scientists think it was about 17 meters (55 feet) in length, or at least twice the size of the living great white shark. This shark was so huge that whales were probably its main prey (along with any smaller animals that it could catch).

There were plenty of whales and other marine mammals for *Carcharocles megalodon* to eat. The toothed whales (odontocetes) and baleen whales (mysticetes) had already appeared in the latest Eocene and were diversifying by the early Oligocene (see chapter 5). In the early and middle Miocene, however, the radiation of whale families accelerated, as a number of odontocete groups arose, including the extinct acrodelphid, rhabdosteid, and kentriodontid dolphins, the beaked whales (ziphiids), the harbor porpoises (phocoenids), true dolphins (delphinids), narwhals (monodontids), the platanistid river dolphins, and the sperm whales (Barnes 1984). Among mysticetes, the middle Miocene saw the evolution of the earliest balaenopterids, or rorquals, which today include the blue whale, humpback whales, finback whales, fin whales, and minke whales. The rorquals have pleats on their throats, so they can gulp a huge amount of water, then close their mouths and squeeze their throat cavity up, forcing water out with their huge tongues and straining krill and other plankton into their screen of baleen in the front of their mouths. All of these

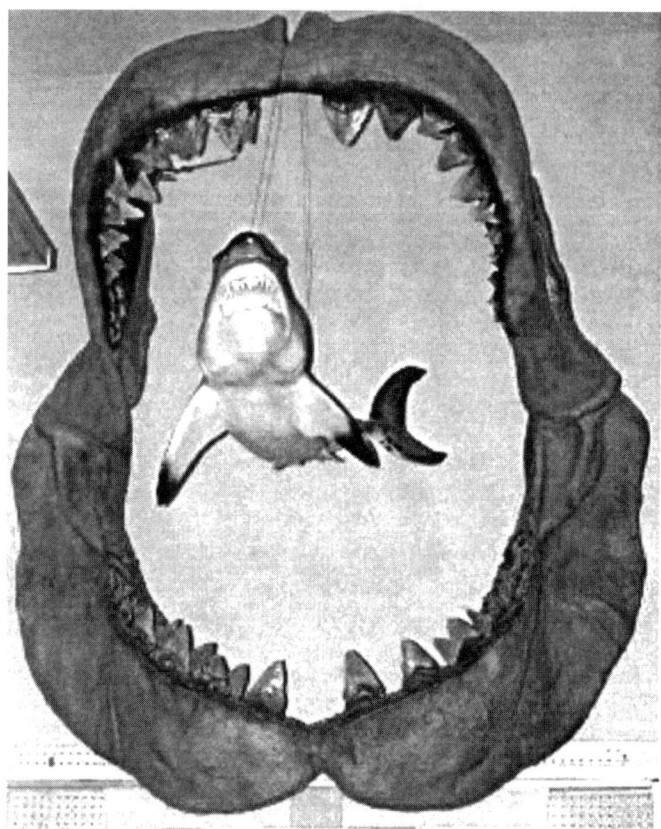

Figure 6.20. Jaws of the gigantic great white shark Carcharocles megalodon, *with a modern great white shark,* Carcharodon carcharias, *in the middle for scale. Photo by the author.*

families of whales that arose in the Miocene coexisted with archaic whale families that had appeared in the Oligocene, including the cetotheres (archaic baleen whales) and the squalodontids (archaic odontocetes), so by the middle Miocene, whale diversity was at an all-time peak.

Joining these whales and dolphins was a new group of marine predators: the pinnipeds, or seals, sea lions and walruses. They arose from a group of terrestrial carnivores most closely related to bears, including *Plesiocyon* and *Amphicynodon* from the Eocene and Oligocene of Europe, and *Parictis, Nothocyon,* and *Drassonax* from the Eocene and Oligocene terrestrial beds of North America (Berta et al. 1989). One of the first semi-aquatic members of the group is the peculiar "beach bear," *Kolponomos,* from the late Oligocene and Miocene marine beds of the Pacific Northwest. Although the skull is bear-like, it had blunt, flat teeth for crushing mollusks, and robust front teeth for prying them loose (Stirton 1960; Tedford et al. 1994). Finally, the lower Miocene beds of California and Oregon yield the enaliarctines, which are the first truly marine relatives of seals and sea lions (Mitchell and Tedford 1973; Barnes 1989;

A

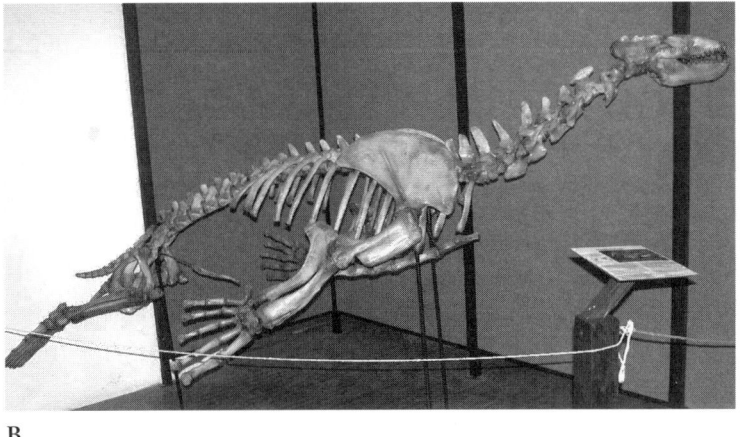

B

Figure 6.21. Early relatives of seals and sea lions. (A) Restoration of the early Miocene pinniped Enaliarctos *from Pyramid Hill, near Bakersfield, California. After Berta et al. 1989. (B) Mounted skeleton of* Allodesmus. *Photo by the author.*

Berta et al. 1989; Berta and Ray 1990). Although they retained many primitive features seen in the bear-like amphicynodontids, they also have specializations of seals and sea lions, including enlarged eyes, an enlarged nasal cavity for regulating the temperature of the blood as they swim, and larger openings for the muscles that control their lips and whiskers. They also have reduced the olfactory lobes of the brain (since the sense of smell is not so important to aquatic mammalian predators) and im-proved the drainage of blood to their brains as an aid to diving. Their bodies (fig. 6.21A) also had well-developed flippers and streamlined shapes, so they would definitely remind us of the living seals. Not long after the enaliarctines, we find the first members of all the living pinniped groups (fig. 6.21B) in the late early to middle Miocene, including the first true seals (*Pontophoca, Praepusa,* and *Crytophoca* in the middle Miocene of Europe and *Leptophoca* in the middle Miocene of North

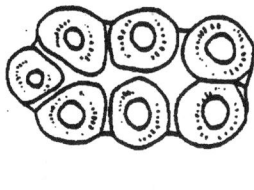

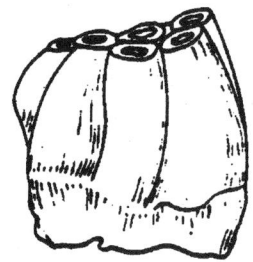

A

*Figure 6.22. Desmostylians, hippo-like mammals
distantly related to elephants, that roamed the shores of
the North Pacific during the Miocene, feeding on aquatic
plants. (A) Restoration of* Desmostylus *feeding along the
rocky shores. From Domning 1994. (B) The distinctive
teeth of desmostylians—composed of a bundle of thick
cylinders of enamel with hollow, dentin-filled centers—
look like a cluster of volcanoes. From Reinhart 1959.*

B

America), the first sea lions (*Pithanotaria* in the middle Miocene of the
Pacific Northwest of North America), and the first walruses (the des-
matophocines of the early Miocene of North America and *Prototaria* of
the early Miocene of Japan).

Finally, the shorelines of the North Pacific were also inhabited by a
semi-aquatic group known as desmostylians (fig. 6.22). These peculiar

beasts looked like a cross between a sea lion and a hippo and were long a mystery in paleontology. Their most bizarre feature was their huge molars, with each cusp developed into a large cylindrical hollow pillar that wore down to look like a cluster of volcanoes. Desmostylians are well known from the late Oligocene and early Miocene of the Pacific Rim (Japan and the Pacific Coast of the United States), where they apparently hunted for food in the tide pools. Their blunt crushing teeth suggested a diet of shellfish, but Clementz et al. (2003) have shown by the chemistry of their bones and teeth that they ate aquatic plants, not mollusks. However, their relationships to other groups of mammals remained elusive, and they were usually placed in isolation in their own order, Desmostylia. But in 1986, Daryl Domning, Clayton Ray, and Malcolm McKenna described an Oligocene form, *Behemotops,* which was much more primitive than any previously known, and they showed that desmostylians were closely related to elephants and manatees.

The Frozen South Pole

Today, we think of Antarctica as a permanently frozen continent (fig. 6.1), weighted down with such a thick ice cap that much of the land is actually below sea level. But as we saw in earlier chapters, for most of the geologic past, it was not frozen. It had been glaciated way back in the Permian (at about 290–250 Ma) but was cool-temperate through most of the Mesozoic, as shown by the lush vegetation that grew there (Kemp 1978). Such mild conditions continued through the early Cenozoic, with no sign of freezing or glaciers until the middle Eocene, at about 49 Ma (Birkenmajer 1987; Birkenmajer et al. 2005). As we saw in chapter 5, the first significant Antarctic glaciers occurred in the early Oligocene, with a renewed pulse of glaciation in the middle Oligocene and one final short pulse at the Oligocene-Miocene boundary. But temperatures warmed up in the early and middle Miocene, so that coral reefs once again grew in New Zealand, and warm-water plankton and mollusks were able to live close to the South Pole.

Those balmy Antarctic days ended forever during the middle Miocene. Oxygen isotope records (figs. 3.22, 6.3) at about 13–14 Ma (Shackleton and Kennett 1975; Savin et al. 1975; Miller et al. 1987, 1991; Zachos et al. 1993, 2001) consistently show a big positive shift in isotopes, indicating a drop in sea-bottom temperatures from about 6°C to about 2°C (43° to 35°F). This means that the bottom waters cooled down to close to their modern near-freezing temperatures. The East Antarctic ice sheet had apparently returned and was probably close to its modern dimensions (Shackleton and Kennett 1975). Ice-rafted sediments are found abundantly in deep-sea cores from all around the Antarctic, showing that there were icebergs floating away in abundance (Craddock and Hollister 1976). Benthic foraminifera from the deep oceans around the world not only recorded this cooling in their shells but also show by their distributions that oceanic circulation had changed as well (Woodruff 1985). The deep waters became so cold that most of the early Miocene benthic species vanished, to be replaced by the groups that are still found in the deep oceans today (Woodruff et al. 1981; Woodruff 1985). Oceanic circulation also became much more vigorous, with deep nutrients brought to the surface both at the Equator and at the poles, where there was a huge expansion of the cold Antarctic water mass, and massive upwelling of siliceous plankton. However, the equatorial surface waters showed no signs of cooling, so the modern extreme temperature difference between poles and Equator first developed in the middle Miocene (Savin et al. 1985). This high Equator-to-pole temperature gradient is one of the main driving forces of modern oceanic circulation, as the heat from the Equator is gradually spread to the cold poles by the movement of ocean currents. When the gradient is extreme, as it has been since 14 Ma, then the circulation is much more active.

What caused this sudden growth of a permanent Antarctic ice sheet, when South Pole glaciers formed and then melted at least three times since the Oligocene began? Clearly, the conditions for growing ice caps (such as the Antarctic Circumpolar Current) had already existed since the Oligocene, but for some reason they were not sufficient to keep the ice sheet permanent. A number of ideas have been proposed to solve this puzzle. According to Schnitker (1980), the key event was flow of the North Atlantic Deep Water (NADW) from its source in the Arctic Ocean and Norwegian Sea to the bottom of the South Atlantic, which brought additional cold to the entire world. This sudden influx of the NADW was apparently caused when the Iceland-Faeroe ridge sank below a critical threshold and allowed this bottom water to escape for the first time. Edwards (1975) also pointed out that the middle Miocene was the time when the Indonesian Archipelago, which arose as Australia began to collide with Asia, blocked the tropical flow between the Indian and Pacific oceans. According to this model, large volumes of water that would have flowed from the Pacific to the Indian Ocean were instead diverted, producing the huge western boundary current that flows north along the Philippines and Japan; this diversion is responsible for the oceanic productivity of the western Pacific. This water eventually returned to the central Pacific as equatorial circulation intensified. Kennett et al. (1985) point out that there were still connections between the Mediterranean and the Indian Ocean until 14 Ma, even though Arabia had begun to collide with Asia as early as 20 Ma. When this connection was finally closed, there was no longer any warm deep saline water produced in the Tethyan region to flow south along the bottom of the Indian Ocean and bring its heat to the Antarctic (Woodruff and Savin 1989).

Currently, the most popular explanation for the mid-Miocene cooling is the Monterey hypothesis of Vincent and Berger (1985). These geologists noticed that the middle and late Miocene is a period of enormous burial of carbon in deep-marine sediments of the Pacific Rim (including the

Monterey Formation of California, which is so rich in organic carbon that it is the major oil-source rock of the region), as well as in limestones in many regions of the tropics, and also in phosphate-rich sediments of the southeastern United States. All of these carbon-rich rocks are products of intense upwelling of carbon from deep in the ocean, which is rapidly consumed by the plankton and then deposited in basins such as those that trapped the shale and diatom-rich rocks such as the Monterey Formation. This hypothesis is confirmed not only by the volumes of carbon-rich sediments but especially by the carbon isotope curve in the middle and late Miocene, which shifts strongly in a negative direction, showing that carbon-12 from the deep ocean was coming back in to circulation (fig. 6.3). If so much carbon had indeed been locked into the shales and limestones of the earth's crust, it would have trapped a lot of atmospheric carbon dioxide and help push the earth further into icehouse conditions. However, the Monterey hypothesis does not explain what triggered the vigorous circulation and upwelling that led to the sequestration of carbon in the earth's crust. Clearly, it must have been some sort of oceanic trigger, such as the loss of the Tethyan warm saline waters, or the flow of the NADW to the South Atlantic, although no one has yet shown which cause is more likely.

Other explanations have been proposed for the long-term cooling of the middle and late Miocene. As has already been discussed, uplift of the Himalayas and Alps had accelerated by the middle Miocene (Rea 1992), so a lot of carbon dioxide could have been taken up with increased weathering due to uplift (Ruddiman and Kutzbach 1991; Raymo and Ruddiman 1992; Ruddiman 1997). Retallack (2001b) proposed that the increase in late Miocene grasslands and grassland soils was an important carbon sink that trapped carbon dioxide and helped end the greenhouse conditions of the early Cenozoic. However, these long-term explanations are relevant only on the scales of the entire Miocene cooling trend. Short-term pulses like the abrupt cooling events of the middle Miocene and the sudden expansion of the Antarctic ice cap must have been largely due to rapid oceanographic changes.

Whatever triggered this global cooling and increased circulation, the effects on marine life through the rest of the Miocene and into the Pliocene were dramatic. The diversity of warm-adapted planktonic foraminifera began to decline (Kennett and Srinivasan 1983; Tappan and Loeblich 1988; MacLeod et al. 2000), whereas the diversity of the phytoplankton (diatoms and coccolithophorids) increased as more nutrients were brought to the surface. The diversity of corals also plummeted as the warm waters they require vanished at middle latitudes, restricting them to the tropics (Coates and Jackson 1985; Budd et al. 1994). Bryozoan diversity also dropped from the middle Miocene peak of about 450 species to fewer than 300 species worldwide (Horowitz and Pachut 1996). Echinoid diversity dropped from about thirty-two to thirty-four species in the shallow-water limy seas in the middle Miocene to only about twenty-three species in the Pliocene (Poddubiuk and Rose 1984; Smith and Jeffrey 2000).

The mollusks responded in dramatic fashion. On the Pacific Coast of North America, where the cold, nutrient-rich waters welled up to chill the shallow seas of central California, warm-water mollusks retreated southward, and cold-tolerant species moved south and took over their ranges in response to the cooling (Addicott 1969, 1970). The diversity of individual molluscan genera also declined. For example, there were eighteen species of the snail *Cancellaria* in the warm middle Miocene, but only seven during the late Miocene, and only three during the Pliocene (Addicott 1970). Similarly, the five species of *Turritella* in the middle Miocene declined to only two by the late Miocene and Pliocene (Addicott 1970). Solitary corals, which had flourished even in the Oligocene and early and middle Miocene, vanished (Durham 1950). Durham (1950) estimated that late Miocene water temperatures in the San Joaquin Basin of central California had dropped to 13°C (55°F), considerably cooler than the 18°C (64°F) estimates for the same region in the middle Miocene.

Clearly, the cooling trend set in motion by the middle Miocene glaciation of Antarctica had global effects on the marine biota. The oceans may have been richer in nutrients and more vigorously circulated (favoring diversification of siliceous plankton, like the diatoms and radiolarians), but for most organisms, the temperate and polar regions became too cold, so their diversity in the nontropical areas declined as species either fled to the tropics or died out. As we shall see in the next chapter, the conditions in the marine realm became even harsher and colder in the Pliocene.

The Growth of the Grasslands

When we visit the savannas of East Africa today, we see extensive areas of grasslands and small patches of forest and scrub, which support a huge diversity of mammals familiar from wildlife documentaries. The modern East African savanna is just a remnant of what was once the prevailing habitat across most temperate and dry tropical regions in the late Miocene. The Great Plains region of North America once supported its own equivalent of the East African savanna (pl. 5), with native mammals playing the roles performed by other mammals in Africa (Webb 1983, 1984, 1985). Instead of elephants, there were mastodonts; instead of hippos, hippo-like rhinos; instead of giraffes, long-necked camels; and so on. This same pattern (but with a different cast of characters) can be seen during the late Miocene in the Pampas of Argentina and on the plains of Ukraine, China, and Pakistan. All of these regions harbored a dense collection of grazing mammals and their predators, all of them dependent on huge seas of grass as their main food resource. Such grasslands still exist today, although most of the incredible diversity of grazing mammals has vanished (for reasons we will discuss in chapter 8).

In North America, this African-style mammalian assemblage was called the Clarendonian Chronofauna by Webb (1969, 1983), because it became a homogeneous established entity starting in the Clarendonian (late Miocene). In North America, the middle Miocene is known as the Barstovian land mammal age (11.5–16 Ma), named after the deposits near Barstow, California (fig. 6.23). It continued through the late Miocene (Clarendonian land mammal age, 8.5–11.5 Ma; and Hemphillian land mammal age, 4.5–8.5 Ma) and then vanished in the early Pliocene (Hemphillian-Blancan boundary in North America, about 4.5 Ma). As mentioned above, the Clarendonian Chronofauna was dominated by at least four genera of gomphothere mastodonts, plus the mammutid (American mastodont) *Zygolophodon*. Camels performed the roles of giraffes, with *Aepycamelus giraffinus* reaching 3.5 meters at the shoulder and 6 meters (19 feet) tall and feeding from the tree canopy; six other genera of camels were built more like antelopes or llamas. The antelope role was also played by a huge diversity (nine genera and dozens of species) of native pronghorns, plus an even greater diversity of horses (fig. 6.24). In one quarry in the Barstovian of Nebraska alone, there are over twelve species of horses that all apparently lived at the same time! Some were browsers that ate leaves with their low-crowned teeth (*Archaeohippus, Hypohippus, Megahippus*), but most had higher-crowned teeth for eating at least some grasses along with leaves (MacFadden 1992). The last of the camel-like protoceratids (fig. 6.8) lived like moose or bushbuck in the thick underbrush, and *Synthetoceras* had a peculiar slingshot-shaped branched horn on its nose, and two horns over the eyes (pl. 7). Teleoceratine rhinos (fig. 6.11B) were the aquatic hippo-like grazer, and aceratherine rhinos (fig. 6.11A) were browsers like black rhinos. Peccaries were the substitute for the true pigs (such as warthogs). The predatory roles were occupied by hyena-like borophagine dogs (pl. 7), bears, mustelids, and beardogs, instead of the cats and hyenas that rule the African savanna today. Although squirrels (sciurids) and beavers (castorids) are common among the rodents, the greatest diversification occurred in the cricetids (New World mice and hamsters), the geomyids (gophers), heteromyids (pocket mice), zapodids (jumping mice), and the lagomorphs (pikas and rabbits).

Although the Clarendonian Chronofauna was a stable entity with most of its genera persisting through the middle and late Miocene, there were changes. The late Barstovian and Clarendonian marked the decline of many groups (especially browsers) that had dominated the earlier Miocene and even Oligocene. The last of the oreodonts (figs. 1.2, 6.9) disappeared in the late Clarendonian. The deer-like blastomerycids and dromomerycids (fig. 6.7) were rare after their Barstovian radiation. The protoceratids (fig. 6.8), too, were rare, and they vanished at the end of the Hemphillian. The last of the horses with low-crowned teeth (anchitherine horses) died out at the end of the Clarendonian too (fig. 6.24). Among carnivorans, the last of the beardogs (amphicyonids) disappeared in the Clarendonian, replaced by both bears and dogs, a great radiation of mustelids (weasels, otters, badgers) and true cats (Felidae), and one last

A

B

Figure 6.23. Formations displaying the middle Miocene Barstovian and late Miocene Clarendonian land mammal ages. (A) The middle Miocene in North America is known as the Barstovian land mammal age (see fig. 6.3), after the rich fossiliferous deposits of the Barstow Formation in the Mojave Desert of California. Here are the typical exposures of the Barstow Formation in Rainbow Basin, folded into a trough-like structure known as a syncline. The resistant ledges in the bedding are volcanic ash layers that have been precisely dated, so that the mammal fossils found there are also well dated. Photo by the author. (B) The middle Miocene Barstovian and late Miocene Clarendonian land mammal ages are well displayed in these deposits of the Ricardo Group in Red Rock Canyon, in the Mojave Desert northwest of Barstow, California. Photo courtesy of L. Spoon.

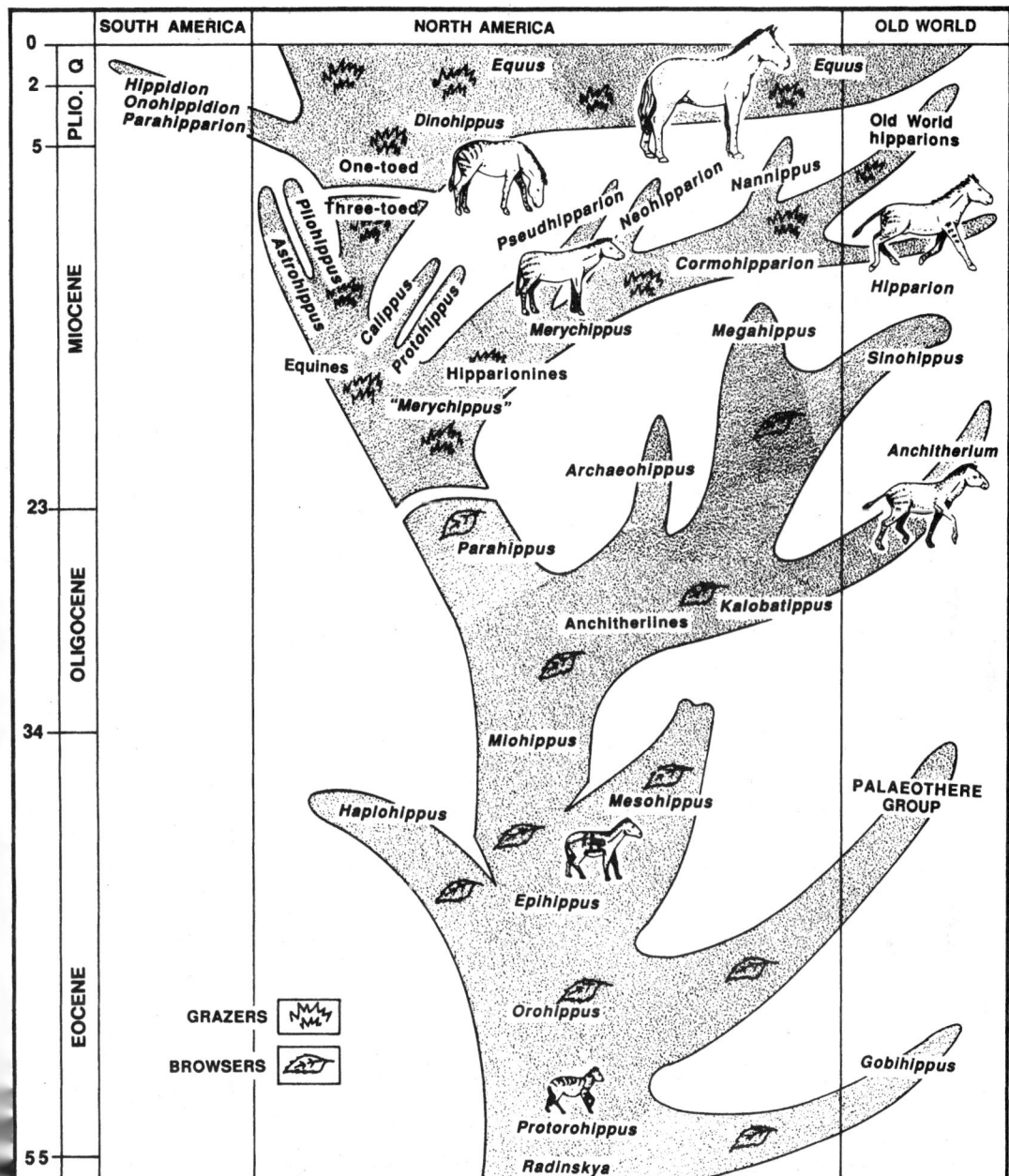

Figure 6.24. Evolution of the horses in North America. Drawing by C. R. Prothero.

surviving cat-like nimravid, the saber-toothed *Barbourofelis* (fig. 6.25), which persisted until the early Hemphillian. Among the small mammals, the hedgehogs vanished in the Clarendonian, along with the once-abundant aplodontids (sewellels), the horned mylagaulids, and the archaic eomyid rodents, relics of the Eocene. As Janis et al. (2000, 2002, 2004) point out, the striking change in the Clarendonian Chronofauna

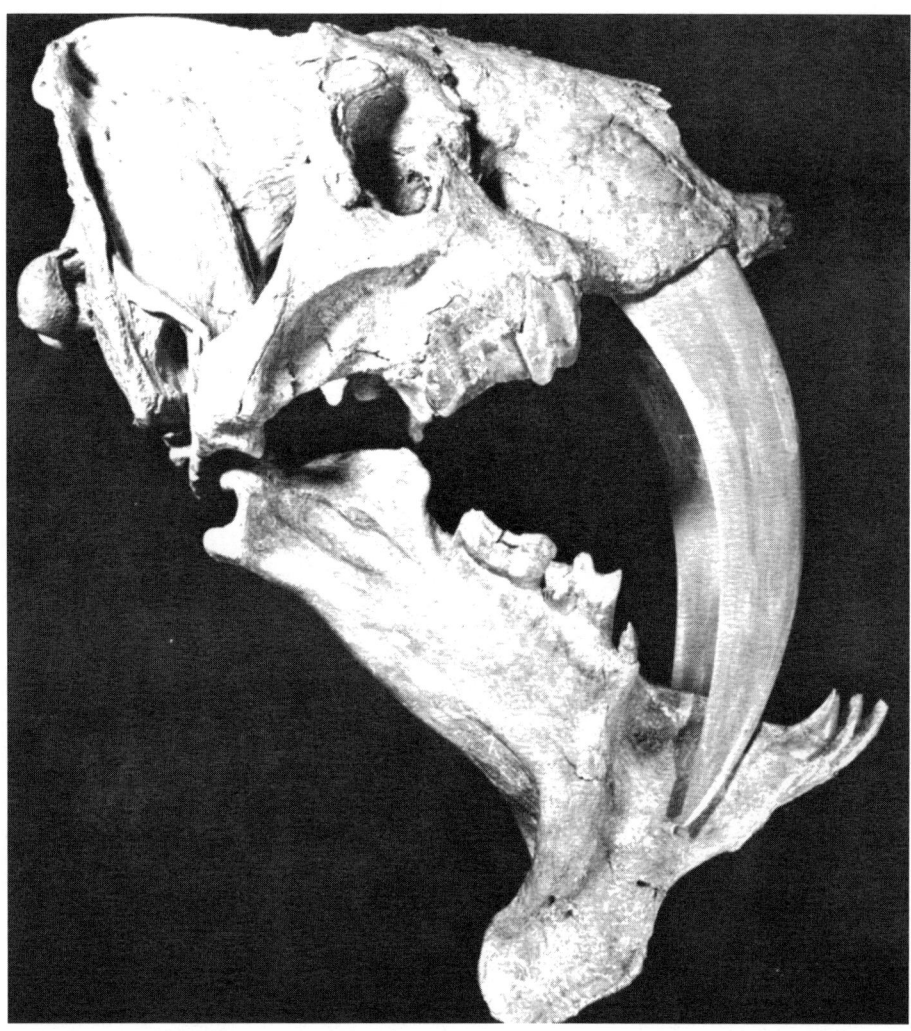

Figure 6.25. Barbourofelis, *the last of the cat-like carnivorans known as nimravids. It had not only large upper canines shaped like sabers but also huge flanges on its lower jaw to sheath them. From Schultz et al. 1970.*

through most of the late Miocene was the loss of browsing (leaf-eating) forms with low-crowned teeth, as grazers (grass eaters) with high-crowned teeth gradually took over. Such browsers included the oreodonts, blastomerycids, dromomerycids, protoceratids, the rhino *Peraceras,* the anchitherine horses, and the archaic rodents (aplodontids, mylagaulids, and eomyids). What remained in the Hemphillian was a fauna dominated by grazers with very high-crowned teeth, including the advanced equine horses, camels, pronghorns, and the later species of the rhinos *Aphelops* and *Teleoceras.*

What kind of landscape and vegetation were home to the Clarendonian Chronofauna? According to Axelrod (1985), Wing (1998), and Graham (1999), there were swamps and dense forests in the humid Atlantic and Gulf Coastal plains, and these regions also supported a peculiar forest-adapted fauna with abundant protoceratids, weird long-nosed camels, and dwarf rhinoceroses (Tedford et al. 1987; Prothero and Sereno 1982). In the Rockies and Great Basin, there were abundant pines and other conifers, as well as trees of the walnut, oak, elm, birch, and willow families, and scrub brush dominated by Mormon tea (*Ephedra*), greasewood, sagebrush, and herbs of the family Chenopodiaceae.

For decades, the classic story alleged that extensive grasslands arose as soon as 16 Ma in the early middle Miocene, as indicated by the increasing hypsodonty of teeth in horses, camels, pronghorns, and rhinoceroses. But the middle Miocene, grass pollen is relatively rare (Leopold and Denton 1987), and only a handful of grass fossils are known (Thomasson 1982, 1985). This striking rarity of grass fossils, which has always been a puzzle, has been further confirmed by the evidence of carbon isotopes. Most plants (including high-latitude and high-altitude grasses) use the Calvin (C3) photosynthetic pathway, which produces a carbon isotope signal of lighter values. But most temperate and tropical tall grasses use another photosynthetic mechanism, the Hatch-Slack pathway (C4), which produces much heavier (richer in carbon-13) isotopic values. Studies of the isotopes of carbonate in ancient soils, and also the carbon isotopes in the fossil teeth of herbivorous mammals, revealed that large areas of grassland-savanna may have originated not as early as 16 Ma but about 7 Ma (Quade et al. 1989; Cerling 1992; Wang et al. 1994; Cerling et al. 1997). If this is true, what were all those mammals with high-crowned teeth eating between 16 and 7 Ma? The flora may have been a mixture of grasses and scrub, often with a lot of dirt and grit on it, which required higher-crowned teeth (Retallack 1997, 2001b; MacFadden 1992). According to Retallack (1997), these sod-forming short grasslands comprised a mixed scrubland-grassland, and not the huge areas of tall grasslands that we now see in Africa and North America. If the carbon isotopes are right, true C4 grasslands forming a tall-grass prairie did not emerge until 7 Ma. Strömberg (2004, 2005), however, has shown that the tiny siliceous fossils of plant cells of grasses (known as phytoliths) show up in the early Miocene, when there were few mammals with hypsodont teeth. Clearly, the puzzle of the origin of grasslands and grazing mammals is not fully resolved yet.

The patterns of the Clarendonian Chronofauna also apply to Eurasia and Africa, albeit with different casts of characters. In Europe (Agusti and Anton 2002), the subtropical evergreen forests of the early Miocene were replaced by seasonal, summer-drought-adapted forests with tough woody vegetation (Axelrod 1975). As the Miocene progressed, this dense forest gave way to more-open habitat, although large grasslands never appeared in Europe during the Miocene. Living in these forests was an assemblage of mammals adapted for eating this tough vegetation, and gradually true grazers appeared by the late Miocene as patches of grassland developed.

They are recorded from localities all over Eurasia, including the famous bone beds in Spain, France, Germany, and Italy, in Pikermi and Samos in Greece, in Maragheh in Iran, in the Siwalik Hills of northern Pakistan (fig. 6.26A), in the Tunggur beds in Mongolia (fig. 6.26B), and in many deposits in China. The hoofed mammals were largely groups that dominate in the Old World today. There was a big radiation of bovids (antelopes and cattle) including huge (300-kilogram) boselaphines (primitive forms related to the modern nilgai antelope), as well as gazelles and several extinct groups. Early Miocene giraffes were built more like antelopes, with forms that had a pair of short unbranched horns, or flat laterally extended horns (fig. 6.27). By the late Miocene, there were huge giraffes (*Brahmatherium*) with thick short necks and moose-like horns. The middle Miocene deer (Cervidae), in contrast, were mostly small primitive forms with no antlers; instead, the males had large canines, like modern musk deer. By the late Miocene, however, deer really began to diversify, with a wide variety of types of antlers. The pigs enjoyed a great radiation, including weird forms like the "unicorn pig" *Kubanochoerus*, which had a long horn pointing out of its forehead.

Both aceratherine and teleoceratine rhinos flourished (as they did in North America), along with the earliest members of the living lineage of Sumatran rhinos (dicerorhinines). There was also a group called elasmotheres that had unusually high-crowned molars with convoluted, crenulated enamel on them. Horses were abundant, primarily the three-toed grazing hipparionine horses, which migrated to Eurasia from North America in several events in the Miocene. Gomphothere mastodonts were the largest herbivores, along with the peculiar deinotheres (fig. 6.14B) with their downward-pointing lower tusks. Another African immigrant, the hyrax *Pliohyrax*, was found in the streams along with pigs. As in North America, the last of the chalicotheres (the gorilla-like *Chalicotherium*; fig. 6.6C) in Eurasia occurs in the middle Miocene, and the beardogs had vanished from Europe by the late Miocene, replaced by true bears and true cats, as well as civets and mustelids. However, true dogs were not found in the Old World yet, and their roles were played largely by a diversification of hyenas, many of which were more like cats and dogs than like modern hyenas. As in North America, squirrels and beavers were still common, but there was a huge radiation of cricetids (New World mice and hamsters), jumping mice, and especially murids, the modern family of Old World rats and mice. Finally, mastodonts and hyraxes were not the only African natives to spread to Eurasia after the Arabian land bridge closed at 18 Ma. By 16 Ma, the primates had made the trip out of Africa too. Starting with the small anthropoid *Pliopithecus* (the first fossil primate described in the history of paleontology, by Lartet in 1834), they soon ranged widely through Europe, and by the late Miocene there was a considerable radiation of dryopithecine apes found in many places in Europe.

A

B

Figure 6.26. Fossiliferous middle to upper Miocene beds, found in many places in Eurasia. (A) Typical exposures of the Siwalik Group in the Himalayan foothills of Pakistan, which are often collected using local camels for transport of heavy fossils. Photo courtesy of L. Flynn. (B) The classic middle Miocene locality of Tunggur in Mongolia, which has produced important mammal fossils for almost a century. Photo courtesy of J. O'Connor.

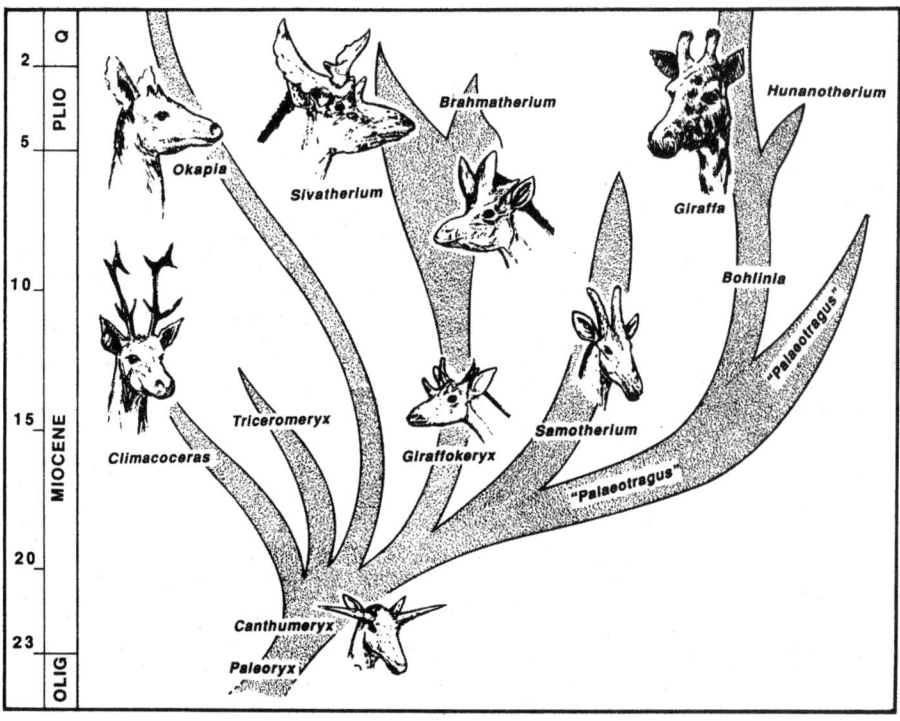

Figure 6.27. *Evolution of the giraffe family during the Miocene. From a primitive antelope-like ancestor in the late Oligocene, giraffes evolved into a variety of shapes and sizes, including the okapi (still alive in the African jungles today), the huge moose-like* Brahmatherium, *and the lineage of* Giraffa, *the only giraffes with long necks. Drawn by C. R. Prothero.*

In North America, the late Miocene faunas gradually saw a shift to herbivores with longer legs and more-hypsodont teeth as tall grasslands expanded in the mid-Hemphillian at around 7 Ma. However, the same did not happen in Europe, since grasslands did not develop until the Pliocene (at around 4 Ma). Instead, there was what European paleomammalogists (Agusti and Anton 2002) call a "Vallesian crisis" at about 9.0–9.5 Ma, during the Vallesian Stage of the European mammalian timescale. According to recent paleobotanical evidence, the persistent subtropical forests of the middle Miocene were abruptly replaced by forests tolerant of more summer drought, with more deciduous trees, such as oaks, maples, alders, hickories, walnuts, and elms, and dense scrub vegetation as well. This Vallesian crisis affected a number of groups of mostly archaic mammals that had survived in the relict subtropical forests and could not tolerate the harsher, woodier deciduous forests that had just developed. A number of browsing rhinos and the last of the European tapirs disappeared, along with many of the more primitive pigs with brachydont teeth. The last of the archaic relict cat-like nimravids and the beardogs also disappeared at this time, while many new bears, cats, and hyenas evolved to replace them. Among rodents, most of the archaic cricetids, glirids (dormice), flying squirrels, and beavers vanished, replaced by a big radiation of advanced hypsodont cricetids and murids. Last but not least, the primate experiment in Europe ended as the dryopithecine hominoids and Old World monkeys that had lived there since 16 Ma vanished at about 9 Ma.

By the late Miocene, Africa was no longer isolated from Eurasia, so most of its mammals were similar to those of Eurasia. Gomphothere mastodonts and deinotheres (fig. 6.14) were common, as were the first true members of the living family Elephantidae, which would soon spread around the world. The evolutionary radiation of antelopes that now dominate the African landscape was in full swing, with many of the modern genera, including kob, reedbucks, gazelles, impalas, antilopines, and true cattle, known by the middle and late Miocene. We also see the first hippopotamuses, along with a huge radiation of pigs (including early warthogs and giant forest hogs), as well as the last of the pig-like anthracotheres. Primitive giraffes were common, most of which still had short, thick necks like the living okapi; but the earliest members of the modern genus *Giraffa* were also present (fig. 6.27). Hipparionine horses were common all over Africa, along with primitive teleoceratine rhinos and the first members of the modern lineage of African rhinos, *Ceratotherium praecox* (related to the living white rhinoceros). Chalicotheres also persisted (as they did in Europe). The predators were similar to those of Europe, consisting mostly of cats, hyenas, and civets—the three predators that predominate in Africa today.

Finally, primates did not vanish from Africa as they did in Europe, but radiated into the Old World monkeys (including the earliest baboons and macaques), numerous genera and species of apes, and the first members of our own family, the Hominidae. Known as *Sahelanthropus tchadensis*, the best specimen is a complete skull (fig. 6.28) from rocks about 6–7 Ma in age from the Sub-Saharan Sahel region of Chad (Brunet et al. 2002). Although the skull is chimp-like with its small size, small brain, and large brow ridges, it had remarkably humanlike features, with a flattened face, reduced canine teeth, enlarged cheek teeth with heavy crown wear, and an upright posture at the very beginning of human evolution. Just slightly younger is the recently discovered *Ororrin tugenensis*, from the upper Miocene Lukeino Formation in the Tugen Hills in Kenya dated between 5.72 and 5.88 Ma. *Ororrin* is known mainly from fragmentary remains, but the teeth have the thick enamel typical of early hominids; and the thighbones and shinbones clearly show that it walked upright. Slightly younger still are the remains of *Ardipithecus ramidus kadabba*, found in Ethiopian rocks dated between 5.2 and 5.8 Ma. These consist of a number of scrappy fossils, but the foot bones show that hominids used the "toe off" manner of upright walking as early as 5.2 Ma. Thus, our human

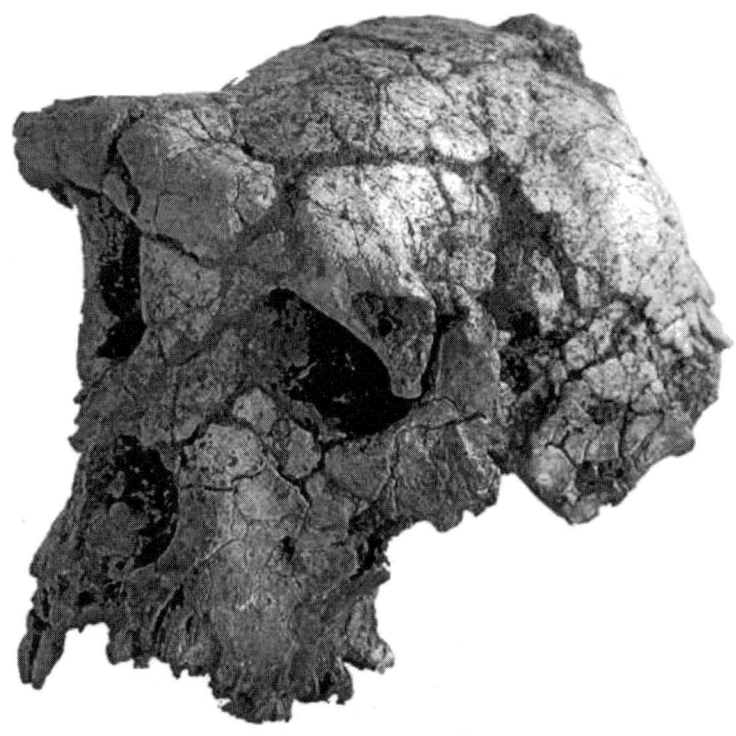

Figure 6.28. The skull of Sahelanthropus tchadensis, *the oldest known fossil of our family Hominidae, from beds 6–7 million years old in Chad. Although its brain was small, and the skull, with its heavy brow ridges, is about the size of a chimpanzee's,* Sahelanthropus tchadensis *had a flat face and the upright posture characteristic of early human evolution. Photo by M. Brunet ©️ MPFT.*

lineage was well established by the latest Miocene, and fully upright in posture, even though our brains were still primitive and our body size was not much different than that of contemporary apes.

In contrast to Africa, South America and Australia continued to be isolated through the rest of the Miocene. In South America, the middle and late Miocene are divided into a number of stages (Flynn and Swisher 1995), which are in order, the Laventan, Mayoan, Huayquerian, and Montehermosan (based largely on faunas from Argentina with a few like La Venta from Colombia). These South American faunas are largely continuations of those found in the early Miocene. The main predators are huge flightless carnivorous birds, the phorusrhacoids (fig. 6.29), which reached up to 3 meters (10 feet) in height (Marshall 1978). As we saw in the Eocene of North America with the diatrymids, when there are no large mammalian predators, there are opportunities for birds to perform the role of top carnivore. The much smaller carnivorous marsupials continued to be important predators (in the absence of bears, cats, or dogs), with some that were dog-like or hyena-like in form. The edentates contin-

Figure 6.29. Phorusrhachids were huge ground birds that were the major predators in South America during the Miocene and Pliocene (in the absence of large mammalian carnivores). They are related to the living cariama bird (shown for comparison) but were much larger and had sharp, hooked beaks for tearing flesh. A few phorusrhachids managed to cross the Panamanian land bridge in the Pliocene and reach North America, where they were truly gigantic. Titanis walleri (found in Florida) reached up to 2 meters (6 feet) in height. From Marshall 1978.

ued to diversify, with many species of ground sloths and anteaters, and the armadillo-like glyptodonts. Most of the native hoofed mammals continued to converge on hippos, rhinos, camels, horses, and other groups found elsewhere in the world. The radiation of caviomorph rodents that had immigrated in the Oligocene was also in full swing, with nine families (including capybaras, agoutis, Guinea pigs, chinchillas, plus several extinct families), as was the radiation of New World monkeys that had also arrived in the Oligocene. However, two immigrant groups clearly suggest that isolation was ending by the late Miocene. One is the raccoon family, represented by the primitive coatimundi *Cyonasua*. The other is the mastodonts, which have recently been reported from late Miocene beds (at least 9 Ma) of the Amazon Basin of Peru (Campbell et al. 2001). These two immigrant groups from North America must have island-hopped across the chain of islands that was Central America at the time, since there was no direct land connection between the continents yet.

Australia, in contrast, continued with its endemic fauna of marsupials and platypuses found at late Miocene localities like Alcoota, with no evidence of immigrant mammals from Asia (other than bats). However, the collision of Australia with Asia to form the volcanic chains of New Guinea and Indonesia-Malaysia had begun, as Australia got nearer to Asia after parting from Antarctica in the Eocene. There is no evidence that any Asian immigrants would cross to Australia for a long time, but New Guinea and other intervening islands have a peculiar mixture of Asian and Australian animals that originated on both continents.

Scylla and Charybdis

The Strait of Messina, between Sicily and Italy, was supposedly the home of the legendary terrors Scylla and Charybdis. In Homer's *Odyssey*, Odysseus and his men were forced to navigate between these two forces

Figure 6.30. (A) Thick deposits of gypsum overlying marine shales at Eraclea Minoa on the island of Sicily. In some places, the Messinian salt and gypsum deposits on Sicily are thousands of meters in thickness. Photo courtesy of R. W. H. Butler. (B) "Swallow-tail" salt crystals. Photo courtesy B. C. Schreiber.

that guarded the narrow straits. The monster Scylla lived in a cave high up on the cliff, from whence she would thrust her long necks (she had six heads), each of her mouths seizing one of the crew of every vessel passing within reach. The other terror, Charybdis, was a gulf nearly on a level with the water. Three times each day the water rushed into a frightful chasm and was then disgorged. Any vessel coming near the whirlpool when the tide was rushing in must inevitably be engulfed; not even Poseidon himself could save it.

Although these mythical monsters no longer guard the strait, the rocks themselves are evidence of an even more terrifying event. Five to six million years ago, the Mediterranean Sea itself dried up and then refilled with a frightening wave of water, over and over again. Near the strait are outcrops that are the basis for the final stage of the Miocene, the Messinian Stage (named after the strait). These outcrops include gravels that were formed in massive floods and enormous thicknesses (up to 2,000 to 3,000 meters!) of rock salt and gypsum (fig. 6.30), which form only when huge bodies of salt water dry up. Geologists had known about this evidence for more than a century but had assumed that the salt and gypsum deposits were produced when a local basin evaporated.

In 1970, Leg 13 of the Deep Sea Drilling Project was removing drill cores from the sea bottom of the western Mediterranean around the Balearic Islands between Spain, France, and Italy. They recovered a variety of puzzling clues (Hsu et al. 1977). As they drilled the shallow margin of the Balearic Basin, they found gravels and other evidence of huge desert flash-flood deposits on the floor of the Mediterranean. Then they drilled in the center of the basin, and underneath the Pliocene and Pleistocene marine deposits, they found thick deposits of gypsum and rock salt, and algal mats that had once formed in shallow pools on the fringe of a giant salty lake, not unlike the Salton Sea of southern California. This was a clinching piece of evidence. The only way that such huge quantities of evaporite minerals (over a million cubic kilometers in total) and algal mats could have formed beneath the center of the deepest Mediterranean was if it had dried up completely, exposing the bottom to sunlight (fig. 6.31). As more and more cores were drilled, the story became even more spectacular. The Mediterranean had dried up a number of times during the latest Miocene, only to refill with catastrophic floods, then dry up again. This event is known as the Messinian salinity crisis.

How could such an amazing event happen? It turns out that the rivers that now feed the Mediterranean, such as the Rhône and the Nile, mostly drain out of the dry "Mediterranean" climates of Italy and Greece and Spain, or out of the Egyptian desert, so their flow is insufficient to keep up with the huge rate of evaporation (3,300 cubic kilometers/year) in this semi-desert latitude. With such high evaporation rates, the Mediterranean Sea would dry up in less than a thousand years if it were not constantly receiving cold water from the Atlantic through the Strait of Gibraltar. If some sort of event cut off this supply of Atlantic Ocean water, the

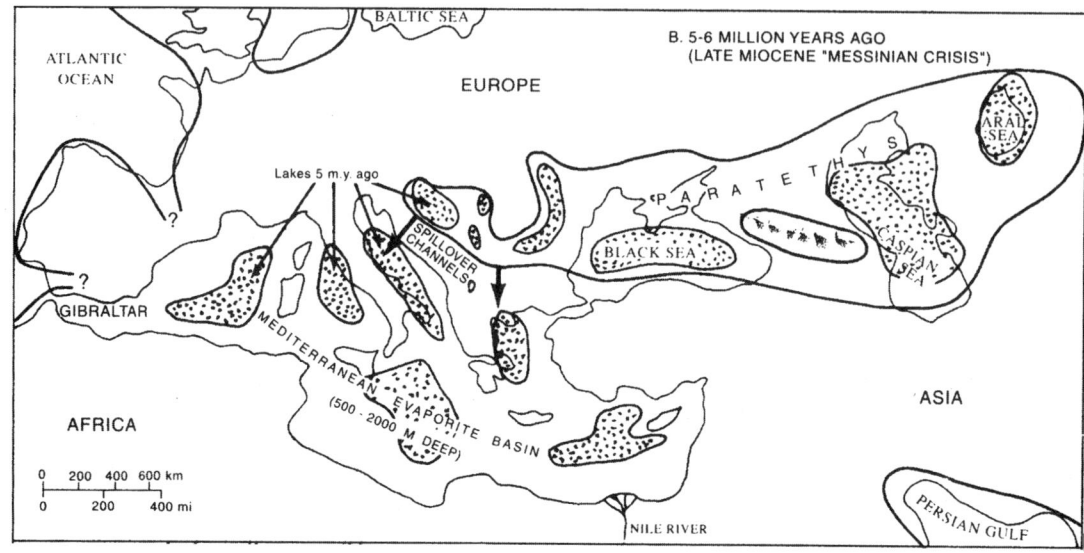

Figure 6.31. During the Messinian salinity crisis at the end of the Miocene, the Mediterranean dried up completely, forming thick salt deposits over 1,000 meters (3,300 feet) below modern sea level and incising a deep canyon beneath the valleys of the Nile and Rhône rivers. The Paratethys Sea also dried up, forming large salt deposits beneath the modern Black, Caspian, and Aral seas. From Prothero and Dott 2003.

Mediterranean would dry up. Indeed, the geological evidence shows that the Atlas Mountains in Spain and Morocco (through which the Strait of Gibraltar is cut) were rising in the late Miocene as Africa collided with Europe, and apparently they began to form a barrier. In addition to the Gibraltar Strait, there were flows through the Betic Corridor in southern Spain and through the Rifian Corridor in northern Morocco. Once these uplifted barriers reached a critical threshold, the Mediterranean was isolated. Each drying episode took about 1,000 years and produced about 70 meters of salt in the bottom of the basin, leading to estimates that at least forty separate episodes must have occurred between 5.6 and 5.8 Ma to produce the 2,000 to 3,000 meters of evaporites now found there (Ryan 1974; Hsu et al. 1977). Once the Mediterranean had dried up, it was only a matter of time before erosion would let the Atlantic water back in, forming a huge waterfall at Gibraltar that was at least ten times the size of Niagara Falls. The water had to be moving so fast through this gap that it broke the sound barrier! Then the mountains would shift again, the drying resume, followed by another flood through Gibraltar—over and over again forty times. Finally, at 5.2 Ma, the Atlantic water rushed in for the last time, the Strait of Gibraltar stayed open, marine waters returned to the Mediterranean permanently, and normal marine sediments were deposited—and the Pliocene had begun.

Further confirmation of this startling hypothesis was discovered when geologists tried to find bedrock beneath the Nile Gorge in southern Egypt so they could provide a firm foundation for the Aswan High Dam. As they drilled and drilled, they found that beneath the thick blanket of Nile sediments was a gorge almost 200 meters below the present level of the Mediterranean. Then surveys were done at the mouth of the Nile, and in the Rhône Valley, and buried gorges almost 1,500 meters deep were found! The only way such deep valleys could have formed so far below the present level of the Mediterranean was if the sea had dried up to 1,500 meters below its present level. Under such conditions, the Nile and Rhône would have cut down to the level of the bottom of the Mediterranean when it was a dry lake bed. Once the Mediterranean refilled in the early Pliocene, the grand canyons of the Nile and the Rhône would have filled with sediment until they matched the modern level of the Mediterranean Sea.

So what triggered this amazing series of events? As we have seen, the collision of Africa with Europe was an ongoing phenomenon in the Miocene, so the rise of the Atlas Mountains was an inevitable result. But what about the late Miocene glaciation? According to data from drill cores around the Antarctic, the West Antarctic ice sheet (which is unstable since it is held in place by floating ice shelves) developed about this time (Shackleton and Kennett 1975; Mercer 1978). In chapter 5, we saw that the first Antarctic ice cap occurred at 33 Ma, and earlier in this chapter we saw that the East Antarctic ice cap became permanent at 14 Ma. But the West Antarctic was still not glaciated in the middle Miocene, and glaciation began only in the latest Miocene. This late Miocene glaciation can be seen in glacial sediments in South America dated at 6.75 Ma (Mercer 1978). In addition, there were oxygen isotope events at 6.3 Ma that indicate a dramatic cooling, and a one-per-mil global carbon isotope shift, which signals a big turnover in the world's oceans (Bender and Keigwin 1979). All of these events seem to be the culmination of a steady global cooling trend through the latest Miocene, as the East Antarctic ice cap got larger and was joined by the West Antarctic ice sheet.

The effects of this global cooling were dramatic. It not only triggered a drop in sea level and cooling global temperatures but also, as we just saw, the Messinian salinity crisis in the Mediterranean region. The Sea of Japan was also cut off from the Pacific, forming a giant freshwater lake (Burckle and Akiba 1978). Huge unconformities were cut in the sedimentary records of most continents and ocean basins at the end of the Miocene, making this global cooling and regression a natural boundary between the Miocene and Pliocene. In addition, the drying of the Mediterranean may have had additional consequences. Ryan et al. (1974) calculated that the drying of the Mediterranean and the withdrawal of all that salt would have lowered global oceanic salinity by 6%. Decreased salinity raises the freezing temperature of water, so ice could form at a higher temperature, which would have helped the Antarctic ice sheets to form more easily once the Mediterranean had dried up. Although Ryan et al. (1974) claim that the reduction in salinity caused the late Miocene glaciation, in fact the evidence shows that the ice sheets were already developing at 6.75 Ma; and because the global oceanic change took place at 6.3 Ma, the much later (5.2–5.6 Ma) drying of the Mediterranean could not have caused the first global cooling, but may have accentuated it.

The latest Miocene global cooling event, and the local drying of the Mediterranean, had some surprising effects on life. On the fringes of the hot dry desert that was once the Mediterranean Sea lived a forest-adapted fauna (Agusti and Anton 2002), including abundant tapirs, pigs, deer with long lyre-like antlers (*Croizetoceros*), diverse antelopes, gomphothere mastodonts with 4.3-meter-long straight tusks (*Anancus*) and the mammutid *Zygolophodon*, sabertoothed cats (*Machairodus* and *Dinofelis*), bears, mustelids, and hyenas. More dramatic was the effect of the drying of the Mediterranean on land corridors. Many African forms, including the hippopotamus *Hexaprotodon*, macaque monkeys, gerbils, and many African antelopes, migrated to Europe around the dry Mediterranean basin during the Messinian. In addition, we see immigrants from even further away, including the first camels and dogs to leave North America and migrate to Eurasia, presumably due to the lowering of the seas across the Bering Strait. A number of typically Miocene mammals also vanished by the Pliocene. These included the giant pig *Microstonyx*, the primitive aceratherine and teleoceratine rhinoceroses (leaving only the ancestors of the modern rhinoceros genera), most of the hipparionine horses, and most of the sabertoothed machairodont cats and Miocene hyenas.

Although the drying of the Mediterranean did not affect life so directly in other parts of the world, the global cooling, rapid change in oceanic circulation, and drop in sea level did. The remaining forests and leafy vegetation were rapidly replaced by dry grasslands and steppe vegetation (Wing 1998; Graham 1999). In North America, many of the animals adapted to the mixed grasslands and woodlands of the Hemphillian vanished (Webb 1984), including the last North American rhinos (both *Teleoceras* and *Aphelops;* fig. 6.11), the last horned camel-like protoceratids (fig. 6.8), the last deer-like dromomerycids (fig. 6.7), the last musk-deer-like blastomerycids, and the last archaic aplodontid, eomyid, and horned mylagaulid rodents (fig. 5.20A). In addition to the loss of these entire families, there were extinctions of species and genera *within* families, including many horses (five genera), peccaries (two genera), pronghorns (seven genera), camels, gomphothere mastodonts (three of the five genera known), mustelids (three genera), bears (two genera), and dogs (three genera). Among small mammals, major extinctions also occurred in the pocket gophers, the jumping mice, the cricetids, the beavers, and the pocket mice and in the shrews and moles as well. According to Webb (1984), this was the largest extinction event in North American land mammals in the entire Cenozoic, dwarfing even those at the end of the Eocene and during the ice ages.

Africa, too, lost some of its archaic groups, such as the last of the teleoceratine rhinoceros *Brachypotherium*, leaving only the modern lineage of

African rhinos. The diversity of antelopes and cattle, pigs, horses, and mastodonts took a hit, requiring the African faunas to evolve new forms in the Pliocene. In Asia, the huge diversity of late Miocene rhinos, horses, mastodonts, pigs, antelopes and cattle, deer, and giraffes was considerably reduced, although all those families persisted. However, the anthracotheres vanished from Asia for good. Likewise, the huge diversity of late Miocene hyenas, mustelids, bears, and cats was greatly reduced in the early Pliocene, but these groups all survived, along with the dogs, which had just immigrated from North America. However, the beardogs, which had straggled on in Asia long after their disappearance elsewhere, finally died out.

In South America, however, the end of the late Miocene (Huayquerian) was not a major extinction. Only a few genera of marsupials, edentates, rodents, primates, and native ungulates vanished (Marshall and Cifelli 1989), and a new diversity arose in the early Pliocene (Montehermosan). Likewise, the Riversleigh localities show that Australia retained the same fauna composition at the Miocene-Pliocene boundary, with only minor adjustments in the genera of marsupials that still ruled the continent (Archer et al. 2001).

Thus, the effects of the Messinian event were highly varied, with dramatic responses in the European continent and in North America, lesser responses in Asia and Africa, and almost no response in South America or Australia. The causes for these varied mammalian faunal changes have yet to be fully explained or even discussed. But the stage was now set for the cool, dry world of the Pliocene and the final development of the ice ages.

For Further Reading

Agusti, J., and M. Anton. 2002. *Mammoths, Sabertooths, and Hominids: 65 Million Years of Mammalian Evolution in Europe*. New York: Columbia University Press.

Kennett, J. P. 1985. *The Miocene Ocean: Paleoceanography and Biogeography*. Geological Society of America Memoir 163.

Hsu, K. J. 1983. *The Mediterranean Was a Desert: A Voyage of the* Glomar Challenger. Princeton, N.J.: Princeton University Press.

Rössner, G. E., and K. Heissig, eds. 1999. *The Miocene Land Mammals of Europe*. Munich: Friedrich Pfeil Verlag.

Figure 7.1. Mollusk fossils from the Pinecrest shell beds. The incredible density of mollusk fossils in the lower Pliocene Pinecrest shell beds in Florida is world famous. Most of these species died out during the glaciation and cooling of the middle Pliocene. Photo courtesy of W. Allmon.

7

The World in Transition: The Pliocene

By the North Gate, the wind blows full of sand,
Lonely from the beginning of time until now!
Trees fall, the grass goes yellow with autumn.
    Li Po, 762 A.D.

Last Gasp of Warmth

The Pliocene is the last epoch of the Tertiary, and the shortest. It began at 5 Ma and ended at 1.8 Ma. Thus, it lasted only 3.2 million years and is shorter than most of the stages within the epochs of the rest of the Cenozoic. Normally, such a short interval of time would not merit a whole chapter, even though it was one of the three original divisions of Lyell's (1833) concept of the Cenozoic (along with the Eocene and Miocene). But the Pliocene is a crucial epoch of transition, from the last of the warmer climates of the Miocene and early Pliocene to the first Arctic glaciers and the beginning of the ice ages in the late Pliocene. A great diversity of warm-water mollusks evolved in the early Pliocene and then were wiped out as glaciation started (fig. 7.1). The savannahs of the Miocene gave way to steppe climates on the northern continents, with abundant mammals adapted to eating grasses and running fast (pl. 8). And because the Pliocene was not long ago, we understand it in much greater detail than we do any previous epoch. We can resolve events in some deep-sea cores down to the level of a century or less, so we can tease out the intricacies of climate change in this crucial transition from the warm Miocene to the ice ages. In fact, whole research groups with their own special acronym, PRISM (Pliocene Research, Interpretation, and Synoptic Mapping), have been funded with millions of dollars in grant money to decipher these complex paleoclimatic events of the Pliocene (PRISM Project Members

1995). Last but not least, the early phases of human evolution happened in the Pliocene, so many more researchers and grant dollars are focused on the details of the Pliocene (especially in Africa) than on any other Tertiary epoch.

The world after the Messinian event (fig. 7.2) had changed in many ways. In most of the world's oceans, temperatures recovered after the latest Miocene ice pulse, and the early Pliocene was a warmer episode on the oxygen isotope curves (Shackleton et al. 1984; Hodell and Kennett 1986; Miller et al. 1987; Kennett 1995; Zachos et al. 2001). Much of the Antarctic ice melted back (Shackleton and Kennett 1975; Berggren and Haq 1976), which has led some researchers to suggest that the entire Antarctic was deglaciated in the early Pliocene (Webb and Harwood 1991; Barrett et al. 1992). However, more-recent detailed analysis (Burckle 1995; Kennett 1995) has shown that the Antarctic ice cap only partially melted back. Nevertheless, there was considerable warming of the southern high latitudes, and early Pliocene ice-rafted sediments retreated to the south, suggesting far fewer icebergs calving off the Antarctic ice sheet (Kennett et al. 1975). All of this melted ice caused global sea level to rise about 25 meters, producing thick lower Pliocene shallow marine beds (fig. 7.1) on many continents (Kennett 1995).

The cause of early Pliocene warming is still controversial. Computer models of climate have tried both elevating the level of carbon dioxide to cause global warming, and also changing the patterns of oceanic circulation to move more heat to the poles. So far, the rise in carbon dioxide does not seem to be a reasonable factor, but the changes in oceanic circulation in the model do fit the predictions of real data (Crowley 1991; Rind and Chandler 1991; Chandler 1993, 1994; Dowsett et al. 1992, 1994; Raymo et al. 1992, 1996; Chandler et al. 1994).

The warming trend can be seen in molluscan and ostracode faunas of the western Atlantic, such as those of the Yorktown Formation of Virginia and North Carolina, the Duplin Formation of Georgia, and the Jackson Bluff Formation of Mississippi (Hazel 1971, 1988; Campbell et al. 1975; Blackwelder 1981; Stanley 1986; Krantz 1990). Diversity of mollusks in the Yorktown Formation was high (about 173 species, according to Stanley 1986), with temperature estimates (based on the isotopic analyses of Krantz 1990) in the mild-temperate range (between 10–22°C, or 50–72°F). Temperatures were seasonal, but the extremes of seasonality seen today, when this region freezes during the winter were not yet occurring. Further south in Florida, the famous Pinecrest shell beds (fig. 7.1) yield well over 200 species of mollusks (Stanley 1986), which lived in subtropical waters along the shoreline of Florida. Early Pliocene faunas are also found around the Caribbean (the Gatun faunal province), where there was apparently strong upwelling as waters of the Atlantic and Caribbean flowed out through the closing gap in Central America into the Pacific Ocean.

On the Pacific side of North America, there is evidence that the steep

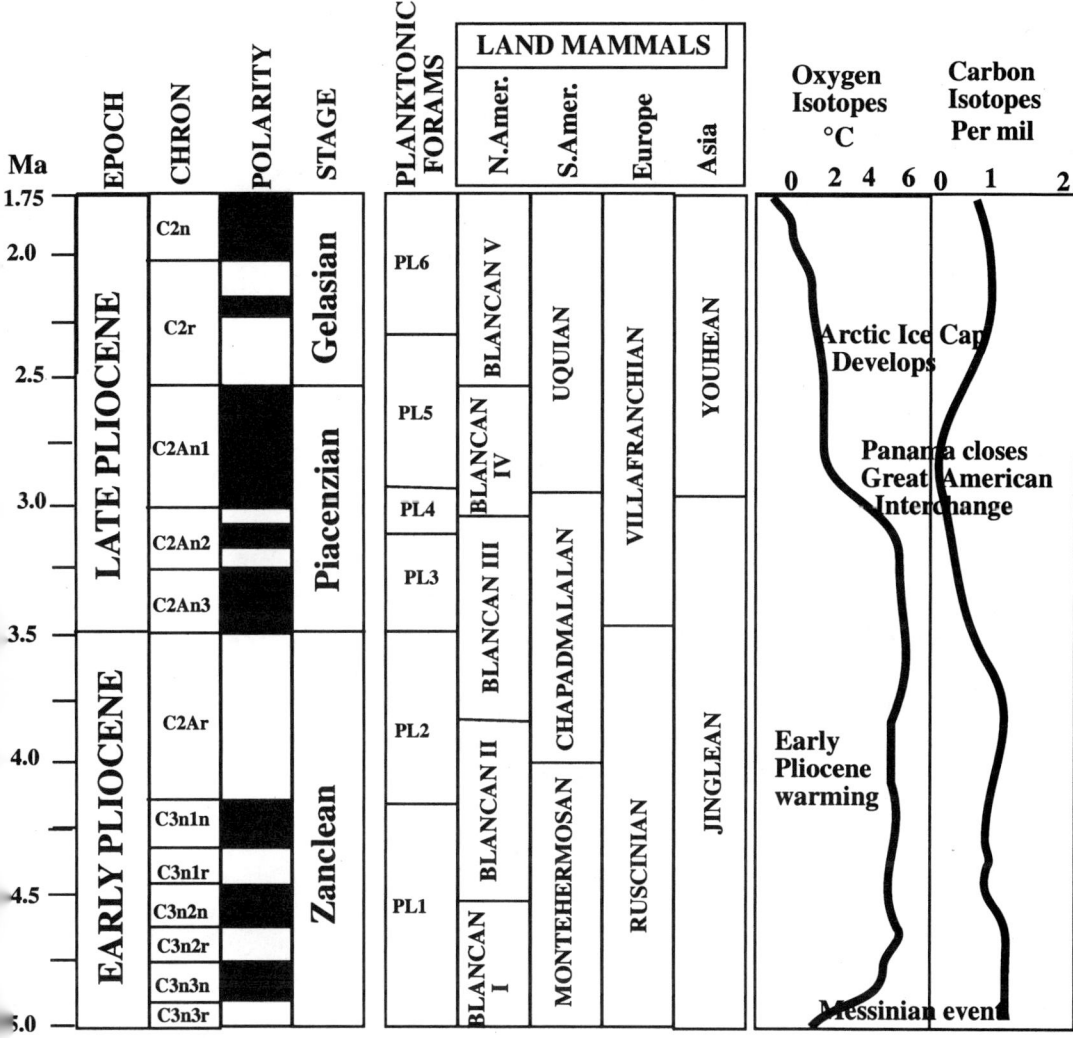

Figure 7.2. The Pliocene timescale (left columns), as recently calibrated by Berggren et al. (1995). The North American land mammal record is after Bell et al. (2004) and Woodburne and Swisher (1995); the South American land mammal ages are after Flynn and Swisher (1995); and the European and Asian record is after Qiu and Qiu (1995). The isotopic records are after Zachos et al. (2001), but they are very generalized and smoothed out.

cooling trend of the late Miocene reversed in the early Pliocene, with a slight warming trend evident in Pliocene molluscan faunas (fig. 7.3) of the Etchegoin and San Joaquin formations of California (Durham 1950). Fossil mollusks from the marine rocks of the Kettleman Hills in the western San Joaquin Basin of California lived in waters with a mean temperature of 13–14°C (55–57°F), with a minimum of 11°C (52°F) and a maximum of 18°C (65°F), so conditions were not highly seasonal. These mollusks are most similar to those living in waters off southern California and Baja

Figure 7.3. Typical slab of fossil mollusks (mainly moon snails) from the lower Pliocene of the Kettleman Hills in central California. From Woodring et al. 1940. Courtesy of the U.S. Geological Survey.

California (Stanton and Dodd 1970). Diversity of the mollusks increased to almost mid-Miocene levels, as a number of subtropical taxa migrated northward again (Addicott 1970).

Although normal marine faunas lived in the warm shallow seas of Europe, the eastern European region was again drowned by an isolated seaway comparable to the Paratethys of the Miocene. It is known as the Pontian Sea, after the old Greco-Roman name of the Black Sea, *Pontus Euxinus,* or "sea that is friendly to strangers," an ironic name for a hostile and treacherous body of water (fig. 7.4). The Pliocene Pontian Sea was an isolated marine basin that covered most of the modern Black and Caspian seas. It had sufficient marine water flowing in from the Mediterranean to keep the salinity of the waters at normal levels most of the time, but there was little exchange of marine faunas. As a result, the marine organisms

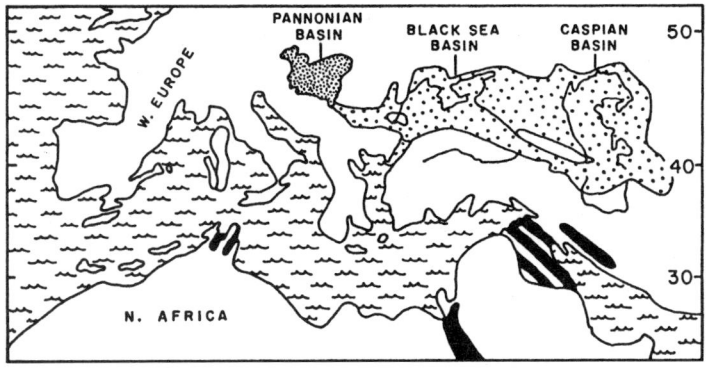

Figure 7.4. *The Pannonian and Pontian basins. During the Miocene and Pliocene, several seas formed as arms of the Mediterranean. The Pannonian Basin covered much of eastern Europe, and the Pontian Basin covered the modern Black and Caspian seas. These isolated bodies of water had their own peculiar water chemistry and evolved their own distinct fauna. From Geary 1990.*

that did inhabit the Pontian Sea evolved quickly in isolation from those in the rest of the ocean. The most remarkable product of this evolutionary radiation is the great diversity of endemic cockles (fig. 7.5), most of which were unique species that are found only in the Pontian Sea during the Pliocene. At least thirty endemic genera are recognized, with many more endemic species, all evolved from the familiar European cockle shell *Cerastoderma*. This genus, which contains only three living species in England and one in the United States, has not changed much since the Oligocene. Clearly, lack of competition from other clams, which never reached the Pontian Sea, allowed these cockles to evolve into many different shell shapes and to occupy niches that are normally inhabited by other kinds of clams. In that sense, they are similar to the Galapagos finches, which have evolved different-shaped beaks for different kinds of food in the absence of other birds that specialized in that food resource.

Steppin' Out

Terrestrial realms also showed the warming trend of the early Pliocene (Agusti and Anton 2002). Warm, humid, nonseasonal climates prevailed all over Europe and Asia, with temperatures about 5°C higher than those of today, and annual precipitation 40–70 centimeters more than the region experiences today. The land vegetation consisted of evergreens and warm mixed forests dominated by bald cypresses. Crocodiles again roamed the shores of Europe, and in southern Spain and Pakistan there were giant tortoises over 2 meters long (fig. 7.9). In the Mediterranean, the climate was drier, so the vegetation was a woody scrubland with many drought-tolerant plants. Living in the early Pliocene (Ruscinian) of Eu-

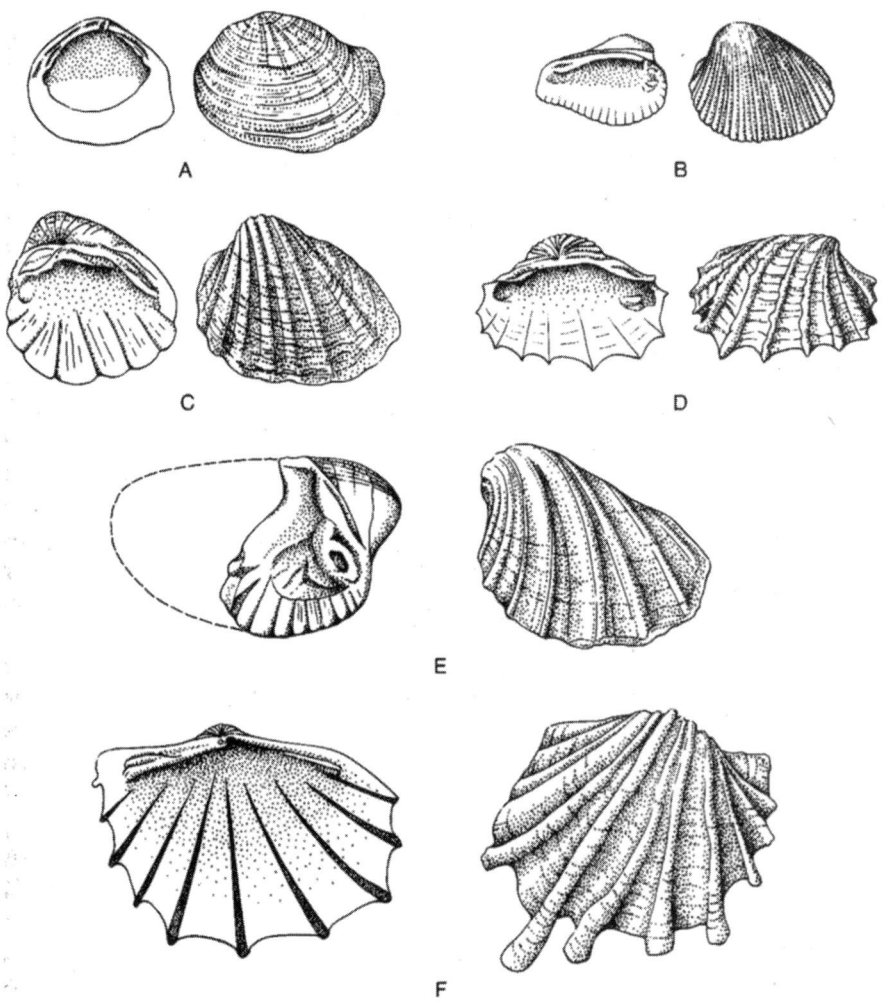

Figure 7.5. Pontian cockle shells. Living in the Pontian Sea (near the modern Black Sea) was a huge radiation of endemic cockle shells, which evolved in isolation to parallel mollusks found in many other part of the world. From Gillet 1946.

rope and Asia was a fauna not much different from that of the late Miocene, except for the loss of the archaic teleoceratine and aceratherine rhinoceroses, hippos, and camels (which had briefly entered during the Messinian salinity crisis). A great diversity of pigs, deer and antelopes, and cattle dominated the herbivores, along with a reduced diversity of hipparionine horses, and relatives of the modern Sumatran rhinoceros, *Dicerorhinus*. The long-tusked gomphothere *Anancus* and the mastodont *Mammut borsoni* (related to the American mastodont, a mammutid) were the largest herbivores in the forests. A few odd stragglers from the Miocene, like eomyid rodents, the hyrax *Postschizotherium chardini*, and the pangolin *Manis hungarica* are found in isolated European localities

Figure 7.6. The running hyena Chasmoporthetes. *Drawn by H. Galiano, in Berta 1981.*

but were not widespread in the Pliocene. Although the diversity of Miocene apes had disappeared, colobus and macaque monkeys were common. Rodents were diverse, with murids (Old World rats and mice) and cricetids (hamsters) dominating the small-mammal fauna. Gerbils were also found, as were flying squirrels and other kinds of squirrels, and beavers. The abundance of arboreal rodents (especially flying squirrels), beavers, and monkeys confirms the dominance of forests in Europe at this time.

Hunting in these Eurasian forests were a broad spectrum of predators. Most of the sabertoothed machairodont cats of the Miocene had vanished and were represented only by *Dinofelis,* plus more-modern cats, such as the first lynxes and lions. The wide diversity of hyenas had also disappeared, replaced by forms that had turned to the modern specialization of bone-cracking and scavenging; although one hyena, *Chasmoporthetes lunensis,* was built like a cheetah (fig. 7.6). Foxes, raccoon dogs, and early pandas were also represented, along with a variety of bears (including the living genus *Ursus* and the huge *Agriotherium*). Mustelids were also common and diverse in the early Pliocene, including martins, weasels, and many extinct groups.

North Africa was much moister than at present, with semi-arid vegetation covering the Sahara Desert region, and tropical forests and moist savanna extending as far north as 21° north latitude (PRISM Project Members 1995). In East Africa, tropical open woodland vegetation was found in areas that are now semi-desert or treeless dry grasslands (Bonnefille 1995). Forests were found only in isolated patches fringing lakes and rivers. Even southern Africa, home of our earliest ancestors, was cloaked in open woodlands and warm-temperate forests where arid scrublands are found today (Scott 1995). In these varied African habitats was a great diversity of African mammals of the Langebaanian land mammal age. They included many living genera of giraffes, antelopes and cattle, and primitive hippos and warthogs, plus hipparionine horses and the ancestral white rhinoceros, *Ceratotherium praecox.* The strange deinotheres, with their downward-pointing lower tusks, lived alongside the long-tusked gomphothere *Anancus,* the stegodont mastodonts, and a radiation of the earliest members of the modern elephant family (fig. 4.16). The rodents included many of the modern African families, such as Old World porcupines, murids, cricetids, and naked mole-rats. Preying on these herbivores was a wide diversity of hyenas, civets, honey badgers, and the earliest lynxes and lions, plus the sabertooths *Machairodus* and *Dinofelis* and the giant bear *Agriotherium.* Last but not least, Pliocene beds in Africa yield abundant primates, including a variety of baboons, macaques, and colobus monkeys and, of course, our own family, the Hominidae.

As we saw in the last chapter, the earliest fossils of the Hominidae occur widely in the late Miocene of northern Africa. These include the recently discovered *Sahelanthropus tchadensis,* from the latest Miocene (6–7 Ma) of Chad in the central Sahara; *Ororrin tugenensis* (5.72–5.88 Ma) from Kenya; and *Ardipithecus ramidus kadabba* (5.2–5.8 Ma) from Ethiopia. All of these taxa are based on fragmentary specimens but show that we hominids were upright and walking from the very beginning of our evolution, even though our brains and bodies were still small. The Pliocene saw an even greater diversity of hominids (fig. 7.7), with a number of archaic species overlapping in time with the radiation of more-advanced hominids. Archaic relics of the Miocene included *Ardipithecus ramidus ramidus,* found in Ethiopia in 1992 from rocks 4.4 Ma in age, which had human-like reduced canine teeth and a U-shaped lower jaw (instead of the V-shaped lower jaw of the apes). Rocks in Kenya about 3.5 Ma in age yield primitive forms like *Kenyapithecus platyops.* By 4.2 Ma, however, the first members of the advanced genus *Australopithecus,* the most diverse member of our family in the Pliocene, are also found. The oldest of these fossils is *Australopithecus anamensis* from rocks dated at 3.9–4.2 Ma near Lake Turkana in Kenya. These creatures were fully bipedal, as shown not only by their bones but also by hominid trackways near Laetoli, Tanzania. The most famous of these early australopithecines is *A. afarensis,* from rocks dated at 3.0–3.4 Ma near Hadar, Ethiopia, better known as "Lucy" by its discoverers, Don Johanson and Tim White.

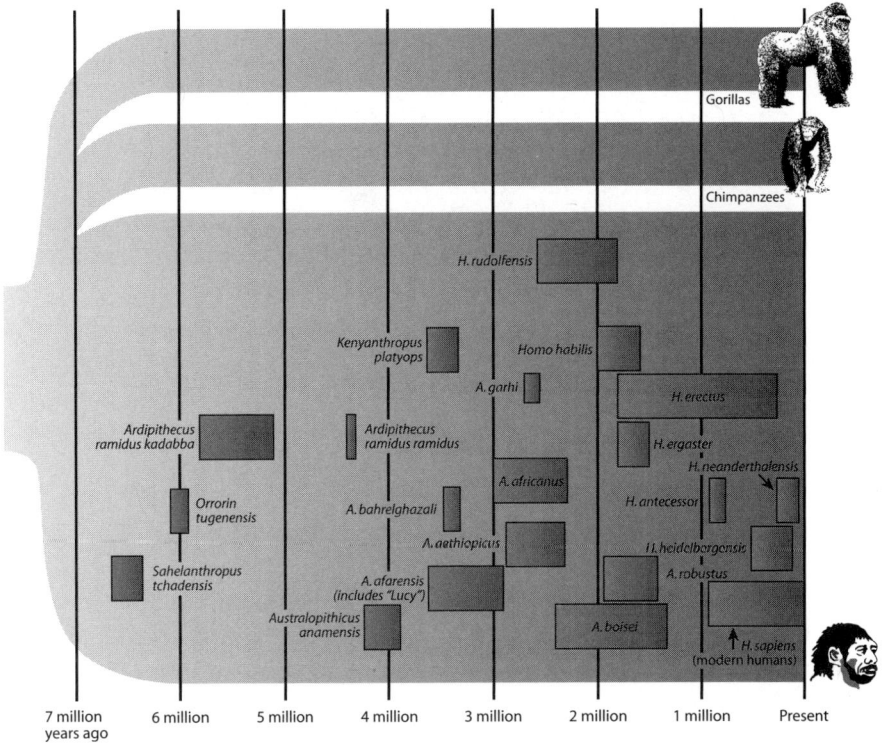

Figure 7.7. Family tree of the hominids. From Prothero and Dott 2003.

When it was discovered in the 1970s, *A. afarensis* was the first early hominid to clearly show a bipedal posture (based on the knee joint and pelvic bones), but it was not as upright as later hominids. These were still small creatures (about 3 feet, or 1 meter tall) with small brains, and they were apelike in having large canine teeth and a large overhung jaw.

By the late Pliocene, hominids had become very diverse in Africa (fig. 7.7). These included not only the primitive forms *A. garhi* (dated at 2.6 Ma) and *A. bahrelghazali* (dated at 3.4 Ma) but also the best-known australopithecine, *A. africanus*. Originally described by Raymond Dart in 1925 based on a juvenile skull (the Taung baby), for decades the Eurocentric anthropology community refused to accept it as ancestral to humans. But as more South African caves yielded better specimens to paleontologists like Robert Broom (especially the adult female skull known as Mrs. Ples), it became clear that *A. africanus* was a bipedal, small-brained African hominid, not an ape. This went contrary to all the accepted notions, which postulated that human evolution was driven by brain size, that bipedalism was secondary, and that evolution had occurred in Europe or Asia. The Piltdown forgery was deliberately set up to reinforce this bias, but by the 1950s, when Piltdown was exposed as a fraud, the evidence from *A. africanus* became undeniable. *Australopithecus africanus* was a rather small, gracile creature, with a dainty jaw, small cheek teeth, no skull crest, and a brain only 450 cubic centimeters in volume. On the basis of its gracile and humanlike features, *A. africanus* is also the best candidate for ancestry of our own genus *Homo*.

In addition to *A. africanus*, the late Pliocene of Africa also yields a number of highly robust australopithecines. These were long lumped into a broad concept of the genus *Australopithecus*, either as distinct species or even as robust males of *A. africanus*. In recent years, however, paleoanthropologists have come to regard them as a separate robust lineage, now placed in the genus *Paranthropus*. The oldest of these is the curious "Black Skull," discovered in 1975 by Alan Walker on the shores of west Lake Turkana, Kenya, from rocks about 2.5 Ma in age. Although it is small in brain and body, the skull is robust with a large skull crest and massive molars, and an advanced dish-shaped face. Currently, scientific opinion places the Black Skull as the most primitive member of *Paranthropus, P. aethiopicus*. It was followed by the most robust of all hominids, *P. boisei,* from rocks in East Africa ranging from 2.2 to 1.2 Ma in age (fig. 7.8A). This form was nicknamed "Nutcracker Man" for its huge thick-enameled molars, robust jaws, wide flaring cheekbones, and the strong crest on the top of its head, indicating a diet of nut-, seed-, or bone-cracking. Originally found by Mary Leakey at Olduvai Gorge (fig. 7.8B) in 1959, it was named *Zinjanthropus boisei* by Louis Leakey, who made his reputation from it. Finally, the rocks of South Africa between 1.6 and 1.9 Ma in age yield the type species of *Paranthropus, P. robustus*. These too had massive jaws, large molars, and large skull crests but were not as robust as *P. boisei. Paranthropus robustus* lived side by side in the same South African caves as *A. africanus*. It is not only more robust but also larger than that species, with some individuals weighing as much as 150 pounds.

Finally, the latest Pliocene saw the first members of our own genus *Homo,* which are easily distinguished from contemporary *Australopithecus* and *Paranthropus* by a larger brain size, flatter face, no skull crest, reduced brow ridges, smaller cheek teeth, and reduced canine teeth. The first of these to be described was *H. habilis* ("handy man"), discovered in the 1960s by Louis and Mary Leakey in Olduvai Gorge, Tanzania, from beds about 1.75 Ma in age (fig. 7.8B). Originally, all of the early *Homo* specimens were shoehorned into the species *H. habilis;* but now paleoanthropologists recognize that this material is too diverse to belong to one species, so several are now recognized. These include the modern-looking *H. rudolfensis* (from beds ranging from 1.9 to 2.4 Ma in age), which also helped to make Richard Leakey's reputation (fig. 7.8C), and the advanced but short-lived *H. ergaster* from beds 1.6–1.8 Ma in age. Finally, African beds about 1.8 Ma in age yield the oldest specimens of the most long-lived hominid ever, *H. erectus,* which will be discussed more in the next chapter.

Although the Pliocene of Africa yields almost a dozen species of hominids, the high diversity of Miocene apes vanished, with only the lineages of the living great apes surviving. The Old World monkeys (especially

A

B

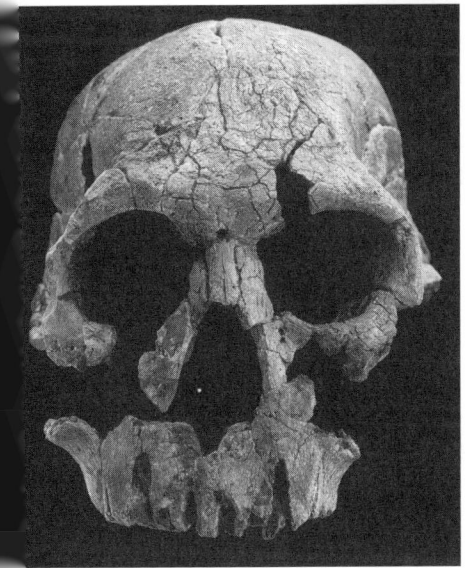

C

Figure 7.8. (A) The famous skull of Nutcracker Man, Paranthropus ("Zinjanthropus") boisei, *discovered by Mary Leakey at Olduvai Gorge in 1959. Photo courtesy of Kenya National Museum. (B) Panorama of Olduvai Gorge in 1973, with the 19-year-old author in the foreground. Photo by the author. (C) Skull of the earliest member of our genus,* Homo rudolfensis, *from Lake Turkana. Photo courtesy of Kenya National Museum.*

Figure 7.9. The gigantic turtle Geochelone atlas. *This turtle was found in the Pliocene of Asia, and gigantic tortoises almost as large were known from South and North America and Europe at the same time. Photo by the author.*

baboons) diversified and largely replaced the apes. Most paleoprimatologists believe that this change is a reflection of the cooling and drying of Africa in the Pliocene, which diminished the forests that sheltered the diversity of apes, replacing them with open grasslands, which favored ground-dwelling baboons, and our walking hominid ancestors.

Other regions likewise showed signs of becoming more arid during the widespread warmth of the early Pliocene. Australia experienced a spread of grasslands and the retreat of the widespread Miocene rain forests, and the first grazing kangaroos (Archer et al. 1995). In South America, the persistently dry climates of the Miocene continued as the Antarctic went through its freeze-and-thaw fluctuations, and early Pliocene (Chapadmalalan) mammals are largely hypsodont grazers (Marshall and Cifelli 1989). However, there were also gigantic turtles (*Stupendemys*) that were over 2 meters (10 feet) long and weighed over 4,000 pounds still surviving in the jungles of South America after having first evolved in the Miocene (fig. 7.9).

Many of the mammals that dominated late Miocene savannas (such as rhinos, protoceratids, dromomerycids, and horned mylagaulid rodents) had vanished from North America by the early Pliocene, or Blancan land mammal age. The diversity of most other groups, such as camels, pronghorns, and horses, was greatly reduced as well (pl. 8). Paleobotanical evidence (Axelrod 1985, 1992; Wolfe 1985; Webb et al. 1995; Wing 1998;

Graham 1999) shows that North America had a slightly warmer climate in the early Pliocene than in the late Miocene; but the early Pliocene was markedly drier in the Great Plains and Rocky Mountains, and extensive treeless grasslands or steppe habitats formed. The Pacific Northwest still had dense forests, but these alternated in drier periods with semi-arid savannas. Even the Gulf Coast, which had been a forested refuge in the late Miocene for the last of the dromomerycids, protoceratids, musk deer, forest-adapted peccaries, giant flying squirrels, and several endemic horses, dried out to a subtropical savanna, and these taxa vanished (Webb 1977; Webb et al. 1995). In their place, the treeless steppes of North America were inhabited by a new assemblage of mammals.

Blancan faunas are known from all over North America, but the localities show a surprising heterogeneity—ranging from Hagerman Fossil Beds in Idaho (fig. 7.10), which were dry and cold, to Mount Blanco in Texas (whence the Blancan got its name), where conditions were milder, to warm subtropical Mexico (Webb et al. 1995). Although the diversity of families of North American mammals had been decimated by the extinction of rhinocerotids, dromomerycids, protoceratids, and mylagaulids, the surviving families diversified into new genera and species (pl. 8). Horses were again common, with many species of both the primitive hipparionine horses (*Nannippus*), as well as the first species of the living genus *Equus*. In Hagerman Horse Quarry in Idaho (figs. 7.10B, C), there were huge numbers of only a few species of grazing horses. One of these species, *Equus simplicidens*, is widely regarded as the earliest zebra. Webb and Opdyke (1995) suggest that these abundant grazers may have accelerated the spread of grasslands by overgrazing and decimating all other forms of vegetation except the hardiest grasses that can regrow quickly. Grazing camels were also abundant, including the huge *Titanotylopus*, which had massive limbs and was 4 meters (12 feet) at the shoulder, and the Blancan llama, *Hemiauchenia blancoensis*. Pronghorns were down to two genera, *Capromeryx* and the four-horned *Tetrameryx*. True deer migrated from the Old World in the latest Miocene, replacing the dromomerycids; and by the Pliocene there were several species, including fossils indistinguishable from modern white-tailed deer. Peccaries (*Mylohyus* and *Platygonus*) were also diverse. Most had relatively large grinding teeth and long limbs for life in the grasslands; their anatomy was comparable to that of the living warthog in East Africa. The gomphothere mastodonts were still abundant, along with the two-tusked stegomastodonts and the first examples of the American mastodont, *Mammut americanum*. Rodents continued to be even more diverse than they were in the Miocene, with many species of squirrels and marmots, pocket gophers, pocket mice, beavers, jumping mice, and a huge diversity of cricetids.

Preying on all these herbivores were diverse carnivores, including many kinds of mustelids (such as the earliest weasels, wolverines, otters, badgers, and skunks), bears, dogs (including early coyotes and the last of the bone-crushing borophagine dogs, *Borophagus diversidens*), raccoons,

A

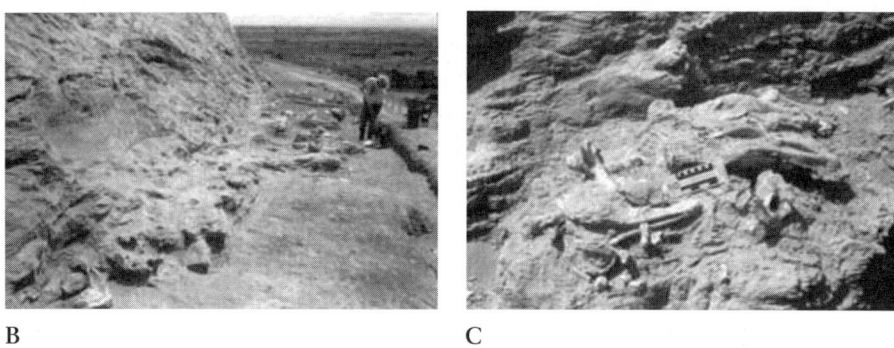

B   C

Figure 7.10. The Hagerman Fossil Beds. These fossil beds in Idaho preserve some of the best records of the early Pliocene (Blancan) in North America. (A) Outcrops of the Glenns Ferry Formation, ancient lake beds exposed along the banks of the Snake River (in the distance). The Hagerman Horse Quarry is the notch in the cliff in the upper left. (B) Close up of recent excavations at the Hagerman Horse Quarry. (C) Block of excavated horse fossils from Hagerman Horse Quarry. Photo A courtesy of D. Ruez; photos B–C courtesy of G. McDonald.

and the last of the North American hyenas, the cheetah-like *Chasmoporthetes* (fig. 7.6). Cats were also widespread, with a variety of sabertoothed forms (*Ischyrosmilus* and *Megantereon*), lynxes, and even a North American cheetah, *Miracinonyx studeri*. This cheetah must have been the fastest predator of the horses and pronghorns on the grassy steppes.

Reptiles of the Blancan included familiar snakes and lizards, plus alligators (consistent with the warming climate) and even giant tortoises twice the size of Galapagos tortoises (almost the size of *Stupendemys* from South America; fig. 7.9), which are found as far north as Nebraska. This is an even stronger indication that the High Plains did not freeze in the early Pliocene.

Of the 106 Blancan genera recognized by Savage and Russell (1983), a third were immigrants from Eurasia or South America. Eurasian immigrants (Webb and Opdyke 1995) included the deer and hyenas mentioned already, as well as bovids (early cattle and antelopes), pandas (*Parailurus*), the grisons (or tayras, a group of Eurasian mustelids), cheetahs, lynxes, the modern bear genus *Ursus*, and the voles and lemmings, which made up the largest part of the cricetid rodent radiation. By the late Blancan, immigrants from South America included a variety of edentates (two families of ground sloths, plus the armadillos and giant armadillo-like glyptodonts), capybaras, porcupines, and the predatory phorusrhachid bird (fig. 6.29) *Titanis walleri*, a huge creature the size of an ostrich but with a thick, sharp beak for ripping its prey apart. These South American immigrants, however, were part of the Great American Interchange, which will be discussed later in the chapter.

The Frigid Arctic

This early Pliocene episode of mild conditions was not long-lived (only about 1–2 million years in duration). By the beginning of the late Pliocene (3.5–3.2 Ma), the earth quickly chilled to its modern ice age conditions with bipolar ice caps. Oxygen isotope data (Shackleton and Opdyke 1977; Miller et al. 1987; Zachos et al. 2001) consistently show a big positive shift of at least 1 per mil (figs. 3.22, 7.2), indicating that bottom-water temperatures had dropped from about 4°C (39°F) to freezing, or below, as they are today. The cooling trend was well developed in the Southern Hemisphere, where the Antarctic ice sheet again expanded to its present dimensions (Kennett and Barker 1990); and ice-rafted sediments show that the volume of icebergs had increased dramatically. Subantarctic surface waters also cooled dramatically, and their effects were felt all the way to temperate regions. Warm, shallow-water benthic faunas in places such as New Zealand were replaced by the cold-tolerant Southern Ocean faunas that live there today (Hornibrook 1992). Glaciers even began to expand on Patagonia (Mercer and Sutter 1982).

But ice-sheet fluctuations in the Antarctic are not a new story—they date back to 33 Ma. What makes the late Pliocene glaciation different is clear evidence that the Arctic was frozen over too. Some glaciers were present in the Arctic even in the late Miocene and early Pliocene (Warnke 1982; Margolis and Herman 1980). But the Arctic didn't completely freeze over and form the Arctic ice cap until the late Pliocene (Berggren 1972; Kennett 1995). This huge expansion of Arctic ice caps accounts for most of the big shift in oxygen isotopes just mentioned. Evidence of late Pliocene glaciers can be seen all over the Arctic region (Thiede et al. 1990), including Iceland (McDougall and Wensink 1996), Alaska (Westgate et al. 1990), and, by 2.7 Ma, the Norwegian Sea (Jansen and Sjoholm 1991) and the open North Atlantic (Shackleton et al. 1984). The expansion of bipolar ice caps pulled a lot of water out of the oceans and helped lower sea level dramatically (Haq et al. 1987), which in turn helped ground the Antarctic ice sheets and increased their stability and promoted their growth. The cooling trend is seen in marine waters everywhere, from the poles to the North Atlantic to the South Pacific and even to the Mediterranean (Keigwin and Thunell 1979).

Why did the Northern Hemisphere finally freeze over almost 30 million years later than the Southern Hemisphere? The peculiarities of this climatic anomaly have long puzzled scientists, but certain things seem clear. The late Pliocene freeze is too rapid and abrupt to be explained by the gradual increase in weathering due to the Himalayan uplift (which may explain the long-term Neogene cooling trend) or by a gradual shift in greenhouse gases. Instead, some sort of rapid cooling due to changes in oceanic circulation best explains the data. For forty years, scientists (Hamilton 1965; Berggren and Hollister 1974, 1978; Keigwin 1976, 1978) have pointed to the coincidence of the closure of the Panamanian seaway at around 3.5 Ma with the synchronous expansion of the Arctic ice cap. Prior to this event, warm waters of the Caribbean flowed freely into the Pacific, and marine organisms passed between the Atlantic-Caribbean and the Pacific as well. But as Central America rose up and began to form a barrier, flow into the Pacific was gradually choked off and eventually shut down altogether. This barrier caused the warm waters of the Caribbean and equatorial Atlantic to be diverted northward along the east coast of North America, forming the Gulf Stream (Brunner 1978). Today, this current brings warm, moist tropical waters from the Caribbean all the way to the North Atlantic, and is partially responsible for the mild climates experienced by the British Isles, which lie at latitudes equivalent to the subarctic regions of Labrador and Finland.

In recent years, our understanding of oceanic circulation has improved dramatically with the discovery of what is known as the "great oceanic thermohaline conveyor belt of currents" (fig. 7.11). Discovered by Wally Broecker (1987, 1991; Broecker and Denton 1989), this flow continuously cycles waters among most of the world's oceans. Shallow waters that are warmed in the tropical Pacific and Indian oceans flow west around the tip of southern Africa, across the tropical Atlantic, and then up the East Coast of North America as the Gulf Stream. As this voluminous warm surface current reaches the North Atlantic, it begins to cool, and its dense salty waters begin to sink north of Iceland. From there, the they continue sinking and then flow south again all the way to the South Atlantic, eastward across the deep Indian Ocean, and eventually across the South Pacific to the North Pacific. There the waters are finally warmed enough as they rise to the surface to rejoin the conveyor belt of surface waters on their path back through the Indian Ocean to the Atlantic (fig. 7.11).

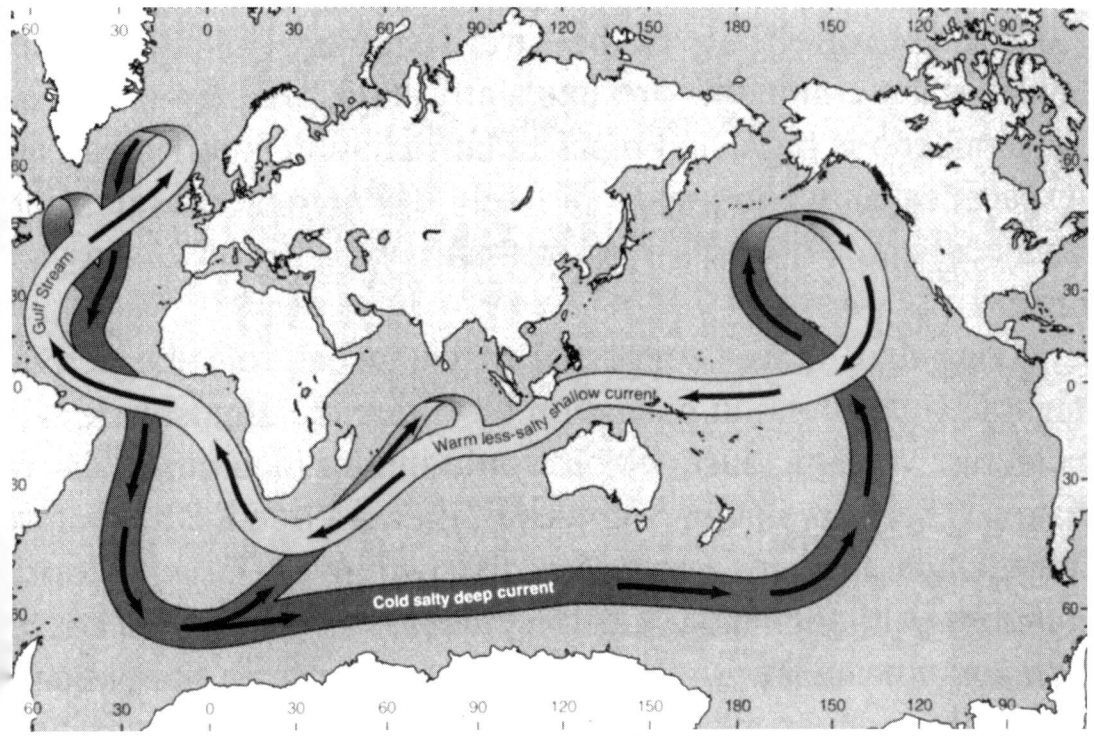

Figure 7.11. Thermohaline conveyer belt of oceanic currents, which links cold deep waters of each ocean with the shallow warm water currents. From Skinner and Porter 1995. Reprinted by permission of John Wiley & Sons, Inc.

Prior to the closure of the Panamanian isthmus, this current did not exist. Instead, low-salinity surface waters of the Caribbean mixed with those of the Pacific. This mixing made the surface waters fresher and more buoyant, so they flowed further north into the Arctic without sinking. The warm current thus kept the Arctic higher in temperature and prevented the development of an Arctic ice cap. Closure of the Panamanian gap shut off this low-salinity flow and forced the much saltier Gulf Stream to sink as it does today, initiating the modern thermohaline conveyor belt among the Atlantic, Indian, and Pacific Oceans. The Gulf Stream would have also brought warm moist air with it to the cold polar regions, providing the moisture needed to generate snow and ice over the Arctic.

The effects of this cooling in the marine realm are dramatic in some places, and not so in others. Plankton in the Antarctic, of course, showed a major effect, as cold water masses around the South Pole expanded and contracted with the cyclic pattern that characterized the ice ages (Fillon 1977). Likewise, North Atlantic planktonic provinces also shifted southward as polar species took over high-latitude waters, and temperate and subtropical species were pushed toward the Equator (Stanley and Ruddiman 1995).

The World in Transition • 249

Along the Atlantic Coast of North America, the rich fauna of more than 361 early Pliocene molluscan species (fig. 7.1) was decimated by the cooling and the shift in currents, causing about 65% of the species to die out (Blackwelder 1981; Stanley and Campbell 1981; Stanley 1986). Most of the species that were affected were purely tropical forms, and almost all the surviving species still live today, and are tolerant of colder conditions. According to Stanley (1986), one of the reasons for this heavy extinction was that marine barriers prevented the tropical species from retreating to the southern Caribbean ("Gatunian") province. Oxygen isotope analyses of shells from the upper Pliocene Chowan River Formation of the Virginia-Carolina coastal plains exhibited a drop of several degrees in mean temperature, from subtropical and warm-temperate conditions to mild-temperate conditions, which exist in these waters today.

By contrast, the extinction of mollusks in the North Sea and eastern Atlantic was much less, as warm-water species migrated south to the tropics and to the Mediterranean and survived (Raffi et al. 1987). Likewise, on the Pacific Rim in California (fig. 7.3) and Japan, most of the mollusk species survived, as the tropical and temperate provinces moved southward in response to glaciation and cooling (Chinzei 1978; Stanley 1987). Extinction hammered reef corals of southern Florida, but the effect on Caribbean coral reefs was much less (Frost 1977). Coral reefs also retreated on the Pacific Rim, with warm-water forms moving to the Gulf of California before dying out in the Pleistocene.

In addition to these indirect effects, closure of the Isthmus of Panama had an even more direct effect on marine life that lived on each side of the barrier. We can see this effect from the changes in oceanic chemistry and temperature. At 3 Ma, the chemistries of the Pacific and Caribbean seawaters were identical, since water flowed freely through the gap. But around 2.4 Ma, oxygen isotope values for Caribbean and Pacific seawaters began to diverge as flow was cut off, and the oceans developed different patterns of temperature and salinity (Keigwin 1978). In addition, molluscan faunas in both oceans were similar in the late Miocene but gradually differentiated during the early Pliocene. By the late Pliocene we begin to see the divergence of species as they became genetically isolated. Today, the Caribbean and Pacific tropical marine faunas are very different (Jones and Hasson 1985; Coates et al. 1992).

The Great American Interchange

Although closure of the Panamanian gateway had a dramatic effect in isolating marine faunas, it had the opposite effect on land faunas in bringing them together. As we have seen in previous chapters, through most of its history South America was an island continent, inhabited mainly by its native fauna of marsupials, edentates, and endemic hoofed mammals that paralleled the forms of mammals on other continents. In the Paleocene and Eocene, the isolation was nearly total, although strange cases like the pantodonts and the arctostylopids managed to make the crossing from China to North America and then to South America (chapter 3). During the late Eocene (chapter 4), African rodents first appeared in South American, generating the huge radiation of South American native caviomorph rodents (agoutis, chinchillas, capybaras, and guinea pigs). During the Oligocene (chapter 5), primates managed to cross to South America, evolving into New World monkeys (spider monkeys, howler monkeys, capuchins, and the like). During the late Miocene, a few more groups managed to hop across the narrowing gap of the Central American archipelago, so South America gained mastodonts and raccoons (Campbell et al. 2001); and a few ground sloths managed to immigrate to North America. The increasing differences between molluscan faunas of the Caribbean and Pacific show that the gap was slowly closing as early as 3.5 Ma and that it continued to close through the late Pliocene. Still, the majority of South American mammals were unable to emigrate from the continent even in the early Pliocene.

The Isthmus of Panama finally closed around 2.7 Ma, and the last barrier to land migration was eliminated. Suddenly, the floodgates burst open, and mammals moved freely between the continents (fig. 7.12). So began the Great American Interchange, one of the most extraordinary evolutionary experiments ever enacted in earth history, when faunas from two continents that had been long isolated suddenly come into contact (Webb 1978; Marshall 1985; Stehli and Webb 1985). Scientists often speculate about what might happen when animals evolve in isolation and then come into contact and competition, but normally we cannot run this experiment without biasing the result through our own interference. However, the Great American Interchange is a natural prehistoric experiment in isolation, contact, competition, and extinction, which reveals much about the nature of competition and adaptation. Not only was a land corridor available for the first time, but apparently there were grasses and savannas along much of the route, so most animals could make the trip.

The effects of the Great American Interchange were highly asymmetrical. Relatively few South American natives made the trip north, with only eight genera (mostly ground sloths, glyptodonts, armadillos, and porcupines and capybaras) representing the "legions of the south" (Webb 1985). Why so few of the South American natives moved north is not clear, but some scientists argue that the mostly tropical South American mammals were not suited to migrating north of the tropics of Central America and Mexico, and were certainly not well adapted to the cold steppes of the late Pliocene of North America. Judging from the results of their interaction with North American mammals, they may not have been able to survive the competition either. For the most revealing statistic is how many of the North American natives successfully invaded South

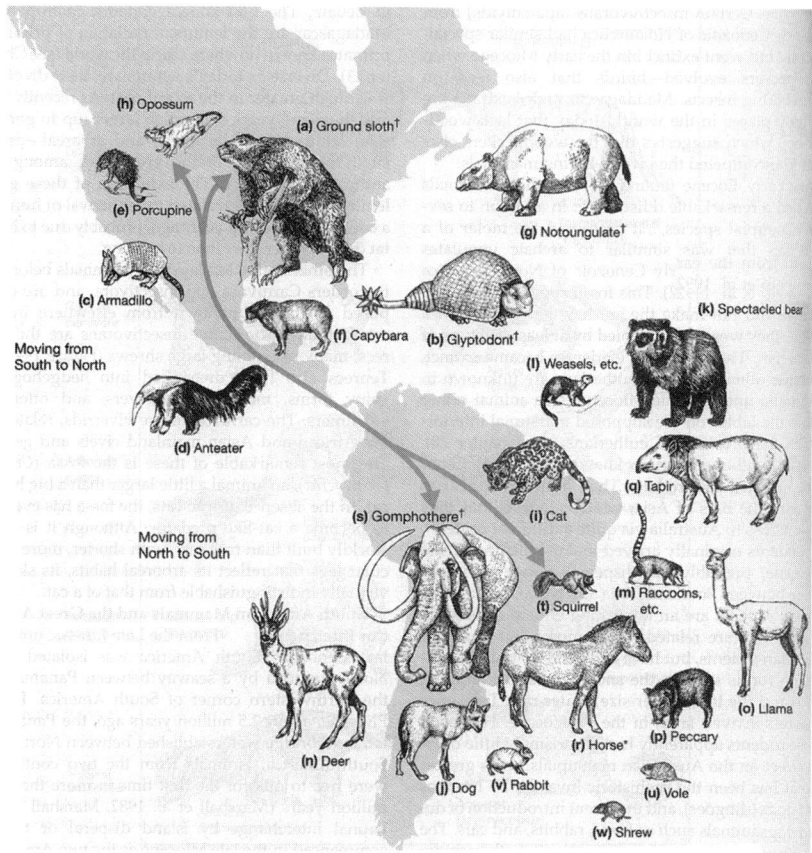

*Figure 7.12. In the middle Pliocene, the Panamanian land bridge connected North and South America in the Great American Interchange. A few groups from South America, such as ground sloths, glyptodonts, armadillos, porcupines, opossums, capybaras, and notoungulates managed to migrate north, but a far greater number of North American mammals moved south and displaced the native South American mammals. These "legions of the north" included mastodonts, mammoths, llamas, deer, horses, peccaries, tapirs, dogs, bears, weasels, raccoons, squirrels, rabbits, shrews, voles, and sabertoothed cats. † = totally extinct animals; * = animals extinct in that area. From Pough et al. 2002. Reprinted by permission of Pearson Education, Inc., Upper Saddle River, N.J.*

America and displaced the native forms. According to Webb (1985), at least twenty-nine North American genera in fifteen families moved south, forming the "legions of the north" that soon came to dominate those of the south. These immigrants included insectivores, rabbits, four families of rodents (pocket mice, field mice, pocket gophers, and squirrels), a huge wave of hoofed mammals (mastodonts, horses, camels, tapirs, peccaries, and deer), and nearly all the North American families of carnivorans (cats, sabertooths, weasels and skunks, dogs, bears, and more raccoon relatives). Today, 50% of the genera and 40% of the mammalian families living in South America are descended from immigrants of the Great American Interchange. Many of these families are so entrenched in South America (and absent in North America) that we hardly realize they were late immigrants to the south. For example, Latin America now houses three species of tapir, but they were native to the Northern Hemisphere until the Pliocene. Tapirs are now extinct in the north and are found only in South America and Southeast Asia. Likewise, camels spent most of their evolution in North America, only to spread to South America and evolve into llamas, alpacas, guanacos, and vicuñas, while they died out in the north. Peccaries were a Northern Hemisphere (mostly North American) group through most of their history, but now they are most common and diverse in Latin America, and they barely make it into the dry deserts of the southwestern states. South America is now the home of many endemic species of deer, descendants of the first immigrants in the Pliocene, and a whole radiation of native dogs that originated when the first canids crossed the land bridge. At one time, South America also had its own endemic species of mastodonts and peculiar hippidion horses that apparently had a short trunk or proboscis.

But the effect on South American natives was the most dramatic of all. At first, late Pliocene faunas of South America (Chapadmalalan land mammal age) had a high diversity as the mix of both immigrants and natives increased the total number of families and genera. But by the latest Pliocene (Uquian) and early Pleistocene (Ensenadan), most of the South American natives had disappeared. According to Marshall and Cifelli (1989), 81% of the families and 83% of the genera that had been native to South America before the Oligocene died out by the Uquian. The vanished species included most of the native South American hoofed mammals, including the horse-like and camel-like litopterns (fig. 7.13A) and the rabbit-like, tapir-like, mastodont-like, and hippo-like notoungulates (fig. 7.13B). They were replaced by rabbits, mastodonts, horses, tapirs, and camels. Edentates, too, were reduced to just a few genera of sloths, anteaters, and armadillos, plus the glyptodonts. The rodents became a mixture of the native caviomorph families plus the large number of immigrant squirrels, pocket mice, pocket gophers, and field mice from North America. Most severely devastated of all, however, were the marsupial carnivores, including the sabertoothed marsupial *Thylacosmilus* (fig. 7.14), the dog-like and hyena-like borhyaenid marsupials, and most of the smaller opossums as well. These were decimated by competition from cats (including sabertoothed cats), dogs, bears, raccoons, and weasels, so that by the Uquian only a handful of native South American opossums remained. Finally, the predatory phorusrhachid birds, the largest predators in South America in the Miocene and early Pliocene, also vanished, even though a few of them, like the giant *Titanis,* managed to reach Florida before the entire group became extinct.

Although it seems like a straightforward case of the northerners outcompeting the southerners, some scientists have argued that rapid uplift

A

B

Figure 7.13. South American native hoofed mammals. Some South American native hoofed mammals managed to survive in the Pliocene and even into the Pleistocene, despite the invasion of the "legions of the north." (A) The litoptern Macrauchenia, *with the camel-like body and long tapir-like proboscis. (B) The huge hippo-like notoungulate* Toxodon. *From Scott 1913.*

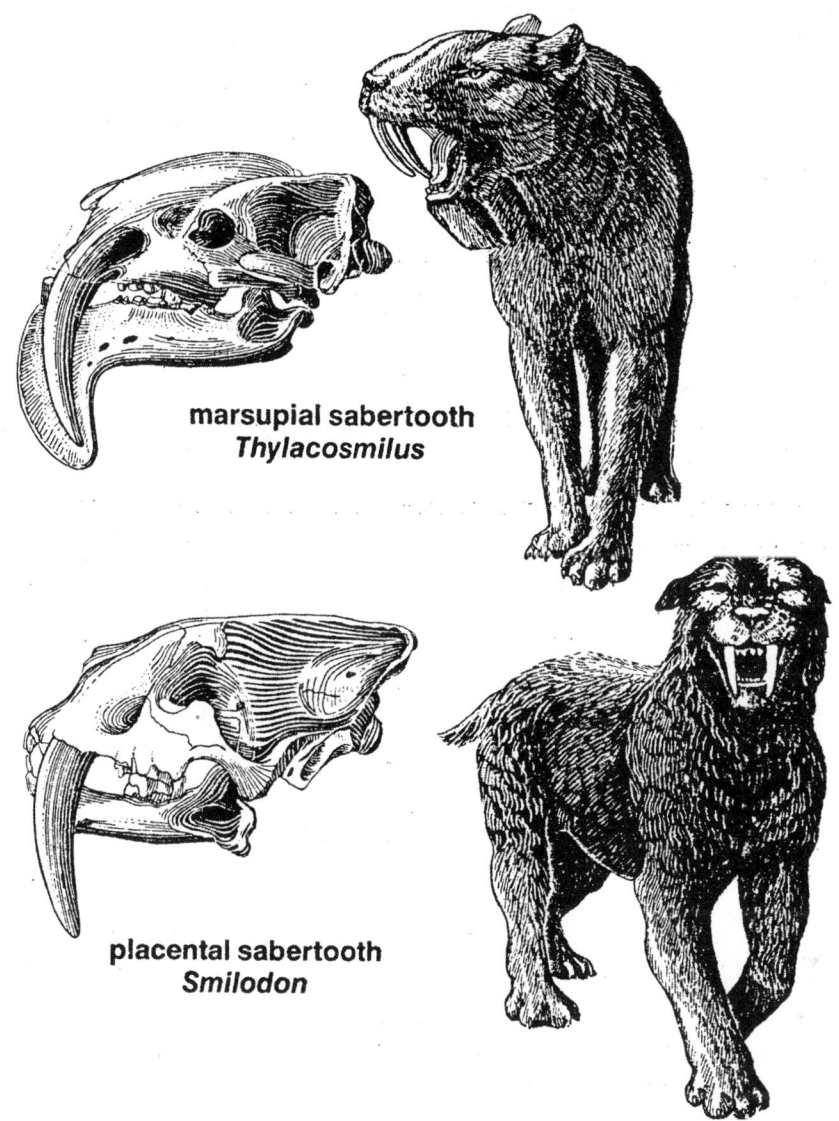

marsupial sabertooth
Thylacosmilus

placental sabertooth
Smilodon

Figure 7.14. *The sabertoothed marsupial* Thylacosmilus. *South America's native marsupial carnivores included the wolf-like borhyaenids and the sabertoothed marsupial* Thylacosmilus, *which showed remarkable convergence on the true cats that also evolved saber-like canines. From Marshall 1981.*

of the Andes and the general drying trend in the late Pliocene and Pleistocene may have also contributed to rapid extinction of the South American natives. This could be true, but the striking fact that so many of the ecologically equivalent South American mammals (e.g., marsupial sabertooths, borhyaenids; horse-like, camel-like, and mastodont-like hoofed mammals) died out so rapidly when their northern equivalents invaded is a strong argument that the northerners outcompeted the natives. In addition, the fact that few of the South American native groups successfully invaded North America is also revealing. Only those groups that had no equivalent in North America (such as ground sloths, armadillos, porcupines, and capybaras) were able to successfully invade the north and become established.

In any case, by the early Pleistocene, South America's fauna was completely modernized and looked much like that of North America during the Pleistocene. The South American fauna differed mainly in the retention of its remaining natives, such as the possums, caviomorph rodents, New World monkeys, and diverse edentates. Ironically, many of the large mammals that once dominated North America (such as camels, peccaries, and tapirs) still survive in South America even though they were eliminated from most of North America in the late Pleistocene. In this way, South America still retains its Pliocene-Pleistocene signature of land animals, and was not nearly as transformed by the late Pleistocene extinctions as was North America.

For Further Reading

Kurtén, B. 1988. *Before the Indians.* New York: Columbia University Press.

Stanley, S. M. 1996. *Children of the Ice Age: How a Global Catastrophe Allowed Humans to Evolve.* New York: Crown Books.

Stehli, F. G., and S. D. Webb, eds. 1985. *The Great American Biotic Interchange.* New York: Plenum Press.

Vrba, E. S., G. H. Denton, T. C. Partridge, and L. H. Burckle, eds. 1995. *Paleoclimate and Evolution, with Emphasis on Human Origins.* New Haven, Conn.: Yale University Press.

Figure 8.1. Zermatt Glacier in the Swiss Alps, as illustrated by Agassiz in his 1840 book on the ice ages. From Agassiz 1840.

8

Ice Time: The Pleistocene

One *Macrauchenia:* Well, why not call it the Big Chill or the
Nippy Era? I'm just saying, how do we know it's an Ice Age?
The other *Macrauchenia:* Because . . . of all . . . THE ICE!!
  Ice Age, Warner Brothers, 2002

Mystery of the Erratics

In the late 1700s and early 1800s, geology was an infant science, prac-
ticed mostly by wealthy gentlemen and clergymen as a hobby. There were
no professional geologists who earned their living at geology, although
mining geology was well advanced. There were no geology professors at
the universities yet, although some European universities taught mineral-
ogy and crystallography in the context of finding ore deposits. Few geologists traveled outside their hometowns in Europe, so they had little experience with glaciers (fig. 8.1) or volcanoes or other phenomena not in their
own backyard. The intellectual framework of geology was also in transition. The literal interpretation of the first chapters of Genesis still shaped
most people's ideas about the age of the earth and the nature of rocks, although pioneers like Scotsman James Hutton were writing about natural,
rather than supernatural, processes and speaking of an earth with "no
vestige of a beginning, no prospect of an end" in 1788. Most geologists
still thought of the layers of strata known as "Tertiary" as having been
laid down by Noah's flood, and they puzzled over huge boulders, some as
big as houses, found widely over northern Europe. They were not made of
any of the local bedrock types but instead came from long distances away
(fig. 8.2). These *erratics,* or "wanderers," were attributed to the immense
powers of Noah's flood. Likewise, the poorly sorted loose gravels, sands,
and muds found just below the surface nearly everywhere in northern Europe were attributed to floodwaters, and called "drift" because they supposedly drifted into place as the floodwaters receded. (We now recognize

Figure 8.2. A large erratic boulder in Scotland, attributed to the movement of ice sheets, and not to Noah's flood. From Geikie 1894.

that this poorly sorted unstratified collection of sediment, called "till," was dumped by the retreat of melting glaciers.)

This "diluvial" school of thought (after the Latin *diluvium,* "flood") was modified and expanded in the early 1800s as geologists with strong religious ties (many of them actually ministers) tried to use the earth to support their religion. However, evidence was beginning to accumulate that contradicted the diluvialist dogma. Sailors had reported seeing huge rocks caught in icebergs as they floated down through the North Atlantic shipping lanes, but no one made the connection to the erratics. As early as 1787, the Swiss minister Bernard Friederich Kuhn interpreted enormous boulders in the Swiss Alps as evidence of ancient glaciation, and in 1794, James Hutton himself visited the Jura Mountains of France and Switzerland and reached the same conclusion. (Ironically, he never recognized the glacial origin of many of the same features in his own backyard in Scotland.) In 1824, the Norwegian Jens Esmark argued that glaciers had once covered most of his homeland and possibly much of Europe; and over the next decade, several other geologists amassed evidence that the Alpine glaciers had once extended far onto the plains, or that the polar ice caps had once spread as far as central Germany.

But the entrenched diluvial explanation of the erratics and the drift was still hard to overcome, both for religious reasons and also simply because of inertia—scientists are human and surrender the notions they learn as students only with reluctance. It took a young Swiss paleontologist, Louis Agassiz, to change ideas about the glacial explanation (fig. 8.3). In 1836, he visited many of the Alpine glaciers with the Swiss geologist Jean de Charpentier, who had been documenting evidence of the expansion and retreat of these glaciers for years. Agassiz was convinced, and at the July 1837 meeting of the Swiss Society of Natural Sciences in Neuchâtel, he presented a paper that startled his audience. Instead of the expected presentation on fossil fishes, Agassiz argued that the erratics

Figure 8.3. A portrait of Louis Agassiz near the Unteraar Glacier, painted by Alfred Berthoud. From Carozzi 1967.

were glacial in origin and that large areas of the globe had once been covered by ice in an *Eiszeit*, or "ice age." The meeting was thrown into turmoil, so that some timid geologists never got around to giving their own papers (including Amanz Gressly, whose idea of "sedimentary facies" later became an important concept in geology). After the meeting, a field trip to the glaciers was hastily put together to convince the most distinguished skeptics in the society, such as Elie de Beaumont and Leopold von Buch—but they remained skeptical. However, Agassiz's 1837 paper and his monumental 1840 book *Studies on Glaciers* were not in vain. The glacial theory had made a splash and now had to be taken seriously. In 1840, Agassiz visited England, where he eventually convinced his two most prominent British critics, William Buckland and Charles Lyell, who then became ardent proselytizers for the glacial theory. But the greatest problem was not religious and intellectual conservatism but the simple fact that most geologists (except those who had visited the Alps) had never seen a glacier in action (fig. 8.1), and no European had any experience with a continental-sized ice sheet at that time. This all changed in 1852, when the first expedition to Greenland by Elisha Kent Kane documented the scale of its continental ice sheets, and Kane's books amazed the Euro-

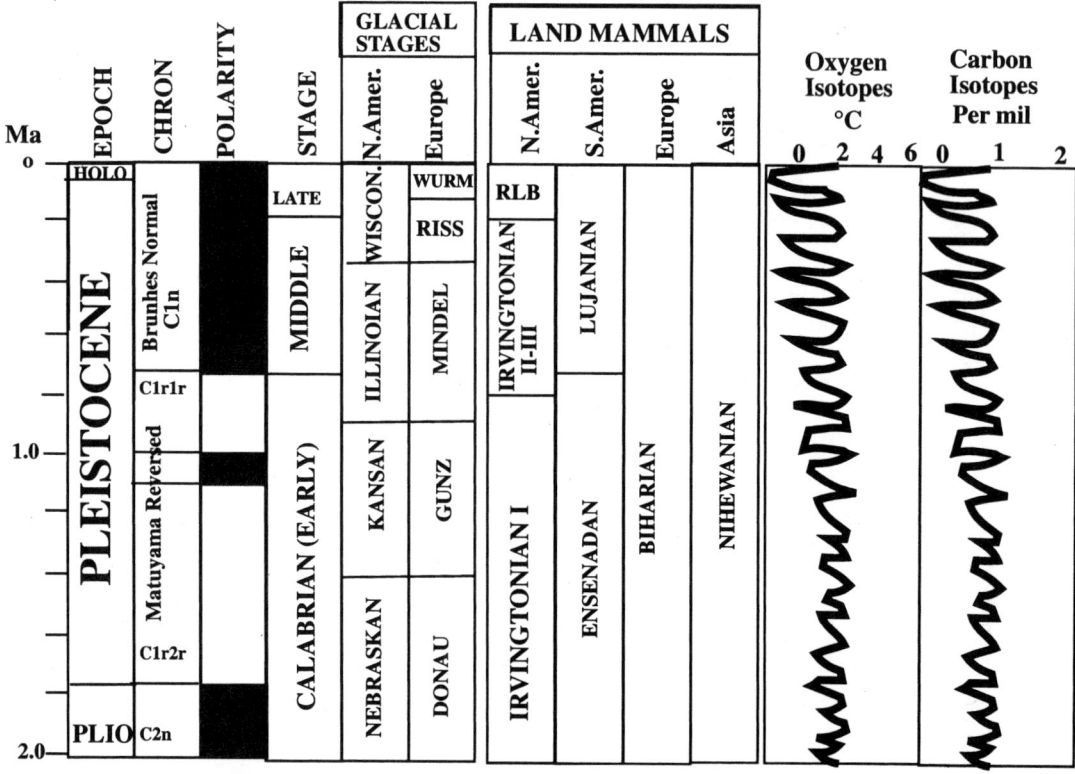

Figure 8.4. The Pleistocene timescale (left columns), as recently calibrated by Berggren et al. (1995). The North American land mammal record is after Bell et al. (2004) and Woodburne and Swisher (1995); the South American land mammal ages are after Flynn and Swisher (1995); and the European and Asian records are after Qiu and Qiu (1995). The isotopic records are after Zachos et al. (2001), but they are very generalized and smoothed out. Abbreviations: RLB = Rancholabrean; WISCON. = Wisconsinan glacial stage.

peans with the sheer magnitude of the Greenland ice (Bolles 1999). Meanwhile, Agassiz left Europe in 1846 and assumed a position at Harvard, where he happily began documenting the evidence for glaciation in North America as well. When he died in 1873, his glacial theory was universally accepted, and geologists had spent over twenty years documenting more and more examples of phenomena that could be explained by the great ice sheets.

As study on the deposits left by glaciers continued, geologists found that there had been not just one *Eiszeit*, as postulated by Agassiz, but several. In some places in Europe, there were at least five major sets of glacial deposits stacked upon one another, suggesting a minimum of five ice ages (fig. 8.4), which became known as the Donau, Gunz, Mindel, Riss, and Würm (from oldest to youngest). North America, too, yielded evidence of multiple major ice advances, which were named for the states that recorded their southernmost extent: Nebraskan, Kansan, Illinoian, and Wisconsinan. Between the glacial tills, there were deposits indicating warm periods, called interglacials, and these too had names. In North America, the glacial retreat between the Wisconsinan and Illinoian was known as the Sangamonian; the one between the Illinoian and Nebraskan, the Yarmouthian interglacial; and that between the Kansan and Nebraskan, the Aftonian interglacial. At the time, geologists were not able to date these deposits precisely, so they counted back from the most recent glacial and equated the Wisconsinan with the Würm, the Riss with the Illinoian, the Mindel with the Kansan, and the Gunz with the Nebraskan. Thanks to recent advances in dating methods, we now know that the Riss and Würm are both equivalent to the Wisconsinan, the Mindel is equivalent to the Illinoian, the Gunz correlates with the Kansan, and the Donau correlates with the Nebraskan (fig. 8.4).

The Scot and the Serb

In addition to the dating problems, geologists were confronted with an even bigger puzzle. What caused the glacial and interglacial cycles? And what caused the earth to slip into the glaciated state in the first place? Many ideas were proposed, but the most influential was the idea first suggested by the Scottish physicist James Croll in 1864. Following calculations of the French astronomer Urbain Leverrier, Croll suggested that the shape of the earth's orbit around the sun (its *eccentricity*) varies from nearly circular to more elliptical on a cycle of about 100,000 years (fig. 8.5). When the orbit is highly elliptical (eccentric), ice age conditions occur, and when it is more circular, interglacials result. Croll reasoned that the amount of sunlight that the earth received in winter is the critical factor: when the earth is in a highly elliptical orbit, it is farther from the sun in the winter and thus cooler, so more snow falls each winter until an ice age results.

Croll also pointed out that the earth cycles in other ways as it travels through space. The tilt (obliquity) of the earth's axis with respect to the plane of its orbit around the sun is currently 23.5°, so that during the Northern Hemisphere winter, the Arctic Circle (the region within 23.5° of latitude from the North Pole) is tilted away from the sun and experiences six months of darkness, and during the summer, it experiences six months of light. (The reverse is true in the Southern Hemisphere.) This angle of tilt of the earth's axis is not constant, and over thousands of years it has been as steep as 24.5° and as shallow as 21.5°. Obviously, when the Northern Hemisphere is tilted at a much steeper angle, the North Pole gets much more sunlight in summer, and the glaciers will melt back. Conversely, when the tilt is much more vertical, the poles get less sunlight, and glaciation is promoted. This cycle of gradual change in the earth's tilt takes about 41,000 years to complete.

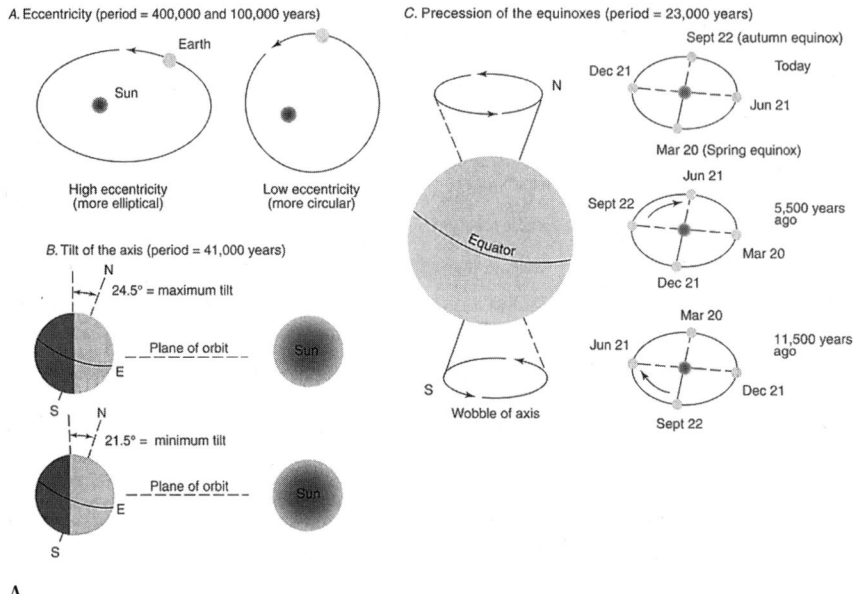

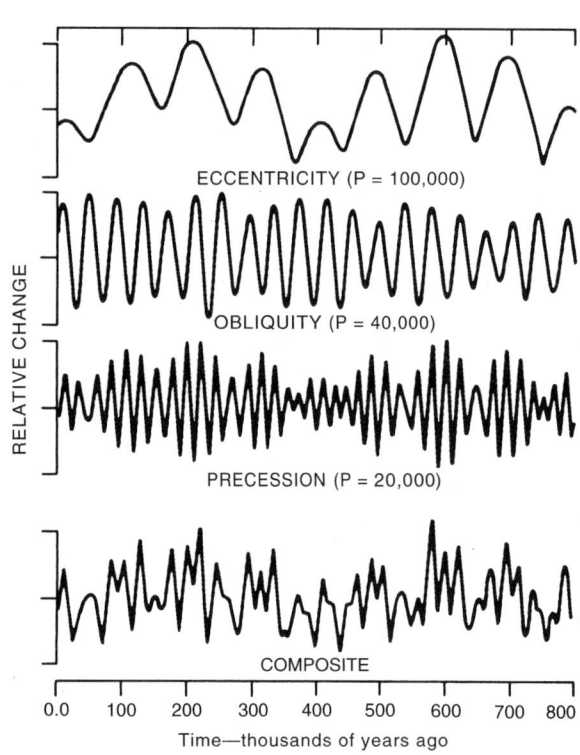

Figure 8.5. The movements of the earth around the sun responsible for the Croll-Milankovitch cycles. (A) The three cycles: the eccentricity of the earth's orbit, the tilt of the earth's axis, and the wobble, or precession, of the earth's axis. (B) The predictable fluctuations of solar energy caused by all three of these cycles combine and interfere to form a complex pattern of incoming solar radiation (bottom). From Prothero and Dott 2003.

In 1842, the French mathematician Joseph Alphonse Adhemar pointed out that the earth's axis not only tilts steeply but also points in different directions as the earth precesses, or wobbles around like a top. The spin axis of the earth, which now points at Polaris as the North Star, pointed to the star Vega as the North Star only 10,000 years ago. In 4000 B.C., the axis pointed to the tip of the handle of the Big Dipper, and as recently as 2000 B.C., it pointed to a spot midway between the Little Dipper and Big Dipper. In 1754, the French mathematician Jean le Rond d'Alembert showed that precession takes place because of the gravitational pull that the sun and the moon exert on the earth's equatorial bulge, and that the precession cycle takes about 22,000 years. These variations of the direction of the earth's axis also affect how much sunlight the poles receive at different times of the cycle, and thus have an effect on the amount of ice cover.

In 1875, Croll wrote a book-length treatment of his theories entitled *Climate and Time*. In 1880, after an accident at the age of 59, he retired to a small house near Perth, Scotland, where he returned to his philosophical writings. His last publication on the topic came in 1890, and he died shortly thereafter. However, by the time of his death, his ideas had not fared too well. Geologists had no way of dating the glacial cycles that were well documented by the 1870s, but what they did know about the age of glaciers was not consistent with Croll's predictions.

The next phase of research occurred a generation later in 1907, when the Serbian astronomer and mathematician Milutin Milankovitch decided to develop a mathematical theory that explained the climates of the earth, today and in the past. Despite the interruptions of two wars (he was even taken prisoner during World War I and did many of his calculations from his prison cell), Milankovitch doggedly persevered. He calculated just how much solar radiation the earth received during Croll's cycles, and he realized that it was not the winter sun but the amount of radiation in the summer when the ice melted that determined whether glaciers melted or retreated. By 1924, he had calculated a solar radiation curve for several latitudes and showed that the peaks and valleys in the predicted solar radiation matched the crude data on glacial advances and retreats that were known at the time. In 1930, he published *Mathematical Climatology and the Astronomical Theory of Climatic Changes,* and his work became known to the rest of the geological community. He then began to calculate how much the ice sheets would respond to changes in solar insolation (the amount of energy received by the earth from the sun), and he predicted how far they would be expected to advance, so geologists could test his theory. In 1941, he published *Canon of Insolation and the Ice Age Problem,* just as the Germans invaded Yugoslavia. His printer's firm in Belgrade was destroyed, and the final page of his book had to be reprinted. He then retired, working on his memoirs and other scientific studies until he died in 1958 at age 79.

Meanwhile, the scientific community had been debating his ideas ever since they were first introduced in 1924. By the late 1950s, they were generally accepted, although the problem was the same one that Croll had faced: dating. Geologists could see there were at least four or five glacial-interglacial cycles, but they could not date the glacial deposits precisely enough to determine whether the cycles matched the Croll-Milankovitch predictions. But in the early 1950s, radiocarbon dating was developed, and at first many of the dates run on late ice age materials did not seem to match Milankovitch's predicted age for the last glacial maximum, or for the warm periods predicted by the astronomical cycles. By the early 1960s, the Milankovitch theory was in disrepute.

The problem, however, was that the dating was still inadequate to test the theory. Radiocarbon dating worked only for objects less than 40,000 years old, which meant that only the end of the last astronomical cycle could be tested; the rest of the events were much older than the limit of the dating method. Moreover, the dates came from isolated specimens in the patchy glacial deposits on land. There was no long-term record anywhere on land that continuously recorded changes in climate that would allow a valid comparison with the astronomical cycle theory. But in the late 1950s and early 1960s, a new source of data emerged that did preserve a continuous record of the past few hundred thousand years. Unlike the episodic, irregular deposition of sediment by glaciers on land, sediments on the deep seafloor are deposited in a nearly continuous rain of muds and planktonic microfossils from the surface. After World War II, the major oceanographic institutions sent out research vessels that began to map and study the ocean floor in detail for the first time. These ships also routinely dropped a long coring device over the side. When retrieved, each core recovered as much as 15 meters of sea-bottom muds, which recorded several hundred thousand years of climatic change in the ocean. Micropaleontologists such as Fred Phleger and David Ericson soon realized that the abundance of temperature-sensitive planktonic foraminifera (such as the warm-water indicator *Globorotalia menardii*) recorded changes in oceanic water temperature, and were good indicators of climatic fluctuations in the ocean. At the same time, Cesare Emiliani developed the use of the oxygen isotope method for measuring the chemistry of the ancient ocean as recorded in the shells of plankton, and soon we had another measure of how earth's climate had changed in the past several hundred thousand years.

Still, dating problems remained. These long cores showed climatic cycles, but how old were they? Only the youngest part of the core was within the range of radiocarbon dating. In the late 1960s, however, the thorium-protactinium dating method was successfully used to date ancient coral reef terraces on Barbados, and this method demonstrated that the ages of the sea level highstands on Barbados matched the Milankovitch prediction for warm interglacials, when the ice sheets melted and sea level rose. In addition, by the mid-1960s, Neil Opdyke had shown that the deep-sea cores could be dated by paleomagnetism. The final clinching research began in the 1970s, when the multimillion-dollar Project CLIMAP (CLImate MApping and Prediction) was launched under the direction of John Imbrie of Brown University. Hundreds of cores from all over the world's oceans were analyzed for both isotopes and changes in the plankton (radiolaria as well as foraminifera) and dated paleomagnetically. By 1972, CLIMAP project members could show that there have been at least nineteen glacial-interglacial cycles in the past 700,000 years (not the four originally discovered on land) and that these cycles were in good agreement with the Croll-Milankovitch theory (fig. 8.6). The final breakthrough was the development of a computerized statistical method called filter analysis, which dissected a complex climatic curve that was a composite of several different cycles into its component cycles. When the best deep-sea climatic records were analyzed in this way, they showed clearly that the complex oceanic climatic signal was a composite of three cycles, which interfered in different ways to produce the highs, lows, and intermediate values of the climatic curve (fig. 8.5). The most prominent cycle was the 100,000-year eccentricity cycle predicted by Croll over a century earlier, and there were weaker signals that matched the 43,000-year tilt (obliquity) cycle, and two peaks at 23,000 and 19,000 years that matched the precession cycle. In 1976, Jim Hays, John Imbrie, and Nick Shackleton published the famous "pacemaker" paper, which finally confirmed that the ice age cycles were caused by the orbital variations first predicted by Croll and Milankovitch. At the end of the decade, this paper was hailed as one of the five most important scientific papers of the 1970s.

Since then, cycles of orbital variation have been sought and often enthusiastically claimed for any cyclical climatic record, even though the evidence and the dating are not always precise enough to establish the 100,000-year, 43,000-year, and 21,000-year periodicities. Bundles of small cycles within larger cycles have been identified as Milankovitch periodicities for cyclic coal deposits of the Pennsylvanian coal beds of the American Midwest (Heckel 1986), which is plausible since there were glaciers in Gondwanaland at that time. Milankovitch periodicities have also been claimed for Triassic lake sediments (Olsen 1984), mid-Cretaceous marine sediments (Fischer 1986), and Devonian limestones (Goodwin and Anderson 1985), although these are slightly less plausible, since they occur when there were no large polar ice sheets. It is not clear how the earth could change sea levels and respond so dramatically to small variations in solar radiation in the absence of ice sheets.

After the 1976 pacemaker paper, many developments occurred in the orbital variation theory. Greenhouse gases, such as carbon dioxide and methane, are now thought to play important roles in modifying climate cycles driven by changes in solar radiation (Ruddiman 2001, 2004). Detailed study of cores covering the Pliocene and Pleistocene has shown that during the Pliocene, the 41,000-year and 23,000-year cycles of tilt and

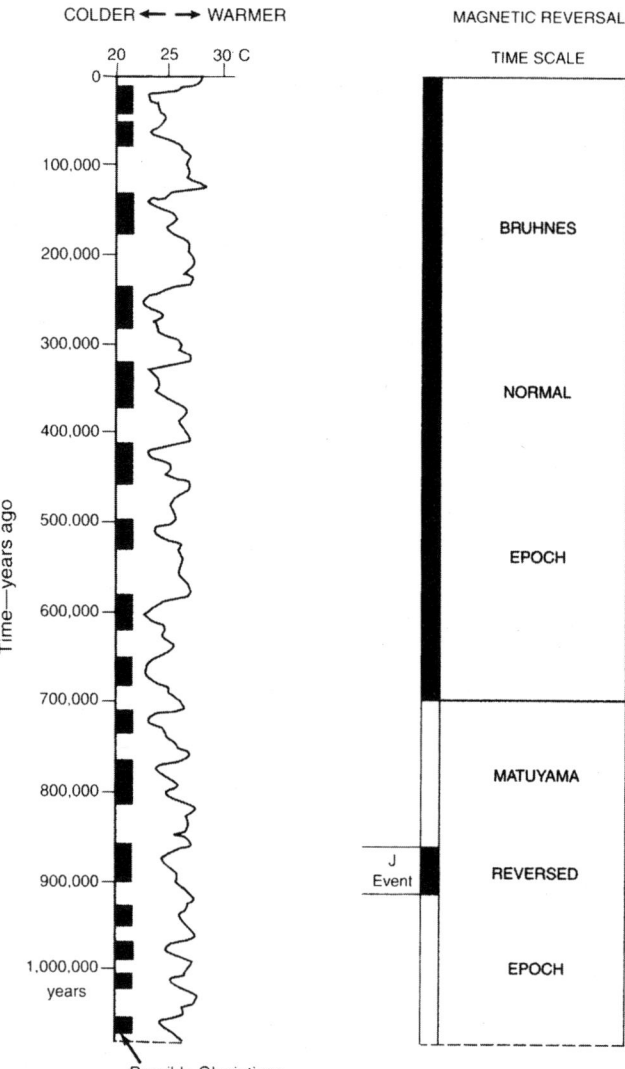

Figure 8.6. The last million years of glacial-interglacial cycles, as recorded in deep-sea cores. From Prothero and Dott 2003.

precession, respectively, were dominant, but about 900,000 years ago, the 100,000-year eccentricity cycle became the more dominant, causing major cycles of 100,000 years between peak warmth of interglacials, or between the coldest part of the glacials. Currently, climatic models suggest that as climate cooled in the Pliocene owing to the changes in continents, oceans, and atmospheric composition, growth of the ice sheets exceeded a critical threshold. Once this had been passed, the effects of the direct differences in solar radiation at 41,000 and 23,000 years became less important, and the huge, slowly changing ice sheets responded mainly to the slower (but less variable in solar radiation) 100,000-year cycle of eccentricity (Imbrie 1992, 1993; Ruddiman 2001).

How did marine life respond to all these rapid changes in oceanic temperature? In most cases, the species of Pleistocene marine animals are still alive today, so we can trace their temperature tolerances. Some of them, in fact, are so temperature-sensitive that they are good paleoclimatic indicators. As mentioned above, planktonic foraminifera have been used in this regard by marine geologists for more than thirty years. For example, the presence of foraminifera with known temperature tolerances tells the micropaleontologist about the water temperature in each centimeter of deep-sea core. The thick-shelled *Neogloboquadrina pachyderma* (fig. 8.7A) is a polar indicator species, whereas *Globigerina bulloides* prefers subpolar waters. *Globorotalia truncatulinoides* (fig. 8.7B) occurs in temperate waters. *Globorotalia menardii* (fig. 8.7D) and the pink-shelled *Globigerinoides ruber* (fig. 8.7C) are found in the tropics and subtropics, whereas *Globigerinoides sacculifer* is strictly tropical. In other cases, individual species show climate in their shell morphology. *Globorotalia menardii*, *Neogloboquadrina pachyderma*, and *Globorotalia truncatulinoides* from deep-sea cores are predominantly right-coiling during warmer interglacial periods but switch to left-coiling when the water cools below a certain threshold during glaciations (fig. 8.7E).

Likewise, changes in the molluscan faunas of the Pleistocene are subtle. Relatively few new species appeared or died out during the glacial and interglacial cycles. Instead, species tended to shift their biogeographic ranges as shallow waters along the coast became warmer or cooler with each glacial advance or retreat.

Land of the Mammoth and Sabertooth

The world of the ice ages is much more familiar to us than any other time in the Cenozoic, since ice age mammals (especially "cavemen," mammoths, and sabertoothed cats) are familiar from many movies, television shows, and even cartoons like the Flintstones and the recent movie *Ice Age*. But the almost 2 million years of the Pleistocene are not one homogeneous interval with a single fauna that persisted through the many ice age fluctuations. Instead, there were clear changes in land mammal faunas on all the continents through the many Pleistocene cycles.

At the peak of the last glacial 20,000 years ago, more than 18 million square kilometers of ice covered northern North America, and about 11 million square kilometers of ice covered most of northern Europe and nearly all of Siberia. By comparison, the modern Antarctic ice sheet covers only 12.6 million square kilometers (which expanded to almost 14 million square kilometers during the ice ages). It is calculated that the North American ice sheet had 27 million cubic kilometers of water frozen into it; the Antarctic, about 26 million cubic kilometers; and the Eurasian, 12 million. The center of the North American ice sheet was over 3 kilometers thick, and its weight depressed the underlying bedrock until it was 370 meters below modern sea level. Some regions, such as Scandinavia, are still rebounding from this pressure 12,000 years after all the ice melted

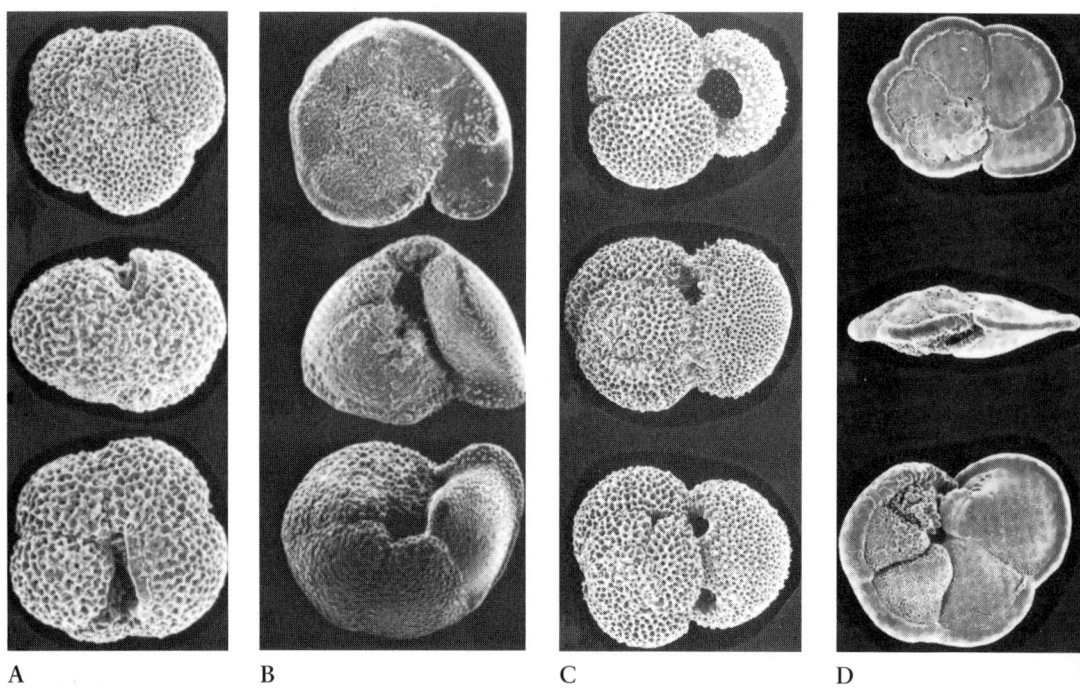

A   B   C   D

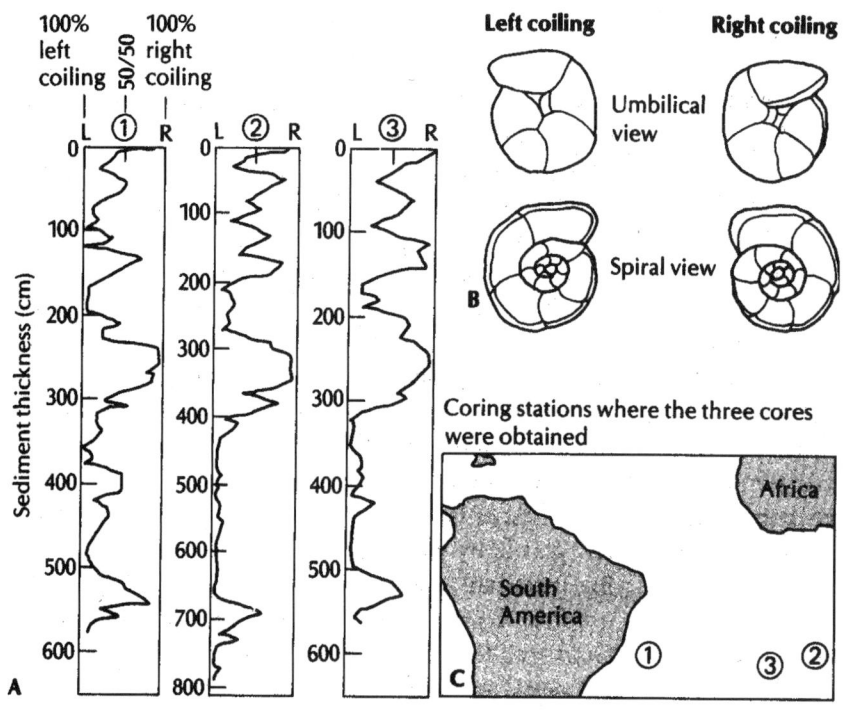

E

Figure 8.7. Planktonic foraminifera, which are sensitive indicators of paleoclimate. (A) Thick-shelled Neogloboquadrina pachyderma *is a polar species. (B)* Globorotalia truncatulinoides *is a temperate species. (C)* Globigerinoides ruber *is a tropical to subtropical species, as is (D)* Globorotalia menardii. *Photos A–D courtesy of J. Kennett. (E) Coiling ratios of planktonic foraminifera (percentage left coiling versus right coiling in each level in a deep-sea core) are sensitive to climate in many species. The populations tend to be right coiling in warmer times and left coiling in colder times. Of the species shown in this figure,* N. pachyderma, G. truncatulinoides, *and* G. menardii *show this behavior. After Eicher 1976.*

away. All of this ice took an enormous amount of water out of the oceans, dropping global sea level more than 120 meters, so that the outer continental shelf 120 meters below present sea level was a dry coastal floodplain.

Meanwhile, the ice sheets radically transformed the land. All of the northern regions of North America as far south as Kansas and Nebraska were scoured flat as the giant ice bulldozer scraped over them again and again. Bedrock in many places has long parallel scratches formed by the rasping of rocks imbedded in the base of the glacier and dragged along. In other places, the glaciers dumped huge erratic boulders and enormous mounds of till. Glacial depressions at the margins of the ice sheets became gigantic lakes, which are now represented by remnants such as Lake Winnipeg (a fraction of the size of glacial Lake Agassiz), the Great Lakes, and the Finger Lakes in New York. In the basins of Nevada and Utah, there were many lakes where there is now desert. These included the immense Lake Bonneville, which once covered most of western Utah but is now represented by the small remnant known as the Great Salt Lake and Bonneville Salt Flats. In northern Idaho, an ice dam trapped glacial meltwater in the narrow valleys of western Montana. Each time the dam broke, huge floods roared across eastern Washington, producing the Channeled Scablands (and Grand Coulee). Giant dune fields of windblown glacial dust, or loess, accumulated in the "breadbasket" regions of the world: the Great Plains, the Ukrainian and Polish plains, and central China. This loess is now a rich soil that supports much of the world's agriculture.

Naturally, the advances and retreats of the ice sheets had enormous effects on the vegetation. During glacial maxima, the ice sheets extended across most of the northern United States and completely covered Canada. South of this region was a broad belt of tundra vegetation in places like Ohio and Iowa, and most of North America was covered by a northern spruce forest like that located today in northern Canada. The southeastern United States had a pine-hardwood forest (like that of the modern Great Lakes), and all the oak forests and prairie vegetation, as well as the swamps of Florida, vanished (fig. 8.8). Tundra covered most of Europe, where today there are mixed deciduous forests and grasses; and steppes covered a much bigger region of Russia and Siberia than they do now.

In North America, most of the Pleistocene is known as the Irvington-

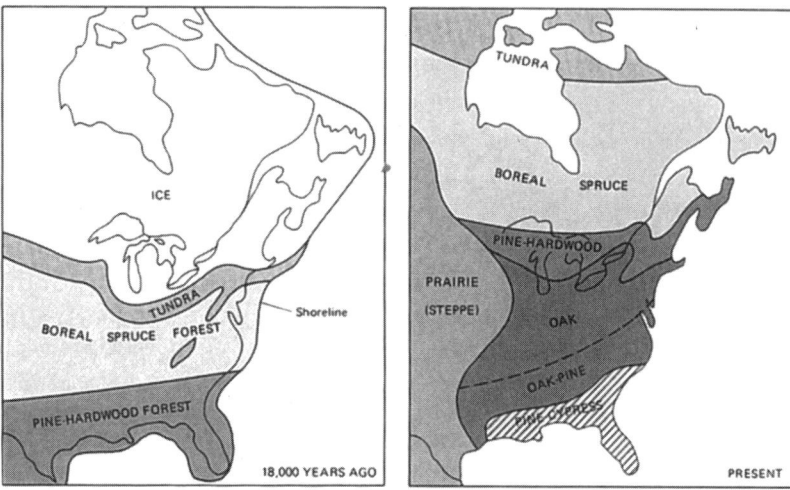

Figure 8.8. Variation in the vegetational zones. During glacial times, the vegetational zones move southward in response to the advance of the ice sheet, but during interglacials, they return to the north as the ice melts. From Prothero and Dott 2003.

ian land mammal age (from 1.9 Ma to about 300,000 years ago). Although most of the fauna looks much like that of the preceding Blancan (Pliocene), there were important changes. The most important was the introduction of mammoths (fig. 8.9A), which apparently migrated over the Bering land bridge from Eurasia at 1.9 Ma. Other mammals first appeared in the Irvingtonian as well, including hares (*Lepus*), jaguars, shrub oxen (*Euceratherium*), and the sabertoothed cat *Smilodon*. With all these new additions, a number of taxa typical of the Miocene and still known from the Blancan disappeared by the early or middle Irvingtonian. These included the last of the North American hyenas (*Chasmoporthetes*; fig. 7.6), the last bone-crushing dog (*Borophagus*), the last primitive sabertooth (*Ischyrosmilus*), the last gigantic camel (*Titanotylopus*), the last primitive proboscidean (*Stegomastodon*), the last American zebra (*Equus simplicidens*), and last of the tiny three-toed horses (*Nannippus*), so that only the modern one-toed horse genus *Equus* persisted. Still, most of the fauna of the Irvingtonian would look familiar; at least 50% of the late Irvingtonian animal species are still alive today. Of the 119 genera of Irvingtonian mammals listed by Savage and Russell (1983), 34% are new appearances, and 19% vanished by the end of the Irvingtonian. About half of the new genera were immigrants from South America, and another 27% were immigrants from Asia.

This mammoth-dominated fauna of the Irvingtonian contained a mixture of both familiar and unfamiliar taxa. Dominating the landscape were the imperial mammoths (*Mammuthus columbi*, descended from the southern mammoth, *M. meridionalis*), which were recent immigrants

A

B

Figure 8.9. Proboscideans of the Pleistocene. (A) The imperial mammoth, Mammuthus imperator, *the largest proboscidean of the ice ages. (B) The American mastodon,* Mammut americanum, *denizen of the Pleistocene forests and bogs. After Scott 1913.*

from Asia at 1.9 Ma. Imperial mammoths (fig. 8.9A) were over 4 meters (13 feet) at the shoulder, with long curving tusks up to 5 meters (16 feet) long. Larger than any elephant that has ever lived, they had huge grinding molars that could crush almost any vegetation, and they thrived in forests and even on the plains and steppes of the Arctic region.

Alongside these elephantids was the last of the primitive proboscideans, the American mastodon, *Mammut americanum* (fig. 8.9B). A relict of the earlier radiation of mastodonts in the Oligocene and Miocene, it had persisted almost unchanged for almost 4 million years, since the early Pliocene, all over Eurasia and North America. The American mastodont was archaic in most of its features: it had simple molars with conical rounded cusps (the name *mastodont* means "breast toothed;" the teeth were the first and only specimens known for a long time), a long flat skull, and a long flat sloping back with no hump on the head or shoulders like that found in mammoths. The American mastodont was only about 3 meters (10 feet) at the shoulder, with slightly curved tusks that reached up to 3 meters (10 feet) in length. One tusk is usually worn down shorter than the other, indicating that individuals were typically right-tusked or left-tusked in their feeding and behavioral preferences. They probably used these tusks to pry up bark and branches and break them into pieces. They lived mostly in the dense forests of the coasts and the interior valley lowlands and bogs (preferring spruce forests) during the ice ages, where they fed on a variety of conifers, leaves, grasses, swamp plants, and mosses (as indicated by stomach contents of mummified specimens). They had coarse brownish outer hairs about 3 centimeters long and a fine woolly undercoat, much like that of aquatic and semi-aquatic mammals such as the moose, so mastodonts were probably fond of bogs and swamps (where many of their skeletons have been fossilized).

In addition to mammoths and mastodonts, the Irvingtonian saw a great diversity of other large herbivores. One-toed horses of the living genus *Equus* were quite diverse, now that the more primitive, three-toed horses of the Miocene were finally extinct. The former included relatives of the modern wild ass, *Equus cumminsi*; the stilt-legged onager *Equus calobatus*; the tiny pygmy onager *Equus tau*; the giant horse *Equus giganteus* (the size of a modern draft horses) from the Plains region; and the small, short-legged *E. conversidens*, which was widespread from Alberta to Mexico. In the late Pleistocene, the large western horse, *E. occidentalis*, was found along the Pacific Coast; it is the horse most commonly found in the Rancho La Brea Tar Pits of Los Angeles. Tapirs, although rare, were also found in more forested parts of the landscape.

Artiodactyls were even more diverse than perissodactyls in the Irvingtonian. They included the long-nosed peccary (*Mylohyus*) and flat-skulled peccary (*Platygonus*), seven species of camels and llamas (from the huge 3.5-meter, or 11-foot, tall *Titanotylopus* to the llama-like *Hemauchenia* and *Palaeolama*), a variety of deer and elk, primitive musk oxen (*So-*

ergilia) and shrub oxen (*Euceratherium*), and a surprising diversity of pronghorns. As discussed in chapter 6, pronghorns (family Antilocapridae) are not true antelopes but members of their own endemic North American family that once performed the roles in North America that different species of true antelopes perform in Africa. At the end of the Miocene, the pronghorns were decimated in North America, but a few lineages survived. By the Irvingtonian, there were at least three genera and five species known, including *Capromeryx, Tetrameryx,* and *Hayoceros.* All of these species had a pair of two-pronged horns, although *Capromeryx* had horns with two equal tines, whereas the horns of *Tetrameryx* had a larger back tine and a short front tine pointed forward.

Along with the immigrant mammoths and bovids from Asia were several immigrants from South America. Sloths and armadillos (fig. 8.10A) had first reached North America during the mid-Pliocene Great American Interchange (see chapter 7), but in the Irvingtonian there were many new immigrant ground sloths, including the huge *Eremotherium* (up to 7 meters, or 21 feet long, and weighing over 3 tons, about the size of an elephant) and the smaller sloth *Nothrotheriops.* The giant armadillo *Holmesina* also spread from the radiation of pampathere armadillos in the Argentine plains. These beasts were the size of a rhinoceros, with a rigid shield of armor over their bodies, only three "bands" to hinge the armor, and a rigid conical tube of armor protecting their tails. Much like the armadillos were the glyptodonts (fig. 8.10B), which had a solid rigid shield over their bodies that had no hinging plates (unlike *Holmesina* and other armadillos). Some of these reached 2 meters (6 feet) in length and weighed over a ton, and many had a bony spike like a mace on the tip of their tails for defense. Although these huge edentates are known from many places, they were abundant mostly in the warmer southern parts of North America (particularly Florida, Texas, and Arizona), with occasional incursions as far north as Nebraska or the Carolinas.

Preying on this great diversity of herbivores was an incredible array of carnivorans. The hyenas and bone-crushing borophagine dogs of the Miocene and Pliocene were gone, but other groups were diverse. These included the living black bear genus *Ursus; Tremarctos,* the ancestor of the spectacled bear; and the huge short-faced bear *Arctodus,* which at 1,500 pounds would dwarf even the largest living bear from Alaska, the Kodiak bear. *Arctodus* was the top predator in North America, and the largest carnivoran that ever lived. Cats, too, were diverse, from the famous saber-toothed cat *Smilodon fatalis* and the dirktoothed cat *Homotherium* to true lions (larger than any living lion), cougars, jaguars, lynxes, and the North American cheetah *Miracinonyx studeri.* Dogs included primitive wolves and coyotes, as well as foxes, which were recent immigrants from Eurasia. The most famous and abundant of these is the dire wolf, *Canis dirus,* a heavily built, short-limbed predator and scavenger (now that hyenas were gone) known from thousands of specimens in the La Brea Tar

A

Figure 8.10. Edentates of the Pleistocene Americas. (A) The gigantic ground sloth Eremotherium, *which reached the size of an elephant. (B) The huge glyptodonts, with their armadillo-like shell and spiked tails. From Scott 1913.*

B

Pits. Primitive raccoons and ringtails were also common. Mustelids were represented by a great variety of primitive weasels, martins, and ermines, as well as wolverines, badgers, otters, and skunks.

Finally, the small mammal fauna in the Irvingtonian was the most familiar of all. The hares (*Lepus*) were late additions to a fauna that included such familiar mammals as pikas and rabbits, shrews and moles, many species of bats, and a tremendous variety of rodents: squirrels, marmots, prairie dogs, pocket gophers, pocket mice, beavers (including the giant beaver *Castoroides ohioensis* (fig. 8.11), which was almost 2.5 meters long and weighed about 200 kilograms, about the size of a bear), jumping mice, porcupines, and the immigrant capybaras from South America. Most diverse of all, however, was the family Cricetidae (New World mice, voles, lemmings, and hamsters), which is represented by dozens of genera and species in the Pleistocene. They evolved so rapidly, and their tiny teeth are so abundant in nearly every Pleistocene deposit,

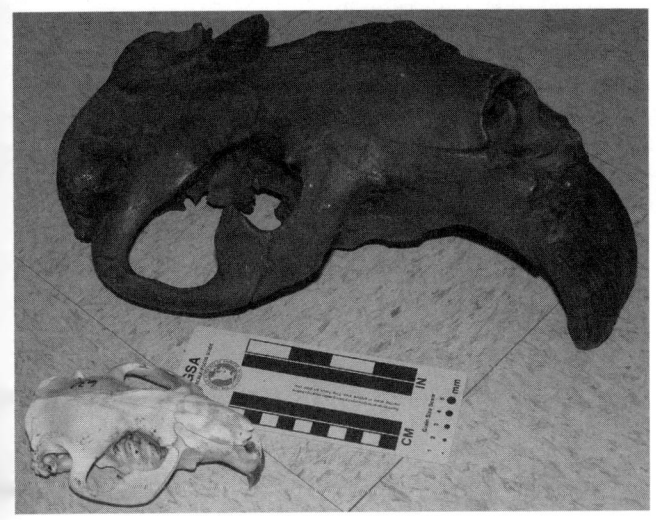

Figure 8.11. Castoroides ohioensis, *the bear-sized ice age beaver, compared to a living beaver skull for scale. Photo by the author.*

that the cricetids are the principal tool for telling time in the Pleistocene in North America. From harvest mice to muskrats to lemmings to field mice, if you want to know the time, you look for cricetid teeth.

This Irvingtonian fauna was remarkably stable for over 1.6 million years. The close of the Pleistocene (300,000 years ago to 10,000 years ago) in North America, however, marks the climax of this ice age mammal assemblage. In glacial terms, the climax began during the Illinoian glaciation and culminated at the end of the Wisconsinan glaciation. It is known as the Rancholabrean, after the spectacular late ice age mammals found at the La Brea Tar Pits (pl. 9, fig. 8.12). In most ways, the Rancholabrean fauna differs only slightly from that of the Irvingtonian. The most striking difference is the immigration of *Bison* from Eurasia, whose first appearance about 300,000 years ago marks the beginning of the Rancholabrean. Some of these bison, like the huge *Bison latifrons,* had horns that spanned over 2 meters in width (fig. 8.13)! But there were also many more immigrant Eurasian species including saiga antelopes, mountain goats and sheep, musk oxen, reindeer, deer and moose, and new species of rodents, many of which were denizens of the rich grassland faunas of the "mammoth steppe" that fringed the ice sheets (Guthrie 1990). Columbian mammoths persisted, but in the north they were competing with the smaller immigrant Eurasian woolly mammoth (*Mammuthus primigenius*), which was only about 3 meters (10 feet) at the shoulder, with a sloping back, long recurved tusks used for scraping away snow to find food, and finger-like projections on its short trunk for grasping. Woolly mammoths had a thick coat of fur that ranged from dark brown to black in color. They were found primarily in Eurasia in the late Pleistocene, and only rarely ventured out of the mammoth steppe of the glacial regions as far south as northern Canada and Alaska.

In Eurasia, the Pleistocene mammal fauna was similar to that of North

Figure 8.12. Typical slab of bones embedded in tar from the Rancho La Brea Tar Pits. Photo courtesy of the Page Museum of La Brea Discoveries.

America, because there was so much travel back and forth across the Bering land bridge. However, for reasons that are not clear, some groups never crossed. Some native North American groups, such as pronghorns and peccaries, never made it to Eurasia, possibly because they were not well adapted to crossing the Arctic, even in its warmest times. The same problem may have prevented true pigs of the family Suidae from leaving Eurasia and invading North America. Other failures to cross are more puzzling. Woolly rhinoceroses (fig. 8.14) were abundant in Eurasia and were well adapted to life in the cold, snowy north, alongside woolly mammoths. Yet although woolly mammoths made the crossing, woolly rhinos never did.

Like the early Pleistocene (Irvingtonian) in North America, the early Pleistocene (Biharian) in Eurasia was dominated mostly by groups that are alive today, albeit largely represented by extinct species (pl. 10). These included primitive mammoths (*M. meridonalis*), the last of the gomphotheres (*Anancus*), and the Etruscan rhino (*Stephanorhinus etruscus*), which was related to the woolly rhino and the living Sumatran rhino. Deer, goats, and cattle were diverse, along with a number of species of archaic horses. The predators included hyenas, a variety of sabertoothed cats, lions, lynxes, and cheetahs, many kinds of wolves and foxes, primitive bears, and a full spectrum of mustelids (weasels, ermines, minks, otters, and wolverines). Around 1.6–1.8 Ma, the Pleistocene in Europe was

Figure 8.13. The enormous horns of the giant ice age Bison latifrons *spanned over 2 meters (6 feet) when the horny sheath covered these bony cores. Photo courtesy of the California Museum of Paleontology.*

marked by the first appearance of the musk ox *Soergilia*, the first *Bison* (long before they appeared in North American in the Rancholabrean, only 300,000 years ago), the goat-like *Hemitragus*, and the giant megacerine deer, which evolved into the giant deer, or "Irish elk" (*Megaloceros giganteus*; fig. 8.15). In truth, it was neither an elk nor exclusively Irish; this huge moose-like deer was found all over Europe and had palmate antlers over 3.6 meters (12 feet) across and weighed over 45 kilograms (100 pounds)! In addition to these Asian immigrants to Europe, there were immigrants from Africa as well. The most impressive of these was the huge 7-ton *Hippopotamus major*, with eye sockets high on the top of its head

Figure 8.14. Reconstruction of the woolly rhinoceros, Coelodonta antiquitatis. *Painting by Z. Burian.*

for better underwater vision. Other African immigrants included a slender giraffid, ostriches, bushbucks, leopards, spotted hyenas, and the first members of our genus to reach Europe, *Homo ergaster.*

The early Pleistocene in Eurasia was dry and cold, but in the late early Pleistocene, which began 1.2 million years ago, warmer interglacials alternated with milder glacial conditions. Forests spread over much of Europe, and temperate elements replaced the steppe fauna of the early Pleistocene. The straight-tusked elephant *Elephas antiquus* (ancestral to modern elephants) dominated the interglacials, whereas the glacial episodes were dominated by the southern mammoth, *Mammuthus meridionalis.* Long-limbed horses descended from *Equus stenonis* were also diverse, and the steppe-adapted grazing Etruscan rhino *Stephanorhinus etruscus* was replaced by the browsing *S. kirchbergensis.* Also indicative of the milder climate in these forests was the appearance of wild boars (*Sus scrofa*) and macaque monkeys. Red deer and roe deer made their first appearance in

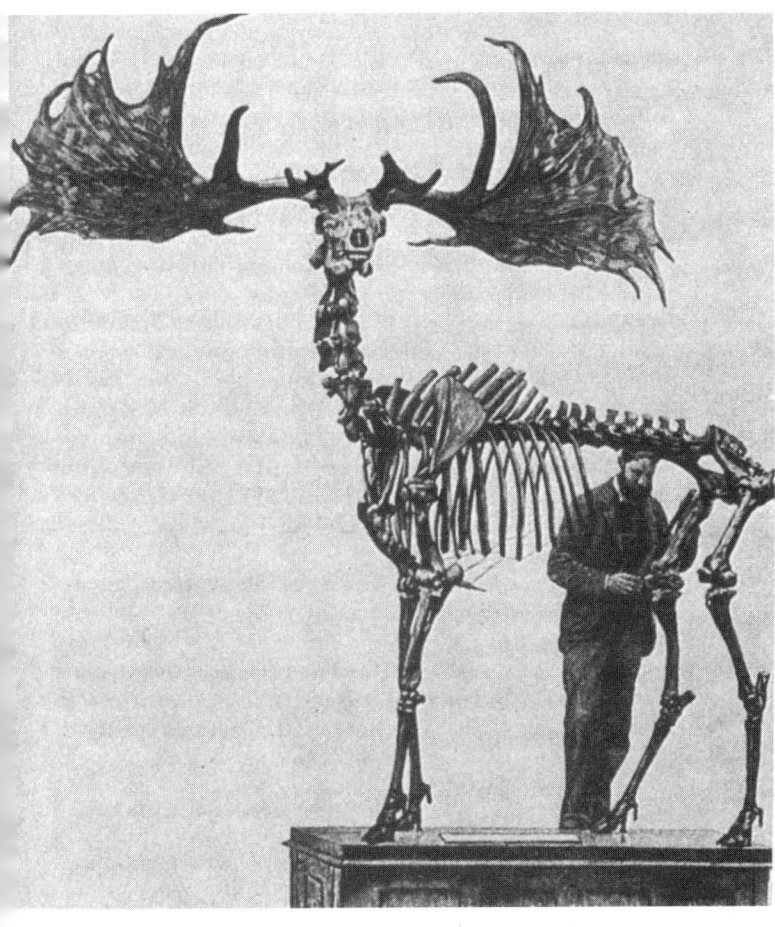

Figure 8.15. The "Irish elk" Megaloceros. The "Irish elk" was neither strictly Irish nor an elk but rather a huge deer with moose-like palmate antlers that were 3.6 meters (12 feet) across and weighed 45 kilograms (100 pounds), 30% larger than those of the living moose. From Millais 1897.

the forests, along with their larger relatives, the megacerine giant deer. More-advanced species of *Bison* emerged, along with the first immigrant mouflon sheep, *Ovis antiqua*.

About 900,000 years ago, the middle Pleistocene began with a dramatic global cooling, and glaciers spread south across Eurasia, covering Scandinavia and most of northern Europe. The glacials became much more severe, with sagebrush-dominated steppes south of the glacial limit and temperate deciduous forests returning during the interglacials. Most of the early Pleistocene mammals survived (except for the sabertooth *Megantereon*), but many more cold-adapted steppe species of goats, musk oxen, sheep, and horses predominated, along with steppe bison and mammoths.

The late Pleistocene in Europe consists of the final glacial-interglacial cycle, known as the Riss and the Würm stages (starting 130,000 years ago). The early late Pleistocene glaciations were not as severe as those of the middle Pleistocene, the interglacials were milder, and the climate was much more like that of our modern world. However, the final late Pleistocene glacial stage (75,000 to 17,000 years ago) was the most severe of all, with nearly all of northern Europe covered by ice sheets; sea level dropped as much as 120

Figure 8.16. The huge rhinoceros Elasmotherium *that roamed the steppes from China and Siberia to Spain. It had a gigantic single horn on its forehead rather than a horn on the nose like all living rhinoceroses. Painting by Z. Burian.*

meters (370 feet), resulting in the formation of land bridges across the Mediterranean islands and between Britain and Europe. Many of the temperate species that had survived in Europe during milder glacial-interglacial conditions finally died out. These included the European hippo, the straight-tusked *Elephas antiquus,* and the forest-adapted rhino *Stephanorhinus kirchbergensis.* In their place were many taxa adapted to the cold sagebrush steppes and snowfields of the ice margins: woolly mammoths, woolly rhinos, steppe bison, saiga antelope, musk oxen, reindeer, ibexes, chamois, and the last and largest of the giant "Irish elk" lineage. In addition, Eurasia saw the spread of the huge elasmothere rhinos, which were almost as heavy as elephants and had a huge horn almost 2 meters (5 feet) long in the middle of their foreheads (fig. 8.16). The predators, too, were large, robust, and adapted to the cold; they are usually preserved in cave deposits, even though they may not have lived in caves most of the time. These include the cave hyena, the cave lion (much larger than the living lion, and reaching 3 meters in length), and the huge flat-skulled cave bears (also about 3 meters long). These animals lived side by side with Neanderthals during the earlier part of the late Pleistocene (pl. 10). By the end of the Pleistocene, they battled early *Homo sapiens*, better known as Cro-Magnons.

Ancestral Homeland

In contrast with Eurasia, the African continent harbored a Pleistocene mammalian fauna that was not much different from that found there today. These fossils have been extensively studied in localities such as Olduvai Gorge in Tanzania (fig. 7.8B) and Lake Turkana in Kenya, as well as many caves in South Africa, because these same faunas also include the earliest record of human evolution. In eastern or southern Africa today, the dominant groups are the huge variety of antelopes and cattle (family Bovidae), and these were also the most common mammals in the Pleistocene. Many are members of living genera, but the diversity of extinct genera and species of antelopes was much greater. Other artiodactyls included a huge variety of pigs, especially the relatives of warthogs. One of these (*Metridiochoerus compactus*) was a giant that stood over 1 meter high at the shoulder and had long curved upper and lower tusks. Each jaw bore a single elongate high-crowned molar with complexly folded enamel, paralleling the condition found in elephants. Hippos were abundant, including the huge *Hippopotamus gorgops*, which had eyes on long stalks above its head. There were also primitive relatives of the giraffes, including the thick-necked, palmate-horned form *Sivatherium*. Although much less common than antelopes, there were a few relatives of the Eurasian deer, and there were camels related to the modern dromedary and Bactrian camels. Among perissodactyls, the primitive ancestors of the modern Grevy's and Burchell's zebra were already present, as were a variety of extinct zebras and asses, and the last of the three-toed horses, *Hipparion*. In the early Pleistocene, the forerunners of the living black and white rhinos still persisted, alongside the Etruscan rhino (related to the living Sumatran rhino) and the huge elasmotheres (fig. 8.16), but late Pleistocene rhino fossils are related to the two species still living in Africa today. There was also a primitive tapir, *Tapirus arvenernesis*, which is presumably related to today's Asian tapir. And the very last of the peculiar clawed perissodactyls known as chalicotheres survived into the early Pleistocene. Some people think they are the legendary "Nandi bears" that are supposedly still alive in the African jungles today. In the early Pleistocene, Africa was still inhabited by a variety of archaic proboscideans, including the long-tusked gomphothere mastodont *Anancus*, the archaic deinotheres with their down-turned lower tusks, and mammoths and the earliest relatives of the living African and Asian elephants. By the late Pleistocene, primitive species of the African elephant (*Loxodonta*) were present, along with *Elephas recki,* which is related to the living Asian elephant.

The small mammals of Africa were mostly groups that are dominant today. There are a number of fossils of primitive aardvarks, and specimens referred to the genus *Procavia,* the modern rock hyrax. Among the rodents were not only the cricetids but also the murid rats and the Old World porcupines, as well as true squirrels, beavers, and kangaroo rats, all no longer found there. Rabbits and pikas were much less common than rodents, as they still are in Africa today. Africa houses a tremendous variety of insectivorous mammals, including shrews, hedgehogs, and moles, as well as elephant shrews, golden moles, and the strange tenrecs of Madagascar. These are nearly all known from Pleistocene fossils as well. And, of course, bats were diverse on all the continents by this time. Fossils of nonhominid primates are quite common in places like Olduvai Gorge and Lake Turkana, including many kinds of macaques, baboons, and mandrills (some of which were gigantic). On Madagascar, there were many extinct relatives of the living lemurs, including *Megaladapis,* which was the size of a gorilla.

Preying upon all these creatures was a great variety of carnivorans. These include not only the fossil relatives of the living great cats, such as lions, leopards, and cheetahs, but also a variety of sabertoothed and dirk-toothed cats (*Megantereon, Dinofelis,* and *Machairodus*). Hyenas were also diverse and included not only the fossil relatives of the living species but also the relict running hyena *Chasmoporthetes.* Today the hunting dog is the only common canid in Africa, but many species of fossil dogs and foxes are known from the Pleistocene. The dominant small carnivorans in Africa today are the viverrids (the mongooses, genets, civets, and their relatives), which were common in the Pleistocene too. However, there were also mustelids, including weasels and otters no longer found in Africa, and the honey badger *Mellivora.*

Last but not least, Africa continued to be the homeland of our own family, the Hominidae. In the previous chapter, we saw that Pliocene rocks of eastern and southern Africa yield a great variety of species of our close relatives, including the earliest members of our own genus *Homo* (*H. rudolfensis, H. ergaster,* and *H. habilis*). These species are known not only from bones but also from their primitive tools, the choppers and hand axes of the "Oldowan" culture. Many of these archaic Pliocene taxa persisted into the early Pleistocene (as recently as 1.6 Ma), including *Paranthropus robustus* and *P. boisei, H. ergaster,* and *H. habilis* (fig. 7.7). By 1.9 Ma, however, a new species had appeared: *H. erectus.* Not only was this creature bipedal and erect (as its species name implies), but its body was almost as large as ours. Some individuals reached 6 feet (2 meters) in height. Its brain capacity was about 1,000 cubic centimeters, only slightly less than ours (fig. 8.17). Like earlier species of *Homo,* they made crude choppers and hand axes ("Acheulian" culture tools), and they certainly knew how to make and use fire and how to construct stone and

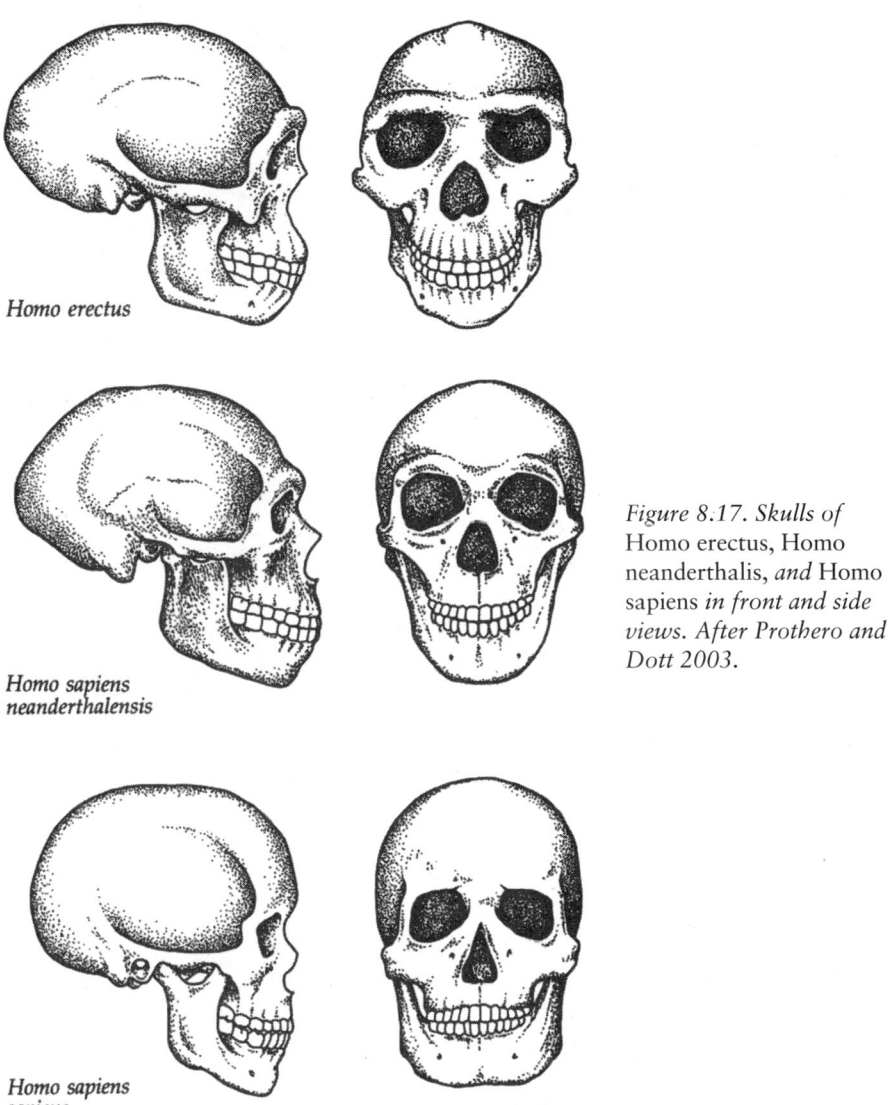

Homo erectus

Homo sapiens neanderthalensis

Homo sapiens sapiens

Figure 8.17. Skulls of Homo erectus, Homo neanderthalis, *and* Homo sapiens *in front and side views. After Prothero and Dott 2003.*

wooden dwellings and small villages. Originally *H. erectus* was confined to Africa, where all of our other ancestors had long lived. The best-known specimen is a nearly complete skeleton found on the shores of western Lake Turkana in 1984 and known as Nariokotome boy (fig. 8.18), which would have reached 2 meters tall when fully grown. By around 1.8 Ma, we have evidence that *H. erectus* finally escaped the African homeland: specimens from Indonesia (originally described as *Pithecanthropus erectus*, or Java Man) have been dated at that age (Swisher et al. 2000), and specimens are also known from elsewhere in Eurasia, such as Romania, that are almost as old. By about 500,000 years ago, we have abundant fossils of *H. erectus* in many parts of Eurasia, including the famous specimens from the Chinese caves at Zhoukoudian, known as Peking Man and

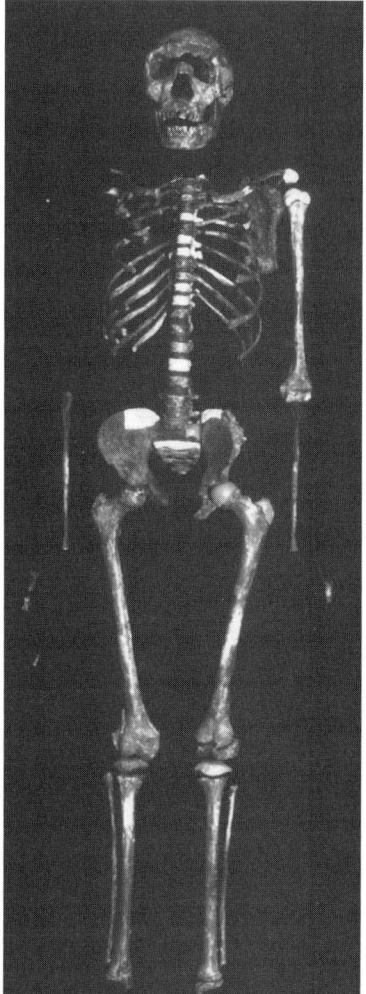

Figure 8.18. The skeleton of the Nariokotome Boy, the most complete Homo erectus *skeleton known. Courtesy of A. Walker.*

dated as old as 460,000 years ago. The latest dating suggests that *H. erectus* may have persisted as late as 27,000 years ago, outlasting even the Neanderthals and overlapping with modern *H. sapiens* (Swisher et al. 2000). *Homo erectus* was thus the first member of our family to live outside Africa, and it roamed through the entire Old World (except Australia and the glaciated regions). *Homo erectus* was not only the first widespread hominid species but also one of the most successful and long-lived species, spanning more than 1.8 million years between 1.9 Ma and 0.03 Ma. During most of that long time, it was the only species of *Homo* on the planet, and it changed very little in brain size or body proportions. If longevity is a measure of success, then one could argue that it was even more successful than we are.

By about 300,000 years ago, another species had become established in western Europe and the Near East: the Neanderthals. These were the first fossil humans to be discovered and were originally dismissed as the

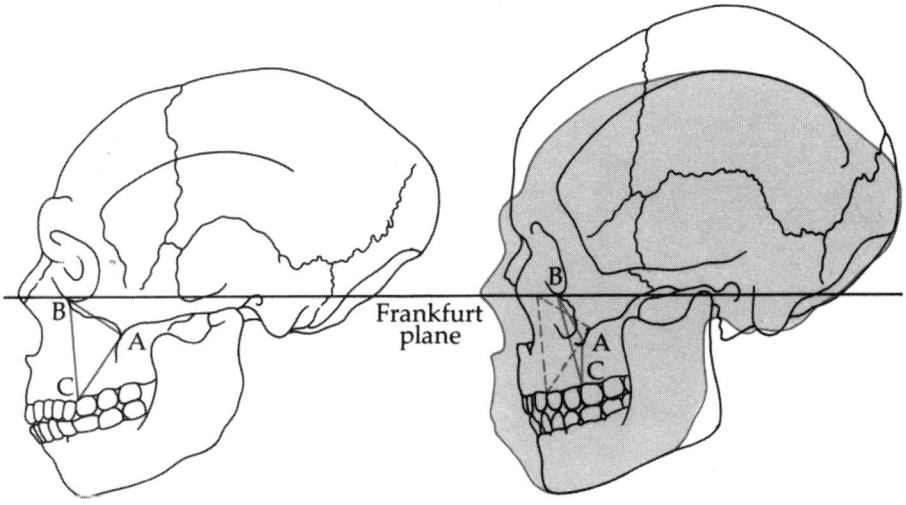

Figure 8.19. Side view of Neanderthal (left) and Homo sapiens (right) skulls, showing how the Neanderthal skull was lower and longer than that of H. sapiens, even though the brain capacities were the same. Neanderthals also had prominent brow ridges and a more prominent face with a weak chin. From Lewin 1984.

remains of Cossacks that had died in caves. The first complete descriptions of skeletons were based on a specimen that suffered from old age and disease, so for decades Neanderthals were thought to be stoop-shouldered and primitive, the classic stereotypical grunting cavemen. Modern research (Stringer and Gamble 1993) has shown that Neanderthals were very different from this stereotype. Although their skulls are distinct from ours in that they had a protruding face, large brow ridges, no chin, and a flatter skull that sticks out in the back (figs. 8.17, 8.19), they had the same brain capacity as we do, and their complex culture included ceremonial burials, which suggests religious beliefs. Their bones (and presumably their bodies) were robust and muscular and slightly shorter than those of the average modern human, perhaps because they lived exclusively in the cold climates of the glacial margin of Europe and the Middle East, where their stocky build (like that of an Inuit or Laplander) would be an advantage. Their tool kits and culture were also more complex: Mousterian hand axes, spearheads, arrowheads, and other complex devices, as well as bone and wooden tools, have been recovered. Some of these tools show complex working and simple carving, so they were artistic as no hominid before had ever been. For decades, anthropologists treated Neanderthals as a subspecies of *H. sapiens,* but recent work suggests that they were a distinct species that probably did not interbreed with, let alone give rise to, our species. The best evidence of their distinctiveness comes from caves in Israel, where layers bearing Neanderthal remains are interleaved with layers containing early modern humans. In addition, Neanderthals appeared later than the earliest archaic *H. sapiens,* so they could not be our ancestors and are instead an extinct European side branch.

At about 100,000 years ago, we see the first fossil skulls (fig. 8.17) and skeletons that look almost indistinguishable from our own species. Some of these "archaic *Homo sapiens*" are known from deposits in Africa dating as old as 500,000 years. About 90,000 years ago, skulls from Africa (such as Klasies Mouth Cave in South Africa) are completely modern in appearance and are universally regarded as *H. sapiens sapiens* (our species and subspecies). Like *H. erectus,* early *H. sapiens* spent most of its history in Africa, finally migrating to Eurasia about 45,000 years ago. There it came into contact with Neanderthals, and for about 9,000 years they co-existed. Mysteriously, Neanderthals vanished 36,000 years ago. There is controversy about whether they were wiped out by *H. sapiens* or by some other cause. Whatever happened, modern *H. sapiens* soon took over the entire Old World, developing complex cultures (Cro-Magnon culture), including the famous cave paintings of Europe and many kinds of weapons and tools. We will look at a bit of the prehistory and history of modern humans in chapter 9.

Giants (and Dwarfs) in the Earth

Many cultures have legends of giants that lived in the mysterious, dark past. These stories often described the giants fighting with the gods or being destroyed by floods. According to Genesis 6:4, "there were giants in the earth in those days" (that is, before Noah's flood). The Greek philosopher Empedocles (ca. 490–430 B.C.) reported huge bones from Sicily, reportedly the remains of a race of giants. Most of these legends were pure mythology, but they were reinforced by the discovery of large bones, some of which were attributed to a huge terrible beast the eastern Europeans called the "mammot" (Tartar for "earth burrower," since they believed the bones discovered in their fields belonged to huge burrowing beasts). For centuries, legends of the "mammoth" grew, and scholars wrestled over the implications of the huge bones and teeth found widely over Europe (Prothero and Schoch 2003). Many could not imagine that they belonged to an animal resembling an elephant (since they were found far from the tropics), nor would they concede that God would allow any creature to die out and break the "great chain of being." But in 1796, the great French naturalist (and founder of comparative anatomy and vertebrate paleontology) Georges Cuvier showed convincingly that the bones and teeth of the "mammoth" belonged to a large beast resembling an elephant, and also that mammoths were extinct. This was the first conclusive proof that extinction had taken place in the past.

Indeed, many mammals of the Pleistocene were giants. Not only were several species of mammoth larger than any living elephant, but the trend toward gigantism is found in many groups. We have already seen how the Pleistocene beaver, *Castoroides ohioensis,* was almost 2.5 meters long and weighed about 200 kilograms, about the size of a bear (fig. 8.11). The giant ground sloth *Eremotherium* was up to 7 meters (21 feet) long, weighing over 3 tons, about the size of a living elephant (fig. 8.10A). The huge camels *Titanotylopus* and *Megatylopus* reached up to 6 meters in height, with necks as long those of giraffes, but they were much more massive animals than giraffes. The "Irish elk" (fig. 8.15) was larger than any living deer or moose. The giant *Bison latifrons* (fig. 8.13) had horns spanning over 2 meters (6 feet), twice the size of those of living *Bison bison*. The huge American lion was almost 3 meters (11 feet) long, larger than any living lion, and the cave bears were as large as or larger than the living Kodiak bear. Africa also had its share of giants: huge baboons the size of gorillas; gigantic warthogs; and the immense ram *Pelorovis,* which weighed more than twice as much as the living Cape buffalo and had long, bow-shaped horns that spanned over 2 meters.

Why was gigantism so widespread among Pleistocene mammals? Perhaps one explanation lies in the fact that many lived in cold climates much of the year, and large body size is advantageous in conserving body heat (especially if the animal also has a thick coat of fur). Certainly, the majority of mammals (from mammoths to deer to musk oxen) in the "mammoth steppe" of Alaska and Canada tended to be larger. However, perhaps another factor is that these animals only seem to be extraordinarily large because today we live in a world where most of the large Pleistocene mammals have vanished (discussed further in the next section). If we consider the whole history of Cenozoic mammals, we find that there were abundant large-bodied animals through most of the last 50 million years, from the brontotheres and uintatheres of the Eocene to the immense indricothere rhinos of the Oligocene, which were the largest land mammals that ever lived. So our impression that Pleistocene mammals are unusually huge is partly biased by the fact that our modern world is virtually depleted in large mammals. We think of gigantism as peculiar, when in fact it was the norm through much of the Cenozoic.

Giant mammals were not restricted to the Northern Hemisphere. In South America, the mammalian fauna after the Pliocene Great American Interchange was dominated by North American immigrants, such as mammoths, mastodonts, horses, camels, and sabertooths. But there were still relicts of the archaic mammals that ruled South America when it was an island continent. These included the hippo-like notoungulate *Toxodon* (fig. 7.13B) and the long-necked camel-like litoptern *Macrauchenia* (fig. 7.13A) which also had a long proboscis, as well as the gigantic ground sloths already mentioned and armadillo-like glyptodonts the size of a Volkswagen (fig. 8.10B). *Toxodon, Macrauchenia,* and the giant glyptodonts and ground sloths were all first discovered when Charles Darwin reached South American during the H.M.S. *Beagle* voyage in the early 1830s. Darwin knew they were extinct beasts unrelated to anything alive today, and puzzling over this problem led to his theory of evolution. He left the description of these strange beasts to Richard Owen, who named and described them

Figure 8.20. Profiles of extinct Pleistocene giant kangaroos (in black silhouette) compared to the smaller living species. (a) Macropus titan *compared to the living* Macropus giganteus. *(b)* Macropus cooperi *compared to the living* Macropus robustus. *After Murray 1984. From* Quaternary Extinctions *edited by P. S. Martin and R. G. Klein © The Arizona Board of Regents. Reprinted by permission of the University of Arizona Press.*

but neither drew evolutionary conclusions from them nor realized their significance as indicators of South America's long isolation.

As the foremost vertebrate paleontologist in Britain in the 1830s and 1840s, Richard Owen also received and described many remarkable fossils from other remote parts of the British Empire. These included the extinct ice age mammals of Australia, which were equally bizarre and confusing. Australia was truly a land of the giants during the ice ages. Almost every lineage of its native marsupial mammals was large in body size: kangaroos 3 meters (11 feet) tall that weighed up to 300 kilograms, twice the size of the largest living kangaroo (fig. 8.20); huge koalas, wombats, bandicoots, and wallabies (Marshall and Corruccini 1978); and the strange tapir-like palorchestids (fig. 8.21). The biggest of all, however, were the diprotodon-tids, wombat-like marsupials (actually related to kangaroos) that were the size of rhinos (fig. 8.22)! Preying on all these marsupials was the peculiar "marsupial lion" *Thylacoleo* (actually related to wombats), which was the size of a leopard but had a peculiar rounded head with a few large blade-like shearing teeth and long sharp protruding incisors in its mouth (fig. 8.23). Along with the thylacines (including the recently extinct "Tasmanian wolf"), these were the dominant mammalian predators in this marsupial-dominated continent. But placental predators were not far off. By the latest Pleistocene, both aboriginal humans and their dingoes had reached the long-isolated island continent. The native marsupials had never been exposed to human hunting or to packs of wild dingoes, so it is likely that most were wiped out when these new placental predators appeared on the scene.

Gigantism was not restricted to mammals in the Pleistocene. Australia harbored a giant monitor lizard, *Megalania* (Molnar 2004), twice the size of the living Komodo dragon (the largest living lizard), huge pythons, and

Figure 8.21. The extinct Pleistocene sloth-like marsupial Palorchestes azeal, *which was about 1.6 meters in length and distantly related to wombats. After Murray 1984. From* Quaternary Extinctions *edited by P. S. Martin and R. G. Klein © The Arizona Board of Regents. Reprinted by permission of the University of Arizona Press.*

Figure 8.22. The extinct Pleistocene wombat Diprotodon optatum, *which was the size of a rhinoceros. After Murray 1984. From* Quaternary Extinctions *edited by P. S. Martin and R. G. Klein © The Arizona Board of Regents. Reprinted by permission of the University of Arizona Press.*

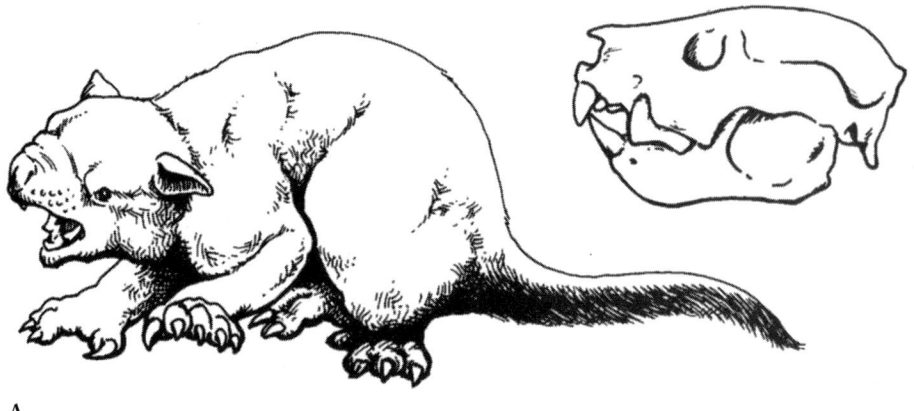

A

Figure 8.23. The extinct Pleistocene "marsupial lion" Thylacoleo carnifex, *which had a single pair of large shearing teeth on each side of its short rounded skull, and prominent protruding incisors. Drawings after Murray 1984. From* Quaternary Extinctions *edited by P. S. Martin and R. G. Klein © The Arizona Board of Regents. Reprinted by permission of the University of Arizona Press. Photo by the author.*

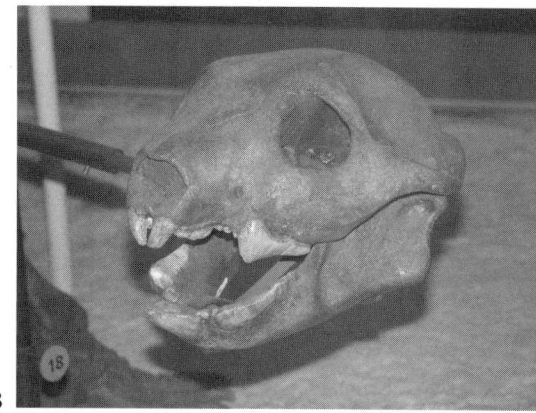

B

giant tortoises. Australia was also the home of huge flightless "thunder birds," or dromornithines (Murray and Rich 2004), which reached 3 meters in height and weighed up to 400 kilograms. By contrast, the living emu and the cassowary are nowhere near as impressive. Madagascar had the immense "elephant bird" *Aepyornis,* and New Zealand was home to the gigantic moa (Worthy and Holdaway 2002). This phenomenon of gigantic birds in isolated regions in the Pleistocene is common and seems to occur mostly on islands or island continents that have no large placental predators (such as large cats or dogs or bears). We saw it with the predatory phorusrhachids of the Miocene and Pliocene of South America and again in Australia, New Zealand, and Madagascar in the Pleistocene. For that matter, the huge predatory diatrymid birds of the early and middle Eocene were comparable, since no large mammalian predators had evolved yet.

Whereas birds tended to become gigantic on islands, large mammals often did the opposite. Dwarfing occurs in many groups of large-bodied mammals found on islands. These include the dwarf mammoths of islands including Sicily and Malta (fig. 8.24) that were only 1 meter high at the shoulder; the dwarfed elephants of the East Indies; and the pygmy mam-

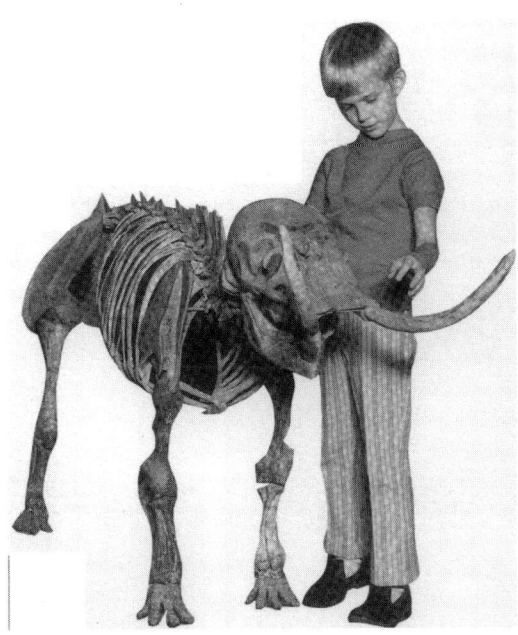

Figure 8.24. The skeleton of the dwarf mammoth from the island of Sicily was dog-sized even when fully grown. Photo courtesy of the University of Nebraska State Museum.

moths of the Channel Islands south of Santa Barbara, California, which were only 2.4 meters (8 feet) at the shoulder. On Wrangel Island in the Bering Sea between Alaska and Siberia, dwarf mammoths survived until 6,000 years ago, long after mammoths had vanished from the rest of the world. Hippos, too, underwent dwarfing. Tiny hippos occurred on several Mediterranean islands, as well as on Madagascar. Why do so many episodes of dwarfed hippos and elephants occur on islands? Most biologists point out that islands are often limited in area and restricted in their food supply. Instead of the broad areas of grasslands that support living elephants and hippos on the mainland, island mammals had to make do with much smaller areas of grasses, and most switched to browsing leaves or mixed diets (as indicated by their teeth). Such a diet cannot support a large body size, but because islands have few large predators, the need for a large body size as protection against predators is reduced. Large body size is costly in terms of the amount of food and energy needed, as well as the longer time required for growth, so it is not surprising that large mammals get smaller when their food resources are limited and there are few large predators.

Death of the Megamammals—Overkill, Climate, or Both?

The world of the late Pleistocene was indeed rich in many kinds of mammals that are no longer with us. The 40,000-year-old fossils of the La Brea Tar Pits include a diverse fauna of extinct mammoths, mastodonts, horses, tapirs, camels, ground sloths, and many individuals of huge short-faced bears, large lions, and sabertoothed cats, as well as living forms such as bison, deer, and pronghorns (pl. 9). Today, only the bison, deer, and pronghorns constitute the "large mammals" of North America, and the cougar and grizzly bear are the largest predators—a pathetic remnant of what lived on this continent only 12,000 years ago. In his *Voyage of the H.M.S. Beagle,* Darwin (1839) wrote, "it is impossible to reflect on the state of the American continent without astonishment. Formerly it must have swarmed with great monsters; now we find mere pygmies compared with the antecedent allied races." Most of the large mammals of the other continents are gone as well, from the mammoths, hippos, rhinos, giant deer, and sabertooths of Eurasia to the mammoths, mastodonts, horses, camels, sabertooths, toxodonts, and litopterns of South America, to the giant marsupials of Australia. The latest Pleistocene was a period of widespread extinction in large mammals, but not so much in the medium-sized and small mammals. The one exception to this pattern is in Africa, which still retains its elephants, rhinos, hippos, giraffes, and a high diversity of antelopes and cattle and gives us a glimpse of the world before the Pleistocene megafaunal extinctions. Darwin's co-discoverer of natural selection, Alfred Russel Wallace (1876), wrote that "we live in a zoologically impoverished world, from which all the hugest, and fiercest, and strangest forms have recently disappeared."

Almost as soon as Cuvier demonstrated that the mammoths and other large Pleistocene mammals were extinct, people have entertained many explanations for their extinction. Most of these explanations have now been discredited by better evidence, but several still remain. They fall largely into two categories: the effects of humans (the "overkill" hypothesis) or the effects of climate. The topic has been extensively debated since the 1950s, with an entire volume published twenty years ago (Martin and Klein 1984), but a consensus is still elusive.

The overkill hypothesis (Martin 1967, 1984) postulates that whenever humans with advanced hunting technologies invade a new area, the large mammals, unwary or insufficiently protected from the voracious hunting techniques of humans, are hunted out of existence. Paul Martin and other overkill advocates point to the coincidence of the arrival of humans with sophisticated Clovis points in North America about 13,000 years ago, and the disappearance of the large mammals between 11,000 and 9,000 years ago. Likewise, the megafauna of South America died out about the same time, supposedly as humans spread south (although some anthropologists argue that humans reached the Americas over 36,000 years ago). The disappearance of mammals in Eurasia and Africa is less clear-cut, since Africa still has its megamammals; but Martin would argue that they evolved side by side with humans, so they were not easily eliminated. In Eurasia, the extinction of the megamammals was also gradual, but Martin argues that it peaked about 32,000 years ago, when sophisticated Paleolithic hunting tools appeared. However, the latest dating shows that mammoths, rhinos, and large deer survived until about 14,000 years ago. Few scientists dispute the overkill hypothesis when it comes to islands like those of the Pacific Ocean (especially New Zealand and Hawaii), the

Caribbean Sea (especially Hispaniola and Cuba), and the Indian Ocean (especially Madagascar and Mauritius). It is well documented that their peculiar endemic animals were wiped out when humans arrived.

But the issue is not so simple as this. Although thirty-two species of large mammals died out in North America between 9,000 and 11,000 years ago, only a few have been shown to have been hunted by humans (mainly mammoths, mastodonts, bison, horses, and camels). For most of the rest of the megamammals, we have no evidence that they were extensively hunted, and critics argue that we should find more such evidence of hunting. Moreover, some of the groups that we know were extensively hunted (such as bison and deer) survived, while others did not. Even more critical for the overkill hypothesis is the fact that many small mammals *did not* go extinct, even though humans find small mammals much easier to catch and kill. Then there are the many bird species that *did* go extinct, even though they were not obvious food for humans. If the Clovis hunters were so voracious and devastating as Martin's "blitzkrieg" model suggests, why did they take several thousand years to wipe out the North American megamammals (and much longer if the 36,000-year-old dates on the earliest sites are correct)? The most devastating argument comes from Australia, where humans were present as early as 40,000 years ago, but the large slow-moving marsupials survived until as recently as 16,000 years ago (Horton 1984). In addition, Australian archeologists have long pointed out that almost none of their sites show much evidence of hunting or butchering of the large Pleistocene marsupials. Finally, the late Pleistocene giant marsupials show a dwarfing trend toward the end of their history, consistent with the idea that they were adapting to climatic change before their final demise.

Beck (1996) tested the overkill hypothesis in North America by examining the date of the last occurrence of each extinct species. According to the prediction, they should have died out first in areas near the Bering land bridge, and last in the most remote areas. In fact, the opposite pattern was found, with the most remote faunas dying out long before those closer to the Bering land bridge.

The alternative school of thought blames most of the late Pleistocene extinctions on climatic change. The period around 10,000 years ago was the transition from the end of the last Pleistocene glacial to the current Holocene interglacial episode, and indeed it was a period of rapid climatic change. Critics argue that there were many previous glacial-interglacial transitions during the Pleistocene, so they cannot see how the events 10,000 years ago were any different from earlier transitions. But recent paleoclimatic studies have shown that the last glacial-interglacial transition was amazingly abrupt, taking place in a few decades or less, and was characterized by extreme swings in climate before the interglacial warming took hold. Apparently, it became much colder and drier than ever before, and there was a great expansion of deserts. Lundelius (1983) and Graham and Lundelius (1984) pointed out that the late Pleistocene fauna consisted of a mosaic of habitats, with a wide variety of animals overlapping in ranges. By contrast, the Holocene is characterized by a few monotonous habitats (desert, grasslands, tundra, taiga) that extend over long distances. Many of today's animals live in narrow niches at the edge of their former range and no longer overlap with animals that once shared their habitat.

Critics of the climate change hypothesis have argued that all of the major extinctions are correlated with the appearance of humans, not with the end of the Pleistocene about 10,000 years ago. But the extinction of North and South American land mammals actually matches the end-Pleistocene climatic signal, and is much later than the arrival 13,000 years ago (or possibly even 36,000 years ago) of humans. Likewise, the disappearances of Australian megamammals 16,000 years ago and the Pleistocene extinctions in Eurasia 14,000 years ago better correspond with the end of the last glacial than with the much earlier arrival of humans on those continents.

Another variation is the "keystone herbivore" hypothesis of Owen-Smith (1987). In his studies of living African elephants, Owen-Smith points out that they are critical to maintaining a diverse habitat and fauna in eastern Africa. Elephants are capable of pruning back and destroying trees and brush at an alarming rate; they act as gardeners, modifying the brush and preventing it from becoming overgrown, so there is a diversity of habitats for many kinds of antelopes and other mammals. When elephants are poached out of an area, the brush quickly becomes overgrown, and diversity drops. If mammoths performed the same role in the Pleistocene, their extinction might have had a critical effect, causing habitat change that affected most other mammals as well. Thus, the extinction of mammoths (either by human overkill, or by abrupt climatic change) may be a key factor.

For the longest time, the debate focused on the mass extinctions at the end of the Pleistocene 10,000 years ago, with the assumption that the megamammals were gone from Eurasia, Africa, and the Americas after that date. But recent discoveries (Stuart et al. 2004) have shown that both the woolly mammoth and giant deer ("Irish elk") survived well into the Holocene (fig. 8.15). The giant deer died out in most places by the beginning of the Holocene but survived in Siberia until about 7,700 years ago. The woolly mammoth, too, survived in Siberia through much of the Holocene and finally vanished from Wrangel Island in the Siberian Arctic as recently as 4,000 years ago, where they had become a dwarfed species. Stuart et al. (2004) point out that these variable last occurrences due to differences in climate and geography do not support the overkill blitzkrieg hypothesis (since human hunters were present in all these regions long before the extinction of giant deer or woolly mammoths), and Stuart et al. show that extinction can be due to a complex of ecological factors that vary from region to region.

In short, there is a tremendous diversity of explanations, with no consensus even after more than fifty years of debate. Larry Marshall (1984), in his paper entitled "Who Killed Cock Robin?" lists these many diverse explanations along a continuous spectrum, with human causes at one extreme and climate at the other. Some scientists have remarked that with so many lethal events, it's a miracle that anything survived. In all likelihood, the end-Pleistocene extinctions were due to a complex series of events, with both human and climatic components. Indeed, that is what a distinguished panel of scientists not affiliated with either the climatic change or the overkill camp concluded: both human hunting and extreme climatic change drove the Pleistocene extinctions (Barnosky et al. 2004). Like most people, some scientists prefer simple answers in black and white; however, nature is rarely so cooperative, and more often the truth lies in a nexus of complex causes.

Whatever the cause, by the beginning of the Holocene about 10,000 years ago, most of the ice age giants were gone, and humans were spread across the entire world, which was largely denuded of its large mammals, and rapidly changing in its environmental hostility. Such changes would continue through the Holocene, and still face us today.

For Further Reading

Barnosky, A. D., P. L. Koch, R. S. Feranec, S. L. Wing, and A. B. Shabel. 2004. Assessing the causes of late Pleistocene extinctions on the continents. *Science* 306:70–75.

Bolles, E. B. 1999. *The Ice Finders: How a Poet, a Professor, and a Politician Discovered the Ice Age*. New York: Counterpoint.

Guthrie, R. D. 1990. *Frozen Fauna of the Mammoth Steppe: The Story of the Blue Babe*. Chicago: University of Chicago Press.

Imbrie, J., and K. P. Imbrie. 1979. *Ice Ages: Solving the Mystery*. Short Hills, N.J.: Enslow.

Johanson, D., and B. Edgar. 1996. *From Lucy to Language*. New York: Simon and Schuster.

Kurtén, B. 1968. *Pleistocene Mammals of Europe*. New York: Columbia University Press.

———. 1988. *Before the Indians*. New York: Columbia University Press.

Kurtén, B., and E. Anderson. 1980. *Pleistocene Mammals of North America*. New York: Columbia University Press.

Macdougall, J. D. 2004. *Frozen Earth: The Once and Future Story of the Ice Ages*. Berkeley: University of California Press.

Martin, P. S., and R. G. Klein, eds. 1984. *Quaternary Extinctions: A Prehistoric Revolution*. Tucson: University of Arizona Press.

Stanley, S. M. 1996. *Children of the Ice Age: How a Global Catastrophe Allowed Humans to Evolve*. New York: Crown Books.

Sutcliffe, A. J. 1985. *On the Track of Ice Age Mammals*. Cambridge: Harvard University Press.

Swisher, C. C., III, G. H. Curtis, and R. Lewin. 2000. *Java Man*. New York: Scribner.

Tattersall, I., and J. Schwartz. 2000. *Extinct Humans*. New York: Westview Press.

Vrba, E. S., G. H. Denton, T. C. Partridge, and L. H. Burckle, eds. 1995. *Paleoclimate and Evolution, with Emphasis on Human Origins*. New Haven, Conn.: Yale University Press.

Figure 9.1. *The Argentière Glacier as portrayed in an etching from 1850 (bottom) and in a photograph taken in 1966 (top). Most of the world's glaciers show a similar pattern of retreat since the end of the Little Ice Age and the beginning of the present period of global warming. From Ladurie 1971.*

9

Our Interglacial: The Holocene

We have met the enemy and he is us.
  Walt Kelly, *Pogo*

Climate and Human History

We have now examined the past 65 million years of earth and life history and have seen how much both have changed. The world went from greenhouse conditions to our present icehouse conditions and now fluctuates between glacial and interglacial conditions. Life on earth has continually changed in response to changing climates. Humans, too, have been affected by climatic change in our distant past. You would not be here to read this book if favorable conditions in Africa in the Pliocene and Pleistocene had not promoted our survival.

But most of this prehistory is the domain of geologists and paleontologists and is rarely read (or considered) by historians or most people who deal with recent human cultures. Our modern human civilization developed in just the past 7,000 to 10,000 years of the Cenozoic. Contrary to popular notions, much of this recent human history was affected by climatic changes, just as major climate changes of the past 65 million years affected all other organisms.

Geologists call the past 10,000 years the Holocene (or Recent) Epoch, and they treat it as a distinct period of the Cenozoic. In reality, this time interval is just another interglacial among the dozens of interglacial-glacial cycles that characterize the entire Pliocene and Pleistocene. In terms of the geological timescale, the Holocene is just the latter part of the Quaternary Period, the final part of the Neogene Subperiod, and the end of the Cenozoic Era (all of which are still continuing as you read this). However, the Holocene interglacial naturally receives much more attention because we are still living in the Holocene, and it encompasses all of human history, and much of our prehistory as well.

The last Pleistocene glaciation peaked about 20,000 years ago, and then there was a 10,000-year interval during which the glaciers receded. For most of the period between 20,000 and 6,000 years ago, there was a rapid rise in sea level of about 1 meter (3 feet) per century, resulting in a total rise in sea level of about 120 meters (380 feet), due to partial melting of the Pleistocene glaciers (fig. 9.1). Then around 6,000 years ago, the rapidly rising trend ended, and the maximum rise until recently has been only 12–15 centimeters per century (a few millimeters per year). The beginning of the Holocene was marked by a rapid rise in temperatures as the last cold spell of the Pleistocene (the "Younger Dryas") ended about 10,000 years ago.

Most of the marine and terrestrial plant and animal species that were alive at the beginning of the Holocene are still living today (except for the many species that have rapidly been exterminated in the past few centuries, as discussed below), so there has been relatively little change in faunas or floras over most of the past 10,000 years. Small changes in land plant communities and in marine faunal communities can be documented over the past few thousand years, but none was as dramatic as those that occurred in the Pleistocene. The most striking difference is between the Pleistocene and the Holocene. After the extinction of the Pleistocene megamammals, the world of the Holocene was depleted of most large mammals on all continents except Africa.

Completely modern *Homo sapiens* had already spread from Africa across Eurasia in the late Pleistocene by about 45,000 years ago, and to Australia by 30,000 years ago or earlier. They may have also spread to North America at this time (according to some archeologists), although the first unquestioned evidence of their passage across the Bering land bridge is dated only about 13,000 years ago. All human cultures were advanced when the Holocene began, with sophisticated tools, shelters, and religious beliefs, many of which still persist in the Aboriginal and Native American cultures in Australia and North America even today. None, however, had developed beyond small hunter-gatherer societies, with small human populations comparable to the size of populations of those that had inhabited the earth for millions of years.

Although an enormous number of events in the past 10,000 years of human and earth history could be discussed, our geological perspective gives us a chance to see a bigger picture amid all the details. Paleoclimatic evidence shows that the typical interglacial lasted only about 10,000 years, between glacial cycles that spanned almost 100,000 years (fig. 9.2). The peak of the last glacial maximum was about 20,000 years ago, and modern climatic conditions and the rapid postglacial rise of sea level occurred about 10,000 years ago, initiating the current Holocene interglacial.

Even the 10,000 years of the Holocene interglacial have been variable in climate as well (fig. 9.3). Climatic cycles of the past 10,000 years include several smaller episodes of warming and cooling that have dramati-

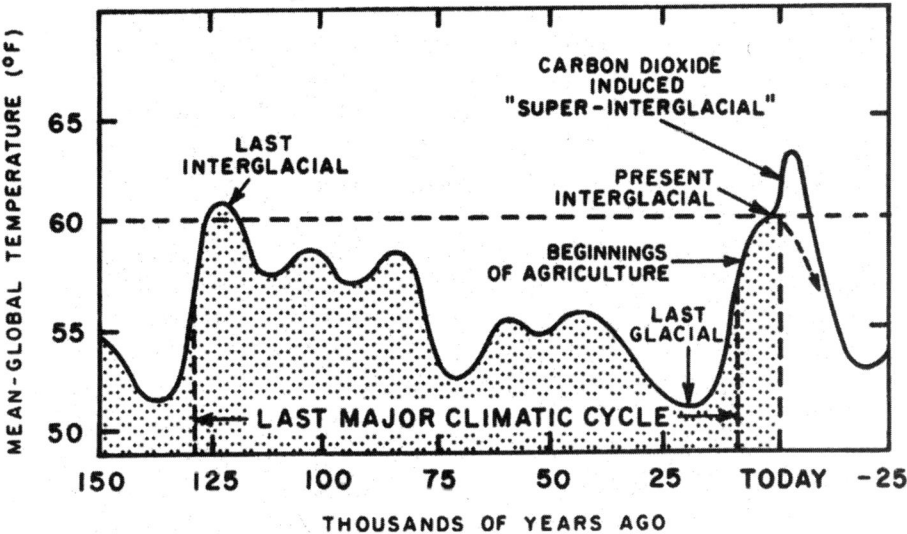

*Figure 9.2. The last 150,000 years of glacial-interglacial cycles.
After Imbrie and Imbrie 1979.*

cally affected human history, although we did not recognize it until recently. The most important event was the warmest period of the
Holocene, known as the Climatic Optimum, about 6,000–7,000 years
ago (5000–4000 B.C.). It is no coincidence that the great civilizations of
Egypt, Mesopotamia, the Indus Valley, and China all originated at this
time. During the Climatic Optimum, all of these regions were wetter and
more fertile, which allowed large-scale agriculture and cities with complex civilizations to grow for the first time. If you visit Egypt or Iraq or Israel today, it is hard to believe that the harsh, dry, forbidding desert was

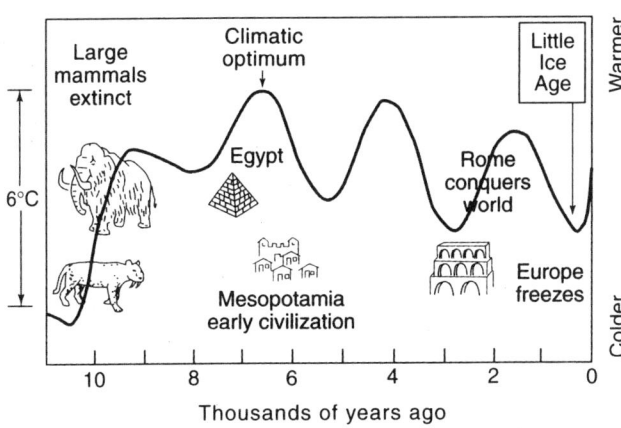

*Figure 9.3. The last
10,000 years of
Holocene climate
variation and landmarks
in human civilization.
After Prothero and Dott
2003.*

once a "promised land" flowing with milk and honey, as described in the Bible and many other ancient accounts. But during the Climatic Optimum, these regions were much wetter and more vegetated—truly the Fertile Crescent described in the history books.

Between 1200 and 800 B.C., climate began to change again, and the Middle East experienced droughts punctuated by annual flooding of the rivers. Under these conditions, the great river valley civilizations needed to contain and distribute the floodwaters for irrigation, store the annual harvests to survive the drought, and distribute their food supplies. These conditions apparently triggered the rise of highly complex, centralized governments (such as those of the Egyptian pharaohs or the Babylonian kings) to coordinate these activities. Civilization might not have arisen in the same way (or perhaps at all) without the push from climatic change for larger, agriculturally based cultures with major cities, complex religions and governments, and the other features we associate with these times.

Other civilizations did not respond to this cooling and drying event so successfully. The great Mycenaean culture of Greece collapsed, shortly after it had conquered Troy. As Herodotus wrote (and archeology has confirmed), Troy and Mycenae were besieged by famine and never recovered. In what is now Turkey, the great Hittite Empire, which had once conquered the Egyptians, also collapsed. The long period of drought in the Middle East weakened many other cultures as well. The warlike Assyrian Empire reached its peak about 800 B.C., subduing nearly every other Mesopotamian culture. The Assyrians even overran Egypt and captured the capital at Memphis. Egyptian unity was not restored until the Assyrians fell to the Babylonians, and the ensuing Pharaonic dynasties were never again as strong.

About 2,500 years ago (500–400 B.C.), another important climatic shift occurred. Known as the Subatlantic Deterioration, it was manifested by changes in wind and moisture patterns that ended the Bronze Age cultures in northern Europe. Consequently, this cold northern climate spurred Germanic invasions of southern Europe and Scandinavia. The world of the classic Greek civilization was considerably colder and wetter than it had been during the time of the Minoans or Mycenaeans. Even the architecture and clothing reflect this climatic change. Minoan paintings show sparsely clothed people living in buildings with flat roofs, suitable for a warm, dry climate, but the classical Greeks wore heavier clothing and had pitched gabled roofs for wet winters. In early Rome, the climate was much colder, and the Tiber River froze often.

About a century later, the climate began to warm in central Europe, and the frozen Alpine passes began to thaw. Hannibal's Carthaginian armies and elephants exploited these conditions to cross the Alps and invade Rome. A few decades later, the Romans conquered what had been the cold regions of northern Europe. Some historians believe that the end of this warm period, around 450 A.D., produced a prolonged freeze and drought in northern Europe, which stimulated the barbarian invasions that ended the Roman Empire.

After this cooling episode, climate began to warm again during the Medieval Warm Period, around 950 A.D. The thaw opened up the icy North Atlantic, and Vikings such as Erik the Red and Leif Eriksson were able to settle Greenland and Labrador. Simultaneously, the great Mayan civilizations of the Yucatan were at their peak in 950 A.D., although they collapsed within a few decades. Historians have long thought that this collapse was due to global climatic shifts, which affected tropical rainfall patterns.

By the 1300s, the Medieval Warm Period was over, and the Viking settlements in Greenland and Labrador disappeared owing to cold and famine. As late as 1200 A.D., Greenland had a population of 4,000 with a successful agricultural economy, but soon the growing season became shorter and shorter, and the northern outposts had to be abandoned by 1350. The coastal seas became choked with ice, cutting off the colonists for months. These hard times can be seen in the archeological remains. Early burials were in deep graves, with coffins made of imported wood. As the colony became impoverished, the people buried their dead using shrouds in shallow graves. Eventually most of the burials were of young people, which suggests a shortened life expectancy. The adults were stunted and misshapen in growth, not tall and robust like their Viking ancestors. They had extraordinary wear on their teeth, suggesting a coarse, gritty vegetable diet. By 1410, the last colonists had abandoned Greenland.

This cooling caused unusually wet and cold years in Europe, which led to flooding and rotting harvests. These events triggered the famines of 1315–1317, leading to an unprecedented number of deaths in what historian Barbara Tuchman (1987) calls the "calamitous fourteenth century." The famine indirectly led to a greater catastrophe, the Black Death, or bubonic plague, which decimated Europe thirty years later. To feed the famine-starved people, huge imports of grain from the Middle East were brought in, and these carried plague-infested rats, which had come from China, where the plague began in 1333. That same year, China experienced huge floods on the Yellow River that killed 7 million Chinese, the largest flood of the Middle Ages. The wet weather also led to the spread of another disease, known as St. Anthony's fire because it produced blackened stunted limbs, as well as convulsions and hallucinations followed by death. It was caused by a fungus that infests rye plants with ergot blight, which could infect a whole sack of flour and sicken hundreds of people. By the 1500s, ergot blight was gone, and later research showed that it could grow only in the cold, damp conditions that existed in the fifteenth century. Today is it virtually unknown.

The changing conditions were also felt in the New World. In the American Southwest, the Anasazi people (ancestors of the Hopi people of Arizona and New Mexico) had built huge pueblos in the cliff sides, with an agricultural economy based on maize. But there were droughts in 1271 to 1285, and the crops failed. Their problem was compounded by clear-cutting of the trees on the plateau above; this in turn promoted flash floods that washed away the crops on the dry creek beds. By 1300, most of the cliff dwellings had been abandoned, and the people moved on to the Hopi mesas, where there was permanent water. Two hundred years later, Coronado searched the area seeking gold from the "Seven Cities of Cibola," but the great pueblos that had spawned this legend were already abandoned.

The climax of the cooling trend of the Middle Ages was known as the Little Ice Age, which lasted from about 1550 to 1850. Old lithographs of the Alps showed that the glaciers had advanced much further than they are now, and paintings by the Dutch and English masters frequently show people skating on frozen canals, lakes, and rivers that do not freeze today. The famous painting of Washington crossing the Delaware River shows large ice floes, but the Delaware rarely freezes now. The Little Ice Age again led to crop failure and great famines in Europe. The great famine of 1594 to 1597 hit Europe severely, causing cannibalism and food riots, and people ate cats, dogs, and even snakes. In 1693, one-third of Finland's population died from another famine. In Scotland, cod fisheries failed, so that by 1691 over 10% of the Scottish population fled to northern Ireland to escape the famine, leading to the conflict that still rocks Ireland today.

The Little Ice Age affected climates all over the world. In northern Africa, preceding wet and fertile years had promoted large civilizations on the fringe of the Sahara in Mali and Ethiopia, but these cultures collapsed during the droughts, floods, and famines of the late 1500s and early 1600s. In 1628 it became so cold in the tropics that snow was reported at low elevations in the equatorial mountains of Kenya and Ethiopia. In India, the great Moghul city of Fatehpur Sikri was abandoned in 1588, only sixteen years after it was finished. This disaster was apparently due to a disruption of the monsoonal circulation pattern that caused severe drought. China experienced a series of severe winters between 1654 and 1676, which led to floods and famines and millions of deaths.

The Little Ice Age peaked with the dismally wet and cold year of 1816, when summer warmth never came, thanks to all the ash in the stratosphere from the Tambora volcanic eruption in Indonesia that year. That summer, nineteen-year-old Mary Shelley wrote *Frankenstein,* and William Polidori wrote *The Vampire.* Both were cooped up by the bad weather with Lord Byron and Percy Shelley in a house near Lake Geneva, Switzerland, so they spent hours writing ghost stories, producing two of the most famous works of horror fiction. The following year, wine grapes were ruined and harvests were poor, so hunger led to more riots and revolution in a world just recovering from the Napoleonic wars. The warming trend at the end of the Little Ice Age was marked by the great Irish potato famine, which was triggered when the wetter and warmer climate promoted conditions conducive to the potato blight fungus. The summer and winter of

1845 were particularly wet and warm, which led to a rapid spread of the blight and the death of more than a million Irish in 1846.

Since the Little Ice Age, the world has experienced almost two centuries of relatively mild warm climates. The warmest periods occurred in the middle of the twentieth century, which also corresponds to the most productive years of modern agriculture. This recent warmth (the warmest time in a millennium) came to an end in the 1940s, and climate has been deteriorating ever since. In the early 1970s, severe droughts in Russia forced the country to buy huge amounts of grain from the United States, and droughts in Africa caused massive starvation seen on television by millions. The past few decades, however, have been hotter and drier than any other time in recorded history, suggesting that climate is changing again. We will look at the predictions for the future later in this chapter.

We humans like to think of ourselves as special and unique, and we like to think we control our destiny. But in fact, our history has been strongly affected by global climatic changes beyond our control. Advanced civilization would never have arisen from the hunter-gatherer societies that characterized most of human prehistory were it not for lucky accidents of climate, such as the Holocene Climatic Optimum. In our cosmic arrogance, we consider our culture to be the inevitable result of our intelligence, but geologic history shows that it was also an accident of favorable conditions as well.

The Sixth Extinction

Paleontologists recognize five major episodes of mass extinction over the past 450 million years, including the late Ordovician extinction 450 million years ago and the "mother of all mass extinctions" 250 million years ago at the end of the Permian, when 95% of life on earth vanished. The Cretaceous-Tertiary mass extinction discussed in chapter 2 was second only to the Permian extinction in intensity. The Eocene-Oligocene extinctions discussed in chapters 4 and 5 were not quite as impressive as the big five mass extinctions, but they are important nevertheless.

But now, in the last few centuries of the Holocene, the earth is experiencing a mass extinction that may be larger than any that has ever occurred on this planet. Richard Leakey and Roger Lewin have dubbed it "the sixth extinction." It became apparent over the past few centuries, as humans wiped out most of the native animals of isolated lands soon after they arrived. In some places, such as Polynesia, humans exterminated whole faunas of South Pacific islands for food. In other cases (especially when European explorers and settlers arrived), they hunted animals more for sport than for food; or the rats, dogs, goats, and cats they introduced wiped out the native species in these fragile habitats. Whatever the motive, the native species of most of the islands of the world are largely extinct now, and pace of extinction has accelerated in the past few centuries.

Human populations have also triggered extinctions on large continents as well. Europe long ago lost many of its native species as farming societies cut down ancient forests after the Middle Ages. As European settlers spread across North America, they felled the native trees and drove many plants and animals to extinction. Other species, like the passenger pigeon, were hunted excessively until they vanished, and the same almost happened to the bison. Now the human scourges of deforestation and excessive hunting are spreading to the great reservoirs of biodiversity, the tropical rain forests of the world. Back in 1979, Norman Myers estimated that 2% of the world's standing forests were being cut down each year, or about 80,000 square miles a year, almost an acre a second. Critics derided his estimates, but most recent surveys using sophisticated satellite imagery have confirmed them. Each year, an area of rain forest the size of Maine is lost. At this rate of deforestation, the tropical forests will lose about 10% of their remaining cover each decade, and they will vanish before the year 2050.

The cutting of these rain forests does even more damage than the cutting of temperate forests did in earlier centuries. For one thing, temperate forests grow back and regain their productivity with other kinds of plants. But all the nutrients in a tropical rain forest are locked up in the trees. Once they are cut down and hauled away, or burned, the tropical soils turn into barren red brick-hard laterites, and nothing can grow on them. The land becomes a wasteland, shedding tremendous amounts of soil into the rivers owing to uncontrolled erosion. The pristine forest cannot grow back, and the subsistence farmers are forced to move on and repeat the slash-and-burn process on another patch of rain forest—all to raise a handful of cattle so that we can have cheap hamburgers.

The secondary effect of the destruction of tropical rain forests is that a disproportionate number of species are driven to extinction, since tropical forests contain more than half of the world's species of plants and animals, although they cover only 7% of the land surface. For example, a mere 25 acres of rain forest may contain over 1,000 species of trees, and E.O. Wilson once counted no fewer than 43 species of ants on a single tree! Thus, the loss of an acre of rain forest may wipe out hundreds of species, whereas the same is not true of temperate forests. Many of these species are insects and other small organisms that have never even been named or studied by scientists, but have vanished from the planet before we even knew they existed. The estimates of how many species are lost each year vary widely depending on the method used, but they are all alarming—from estimates as low as 17,000 species per year to some as high as 100,000 species per year. Even moderate estimates suggest that one species is lost every five minutes! By contrast, paleontologists estimate that normal "background" rates of extinction are one species lost every four years, and even the big five mass extinctions did not produce rates much more than ten times this (i.e., ten species every four years). By contrast, the current rate of about 30,000 species lost per year is 120,000 times the normal background rate of extinction!

The effects are obvious when you travel to the tropics. In his book

African Silences, Peter Matthiessen surveyed tropical Africa, which he had visited many times in his younger years. Most of the great forests of West Africa (and their wildlife) are gone, replaced by barren wasteland and starving people. Even in the national parks and refuges, most of the wildlife is gone, hunted for "bush meat" by poachers. As Matthiessen put it, "the great silence that resounds from a wild land without sign of human life, from which all the great animals are gone, is something ominous. Mile after mile, we stare down in disbelief." The same is happening to the Congo jungle, the true heart of Africa's wildlife. The exotic African jungles of Stanley and Livingstone, of Joy Adamson and Dian Fossey and Jane Goodall, full of jungle beasts we know from the Tarzan movies and *Daktari,* have nearly vanished. Similar stories can be told of the loss of game in the East African savanna (even in the game parks) as rhinos and elephants are poached for their horns and ivory, and other animals are killed for meat or to protect cattle and crops. The East Asian rain forests are vanishing at even a greater rate, as is the Amazon rain forest, the largest reservoir of life on earth.

Most people do not care if a few unknown insect species vanish, but the effects reach throughout the food chain. Recent estimates (see Web links at http://www.well.com/user/davidu/extinction.html) show that 25% of all mammal species on earth, from the cute and cuddly pandas and chimpanzees and orangutans to the elegant tigers, giraffes, cheetahs, whales, hippos, elephants, and rhinos to many more species of rodents and monkeys that are not so popular, are threatened with extinction in the next few decades. Similar numbers can be seen in every other group: 33% of bird species, 20% of reptile species, 25% of amphibian species, and 35% of fish species. The devastation of marine life is equally frightening. Overfishing and destruction of the delicate coral reef habitat (typically by throwing dynamite overboard and catching whatever floats up dead) has endangered thousands of species of marine fishes, including sharks, tuna, many coral reef fishes, and even sea horses. The ruthless efficiency of modern industrial fishing techniques means that only 10% of the large oceanic fishes remain, and most of the world's fisheries are crashing at an alarming rate, even as fish becomes an even more important part of the human diet. In addition to animals, plants are suffering from the direct effects of being cut down and burned. The estimates are shocking: half of the earth's plant species face extinction, and even in relatively low-diversity areas like the United States, 29% of plant species are threatened. The list of potential losses of plants, from most species of conifers to palm trees and bamboo and most tropical trees, means that a tremendous number of useful and important species will be gone. In the near future most plants and animals will be known only from zoo and botanical garden specimens—or from pickled or dried specimens in jars and cabinets in museums.

Why should we care if we exterminate most of the world's wild plants and animals? Most biologists argue that these organisms have as much right to exist as we do and, for moral and philosophical reasons, deserve to share this planet with us. But there are practical reasons as well. Every year we discover new medicines and other important uses for wild plants and animals, many of them from the tropical rain forests. Yet most of these species are destroyed before we can study them and find out if they might be useful to us. Some people may not find the loss of a few species here and there troubling (although it is not a few species but wholesale decimation). But Paul and Anne Ehrlich posed an interesting analogy as a counterargument. Imagine that you are on an airplane, and you see one or two rivets fall out of the wing. One or two rivets may not make much difference, but as each additional rivet is lost, the wing becomes more fragile, until the plane (and our planet) crashes. As Doug Adams and Mark Carwardine wrote in *Last Chance to See,* "Even so, the loss of a few species may seem almost irrelevant compared to major environmental problems such as global warming or the destruction of the ozone layer. But while nature has considerable resilience, there is a limit to how far that resilience can be stretched. No one knows how close to the limit we're getting. The darker it gets, the faster we're driving. There is one last reason for caring, and I believe no other is necessary. It is certainly the reason why so many people have devoted their lives to protecting the likes of rhinos, parakeets, kakapos, and dolphins. And it is simply this: the world would be a poorer, darker, lonelier place without them" (Adams and Carwardine 1990: p. 211).

The Future?

The culprit in this horrendous mass extinction is not hard to identify. As the comic-page opossum Pogo once said, "We have met the enemy and he is us." Humans (or their domesticated animal proxies) are the direct or indirect cause of nearly all this global ecological catastrophe, and the root of the problem comes down to a simple dynamic: human population growth (fig. 9.4).

For the first 3 million years that *Homo* has lived on this planet, human populations were small and sustainable by the local resources that hunter-gatherers could provide. Even when *Homo* spread out of Africa and across Eurasia, there were never more than a few million humans on this planet. The beginnings of agriculture and civilization allowed a great expansion in human populations by increasing the food base, so that by 1 A.D., the world's population is estimated at about 300 million. But in the 1700s, the Industrial Revolution raised living standards and began to diminish the effects of famines and epidemics (at least in Europe), and population growth accelerated. By 1750, there were about 760 million people on this planet, and there were over 1 billion by 1800. About 86% of those people lived in Asia and Europe (65% in Asia alone). Then the expansion of industrialization increased Europe's share, so that by 1900, there were 1.6 billion people, 25% of which were in Europe. After the devastation of

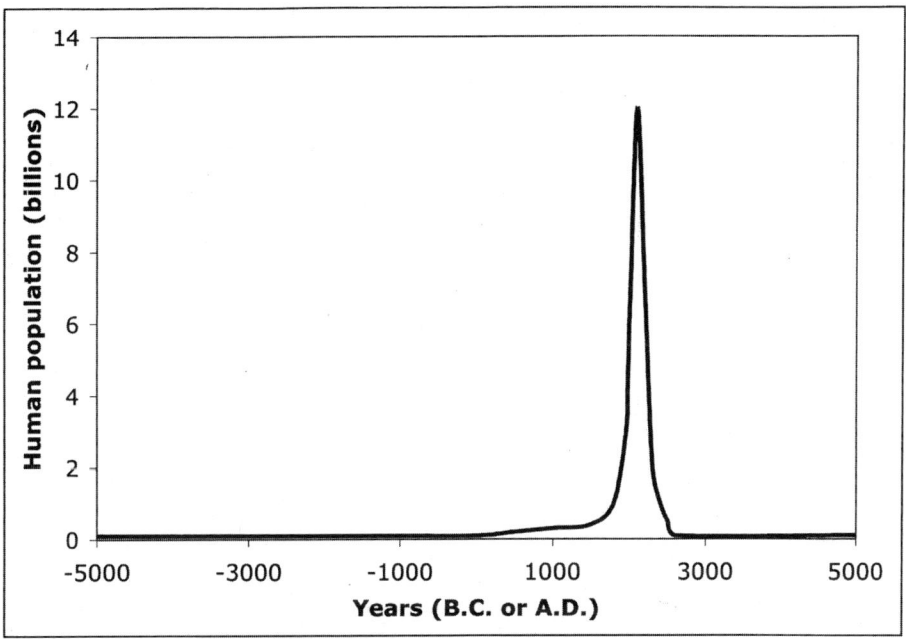

Figure 9.4. Human population growth. Drawn by the author.

World War II, population growth again expanded dramatically, this time in the underdeveloped countries of the world (especially those in Africa and Asia), as modern medicines and health care depressed the mortality rate. The third billion was reached in 1960, only 30 years after we reached 2 billion. The fourth billion came between 1960 and 1974 in only 14 years. The fifth billion was added by 1987, just 13 years later. By 1999, just 12 more years later, world population exceeded 6 billion. As I write this in 2004, we have already passed 6.4 billion, and we may exceed 7 billion before 2010, only 6 years later. As you read this, 14 to 15 babies are born every 6 seconds! Each hour there are 8,800 more mouths to feed; each year, there are more than 80 million more people than the year before!

For most of human history, population growth was impossible because of natural constraints on human life spans: war, famine, infant mortality, and disease. But as Western medicine has decreased mortality due to disease and as abundant food has reduced famine, populations have exploded. Unfortunately, we have not adopted the necessary birth controls to hold the growth in check, so the planet is becoming terribly overcrowded, with little or no real effort to stop it. The sad irony is that most of this overpopulation is occurring in underdeveloped countries that are least able to feed and support their people, and so their poverty increases. This situation leads to a moral dilemma: about 30% of the world's resources are used in the underdeveloped countries, which have 70% of the world's population, while our wasteful industrialized society has only 30% of the population but uses 70% of the resources. We cruelly tantalize developing countries with cars and televisions and other amenities that they cannot possibly attain with their population and economic problems. We would not have such wealth if we had not lived so wastefully at their expense (and at the expense of future generations) and exploited the bulk of the earth's resources before the developing world (and our children) could use them.

The population bomb is already turning this planet into a devastated wasteland, overrun by humans, cattle, and rats, with few native wild animals or plants. Whether we want this trend to continue is partly up to us, and whether we can practice birth control and restrain population growth before the traditional tools of the grim reaper (war, famine, and pestilence) do it for us. But it is also controlled by outside factors. Geologists and paleontologists take the long view on such things, noting that most Pleistocene interglacial episodes lasted only 10,000 years (fig. 9.2). Our present Holocene interglacial has already lasted that long, so if nothing else changes, we should be headed into the next glacial in the next century or so. This may be a gradual process, or if recent research on ice cores is correct, it may take place abruptly in a few decades—much faster than human societies will be able to cope.

However, natural processes may be trumped by our own interference again. Instead of heading into the next ice age, many scientists now think that we may instead overheat the planet and move into a "super-interglacial" thanks to global warming caused by our emission of greenhouse gases (fig. 9.2). Since the 1800s, atmospheric carbon dioxide levels have risen from about 280 ppm to over 345 ppm today, and they have increased over 9% just since 1958, when they were only 315 ppm! Yet even during the warmest interglacials of the Pleistocene, carbon dioxide levels did not exceed 300 ppm (fig. 9.5). Global temperatures have already increased by more than 1°C in the past century, yet as we saw above (fig. 9.3), the fluctuation between extreme glacial and interglacial conditions was no more than 5–6°C worldwide. Since we are already in an interglacial and near maximum warmth for the planet, the addition of more greenhouse gases will push us into a "super-greenhouse" as warm as the Cretaceous world, when sea levels rose and drowned the continents, and there were no polar ice caps (see chapter 3).

A warmer planet may not seem like such a bad thing at first glance, but there are plenty of negative consequences. Melting the polar ice caps causes a global rise in sea level. Even a 1-meter rise would drown some coastal cities and lowlands, and a 6-meter rise would drown all of New York and London and most of the world's coastal plains. Complete melting of the ice caps would cause a sea level rise of over 150 meters, which would drown nearly all the world's lowlands and coasts. St. Louis would become the mouth of the Mississippi as New Orleans vanished beneath the sea. Paleoclimatologists also predict that global weather patterns

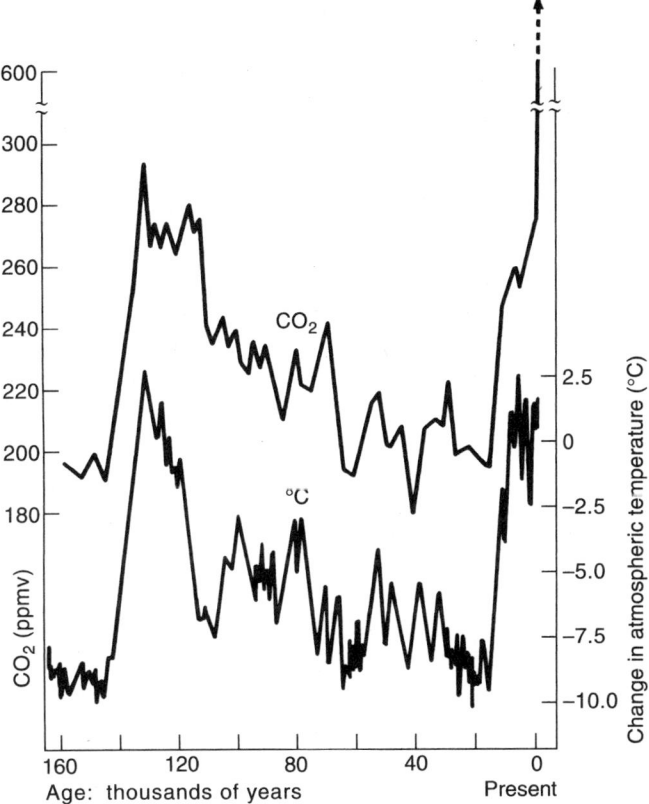

Figure 9.5. The last 160,000 years of global temperature and atmospheric carbon dioxide, as recorded by gas bubbles frozen into the Antarctic ice cap. The peak of the last interglacial (about 130,000 years ago) shows up clearly, as does the decline into the coldest period of the last glacial maximum about 20,000 years ago. Global temperatures then warmed steadily, reaching a stable level of the past few thousand years but marching upward in this century. Even more dramatic is the increase in carbon dioxide due to human burning of fossil fuels. Prior to the past two centuries, the concentration was only about 280 ppm (parts per million), but as of 1992, it had already reached 355 ppm, higher than any previous interglacial warm period. Many scientists project that the level will reach 600 ppm in less than 40 years, even with major efforts to cut back on our burning of fossil fuels. After Barnola et al. 1991.

would change, so that the "breadbasket" farm belts of the Northern Hemisphere would become too hot and dry for agriculture. Other areas might become wetter, but the balance of agricultural power would certainly change, and food production would not be able to keep up with human population growth.

There is another possibility: global warming may actually trigger the next ice age. As I pointed out in chapter 7, the modern global conveyor belt (fig. 7.11) of ocean currents brings warm waters of the Gulf Stream to the North Atlantic, keeping England's climate mild and preventing the ice sheets from advancing. But during the very end of the last glacial (the Younger Dryas, about 11,000 years ago), ice-core and deep-sea sediment records reveal that there was a rapid ice advance before the final retreat of the glaciers. Climatic modeling (Stocker and Schmittner 1997; Broecker 1999a,b) has shown that an episode of melting released huge volumes of fresh water from Greenland and Canada and Siberia, which formed a freshwater lid on the North Atlantic. This, in turn, deflected the denser warm salty Gulf Stream current into deeper waters and prevented its warmth or moisture from reaching the North Atlantic. A rapid glaciation ensued, developing in just a few years or decades. Many climatologists think that this possibility is realistic for our own future, since our global interference with greenhouse gases has rapidly melted many of the Northern Hemisphere glaciers and is now releasing huge volumes of fresh water into the North Atlantic.

We could go on and on discussing other disturbing trends of the Holocene, from the other types of pollution and environmental damage we are causing, to the limits of growth of societies as our resources are exhausted (see Deffeyes 2001). The limits of oil and gas are already in sight as I write this in 2005. But the overall message is clear. Humans are not independent of this planet, nor are they entitled to do anything they want to it. Instead, we are one species among millions that have lived during the Cenozoic. We have not even been around more than 100,000 years, whereas most mammalian species survive much longer. We would not be here if it were not for a long and fortunate chain of contingent circumstances: the extinction of the non-avian dinosaurs, the favorable changes in climate, and the events that allowed our primate ancestors to survive when most other species did not. Our entire civilization is a product of a short-term warming called the Holocene, and now we are looking at the end of the Holocene becoming either too warm or too cold. We can no longer assume that the world will be hospitable to our excessive demands on it. We have pushed the planet to its limits and polluted our nest by overpopulation and overexploitation of its resources, and we are already paying the price. Whether we will survive much longer will be partially depend not only on whether we make wise decisions about population control and economics but also on the earth's own changes, which we do not control.

From the geological perspective, humans are but a brief blip (fig. 9.4) in the billions of years of geological time. In 1985, Prosh and McCracken published a satirical article in the normally staid *Geology* magazine, speculating on what the post-Holocene epoch of the Cenozoic (following the extinction of humans) might be called. Following the trend established by the Paleocene through Holocene, they playfully suggested such names as "Nothingcene," "Changeofcene," and "Weshouldhavecene."

For Further Reading

Adams, D., and M. Carwardine. 1990. *Last Chance to See*. New York: Harmony Books.

Bryson, R. A., and T. J. Murray. 1977. *Climates of Hunger: Mankind and the World's Changing Weather*. Madison: University of Wisconsin Press.

Deffeyes, K. F. 2001. *Hubbert's Curve: The Impending World Energy Shortage*. Princeton, N.J.: Princeton University Press.

Ehrlich, P., and A. Ehrlich. 1990. *The Population Explosion*. New York: Simon and Schuster.

Gore, A. 1992. *Earth in the Balance: Ecology and Human Spirit*. New York: Penguin.

Gribben, J. 1982. *Future Weather and the Greenhouse Effect*. New York: Delacorte.

Ladurie, E. L. R. 1971. *Times of Feast, Times of Famine*. New York: Doubleday.

Lamb, H. H. 1972. *Climate: Past, Present, and Future*. London: Methuen.

———. 1982. *Climate History and the Modern World*. London: Methuen.

Leakey, R., and R. Lewin. 1995. *The Sixth Extinction*. New York: Doubleday.

Matthiessen, P. 1991. *African Silences*. New York: Random House.

Mithen, S. 2004. *After the Ice: A Global Human History*. Cambridge, Mass.: Harvard University Press.

Oppenheimer, J., and R. Boyle. 1990. *Dead Heat: The Race against the Greenhouse Effect*. New York: Basic Books.

Pielou, E. C. 1991. *After the Ice Age: The Return of Life to Glaciated North America*. Chicago: University of Chicago Press.

Prothero, D. R., and R. H. Dott Jr. 2003. *Evolution of the Earth*. 7th ed. New York: McGraw-Hill.

Schneider, S. H. 1990. *Global Warming: Are We Entering the Greenhouse Century?* San Francisco: Sierra Club.

Weiner, J. 1990. *The Next One Hundred Years: Shaping the Future of Our Living Earth*. New York: Bantam Books.

Bibliography

Adams, C. G., D. E. Lee, and B. R. Rosen. 1990. Conflicting isotopic and biotic evidence for tropical sea-surface temperatures during the Tertiary. Palaeogeography, Palaeoclimatology, Palaeoecology 77:289–313.

Adams, D., and M. Carwardine. 1990. *Last Chance to See.* New York: Harmony Books.

Addicott, W. O. 1969. Tertiary climatic change in the marginal northeastern Pacific Ocean. *Science* 165:583–586.

———. 1970. Tertiary paleoclimatic trends in the San Joaquin Basin, California. *U.S. Geological Survey Professional Paper* 644-D:1–19.

———. 1976. Neogene molluscan stages of Oregon and Washington. In A. E. Fritsche, H. TerBest Jr., and W. W. Wornardt, eds., *The Neogene Symposium,* pp. 95–115. San Francisco: SEPM, Pacific Section.

Agassiz, L. 1840. *Etudes sur les glaciers.* Neuchâtel: Privately published.

Agusti, J., and M. Anton. 2002. *Mammoths, Sabertooths, and Hominids: 65 Million Years of Mammalian Evolution in Europe.* New York: Columbia University Press.

Alroy, J. 2002. Extraterrestrial bolide impacts and biotic change in North American mammals. *Journal of Vertebrate Paleontology* 22 (3): 32A.

Alvarez, L. W., W. Alvarez, F. Asaro, and H. V. Michel. 1980. Extraterrestrial cause for the Cretaceous-Tertiary extinction. *Science* 208:1095–1108.

Alvarez, W., F. Asaro, H. V. Michel, and L. W. Alvarez. 1982. Iridium anomaly approximately synchronous with terminal Eocene extinctions. *Science* 216:886–888.

Archer, M., S. J. Hand, and H. Godthelp. 1995. Tertiary environmental and biotic change in Australia. In E. S. Vrba, G. H. Denton, T. C. Partridge, and L. H. Burckle, eds., *Paleoclimate and Evolution, with Emphasis on Human Origins,* pp. 77–90. New Haven, Conn.: Yale University Press.

———. 2001. *Australia's Lost World: Prehistoric Animals of Riversleigh.* Bloomington: Indiana University Press.

Archibald, J. D. 1982. A study of Mammalia and geology across the Cretaceous-Tertiary boundary in Garfield County, Montana. *University of California Publications in Geological Sciences* 122:1–286.

———. 1993. The importance of phylogenetic analysis for the assessment of species turnover: A case history of Paleocene mammals in North America. *Paleobiology* 19:1–27.

———. 1996. *Dinosaur Extinctions and the End of an Era: What the Fossils Say.* New York: Columbia University Press.

Archibald, J. D., and L. J. Bryant. 1990. Differential Cretaceous-Tertiary extinctions of non-marine vertebrates: Evidence from northeastern Montana. *Geological Society of America Special Paper* 247:549–562.

Archibald, J. D., W. A. Clemens, P. D. Gingerich, D. W. Krause, E. H. Lindsay, and K. D. Rose. 1987. First North American land mammal ages of the Cenozoic Era. In M. O. Woodburne, ed., *Cenozoic Mammals of North America: Geochronology and Biostratigraphy*, pp. 24–76. Berkeley: University of California Press.

Asaro, F., L. W. Alvarez, W. Alvarez, and H. V. Michel. 1982. Geochemical anomalies near the Eocene/Oligocene and Permian/ Triassic boundaries. *Geological Society of America Special Paper* 190:517–528.

Askin, R. A. 1992. Late Cretaceous–early Tertiary Antarctic outcrop evidence for past vegetation and climates. *Antarctic Research Series* 56:61–73.

Aubry, M.-P. 1992. Late Paleogene calcareous nannoplankton evolution: A tale of climatic deterioration. In D. R. Prothero and W. A. Berggren, eds., *Eocene-Oligocene Climatic and Biotic Evolution*, pp. 272–309. Princeton, N.J.: Princeton University Press.

———. 1998. Early Paleogene calcareous nannoplankton evolution: A tale of climatic amelioration. In M.-P. Aubry, S. G. Lucas, and W. A. Berggren, eds., *Late Paleocene–Early Eocene Climatic and Biotic Events in the Marine and Terrestrial Records*, pp. 158–203. New York: Columbia University Press.

Aubry, M.-P., F. M. Gradstein, and L. F. Jansa. 1990. The late early Eocene Montagnais meteorite: No impact on biotic diversity. *Micropaleontology* 36 (2): 164–172.

Aubry, M.-P., S. G. Lucas, and W. A. Berggren, eds. 1998. *Late Paleocene–Early Eocene Climatic and Biotic Events in the Marine and Terrestrial Records*. New York: Columbia University Press.

Axelrod, D. I. 1985. Rise of the grassland biome, central North America. *Botanical Review* 51:163–201.

———. 1992. Climatic pulses, a major factor in legume evolution. In P. S. Herendeenand and D. L. Dilcher, eds., *Advances in Legume Systematics*, pt. 4, *The Fossil Record*, pp. 259–279. Kew, England: Royal Botanic Gardens.

Baldauf, J. G. 1992. Middle Eocene through early Miocene diatom floral turnover. In D. R. Prothero and W. A. Berggren, eds., *Eocene-Oligocene Climatic and Biotic Evolution*, pp. 310–326. Princeton, N.J.: Princeton University Press.

Baldauf, J. G., and J. A. Barron. 1990. Evolution of biosiliceous sedimentation patterns—Eocene through Quaternary: paleoceanographic response to polar cooling. In U. Bleil and J. Thiede, eds., *Geological History of the Polar Oceans: Arctic Versus Antarctic*, pp. 575–607. Amsterdam: Kluwer Academic Publishers.

Barker, P. F., and J. Burrell. 1977. The opening of the Drake Passage. *Marine Geology* 25:15–34.

———. 1982. The influence upon Southern Ocean circulation sedimentation and climate of the opening of Drake Passage. In C. Craddock, ed., *Antarctic Geoscience*, pp. 377–385. Madison: University of Wisconsin Press.

Barnes, L. G. 1984. Whales, dolphins and porpoises—origin and evolution of the Cetacea. *University of Tennessee, Department of Geological Sciences, Studies in Geology* 8:139–154.

———. 1989. A new enaliarctine pinniped from the Astoria Formation, Oregon, and a classification of the Otariidae (Mammalia: Carnivora). *Natural History Museum of Los Angeles County Contributions in Science* 403:1–26.

Barnola, J. M., M. Bender, and T. Sowers. 1991. *Trends '91: A Compendium of Data on Global Change.* ORNL/CDIAC-46, Carbon Dioxide Information Analysis Center, Oak Ridge National Laboratory, Tennessee.

Barnosky, A. D., P. L. Koch, R. S. Feranec, S. L. Wing, and A. B. Shabel. 2004. Assessing the causes of late Pleistocene extinctions on the continents. *Science* 306:70–75.

Barrera, E., and G. Keller. 1990. Foraminiferal stable isotope evidence for gradual decrease of marine productivity and Cretaceous species survivorship in the earliest Danian. *Paleoceanography* 5:867–870.

Barrett, P.J., C. J. Adams, W. C. McIntosh, C. C. Swisher III, and G. S. Wilson. 1992. Geochronological evidence supporting Antarctic deglaciation three million years ago. *Nature* 359:816–818.

Barron, E. J. 1984. Climatic implications of the variable obliquity explanations of Cretaceous-Paleogene high-latitude floras. *Geology* 12:595–598.

———. 1987. Eocene equator-to-pole surface ocean temperatures: A significant climate problem? *Paleoceanography* 2:729–732.

Barron, J., B. Larsen, and the Leg 119 Shipboard Scientific Party. 1989. *Proceedings of the Ocean Drilling Program* 119:1–939.

Beard, K. C. 1998. East of Eden: Asia as an important center of taxonomic origination in mammalian evolution. *Bulletin of the Carnegie Museum of Natural History* 34:5–39.

———. 2004. *The Hunt for the Dawn Monkey: Unearthing the Origin of Monkeys, Apes, and Humans.* Berkeley: University of California Press.

Beck, M. W., 1996. On discerning the causes of late Pleistocene megafaunal extinctions. *Paleobiology* 22:91–103.

Bell, C. J., E. L. Lundelius Jr., A. D. Barnosky, R. W. Graham, E. H. Lindsay, D. R. Ruez Jr., H. A. Semken Jr., S. D. Webb, and R. J. Zakrzewski. 2004. The Blancan, Irvingtonian, and Rancholabrean land mammal ages. In M. O. Woodburne, ed., *Late Cretaceous and Cenozoic Mammals of North America: Biostratigraphy and Geochronology,* pp. 377–385. Berkeley: University of California Press.

Bender, M. L., and L. D. Keigwin. 1979. Speculations about the upper Miocene change in abyssal Pacific dissolved bicarbonate ^{13}C. *Earth and Planetary Sciences Letters* 45:383–393.

Berggren, W. A. 1971. Tertiary boundaries and correlations. In B. M. Funnell and W. R. Riedel, eds., *The Micropaleontology of Oceans,* pp. 693–809. Cambridge: Cambridge University Press.

———. 1972. Late Pliocene–Pleistocene glaciation. *Initial Reports of the Deep Sea Drilling Project* 12:953–963.

———. 1982. Role of ocean gateways in climatic change. In W. Berger and J. C. Crowell, eds., *Climate in Earth History,* pp. 118–285. Washington, D.C.: National Academy of Sciences.

Berggren, W. A., and B. U. Haq. 1976. The Andalusian Stage (late Miocene); biostratigraphy, biochronology and paleoecology. *Palaeogeography, Palaeoclimatology, Palaeoecology* 20:67–129.

Berggren, W. A., and C. D. Hollister. 1974. Paleogeography, paleobiogeography and the history of circulation in the Atlantic Ocean. *SEPM Special Publication* 20:126–176.

———. 1978. Plate tectonics and paleocirculation; commotion in the ocean. *Tectonophysics* 38:11–48.

Berggren, W. A., D. V. Kent, and J. Hardenbol, eds. 1995. *Geochronology, Time Scales, and Stratigraphic Correlation.* SEPM Special Publication 54. Tulsa, Okla.: SEPM.

Berner, R. A., A. C. Lasaga, and R. M. Garrels. 1983. The carbonate-silicate geochemical cycle and its effect on atmospheric carbon dioxide over the past 100 million years. *American Journal of Science* 283:641–683.

Berta, A. 1981. The Plio-Pleistocene hyaena *Chasmoporthetes ossifragus* from Florida. *Journal of Vertebrate Paleontology* 1:341–356.

Berta, A., and C. E. Ray. 1990. Skeletal morphology and locomotor capabilities of the archaic pinniped *Enaliarctos mealsi. Journal of Vertebrate Paleontology* 10:141–157.

Berta, A., C. E. Ray, and A. R. Wyss. 1989. Skeleton of the oldest known pinniped, *Enaliarctos. Science* 244:60–62.

Beyrich, H. E. von. 1954. Über die Stellung die hessischen Tertiar-bildungen. *Konigliche Preussische Akademie Wissenschaften Berlin Monatsheft* 1854:664–666.

Birkelund, T., and E. Hakansson. 1982. The terminal Cretaceous in Boreal shelf seas: A multicausal event. *Geological Society of America Special Paper* 190:373–384.

Birkenmajer, K. 1987. Tertiary glacial and interglacial deposits, South Shetland Islands, Antarctica: Geochronology versus biostratigraphy. *Bulletin of the Polish Academy of Science, Earth Science* 36:133–145.

Birkenmajer, K., A. Gazdzicki, K. P. Krajewski, A. Przybycin, A. Solecki, A. Tatur, and H. I. Yoon. 2005. First Cenozoic glaciers in West Antarctica. *Polish Polar Research* 26:3–12.

Blackwelder, B. W. 1981. Late Cenozoic stages and molluscan zones of the U.S. Middle Atlantic Coastal Plain. *Journal of Paleontology Memoir* 12:1–34.

Boersma, A., I. Premoli-Silva, and N. J. Shackleton. 1987. Atlantic Eocene planktonic foraminiferal paleohydrographic indicators and stable isotope paleoceanography. *Paleoceanography* 2 (3): 287–331.

Boersma, A., I. Premoli Silva, and P. Hallock. 1998. Trophic models for the well-mixed and poorly mixed warm oceans across the Paleocene/Eocene boundary. In M.-P. Aubry, S. G. Lucas, and W. A. Berggren, eds., *Late Paleocene–Early Eocene Climatic and Biotic Events in the Marine and Terrestrial Records,* pp. 118–285. New York: Columbia University Press.

Bolles, E. B. 1999. *The Ice Finders: How a Poet, a Professor, and a Politician Discovered the Ice Age.* New York: Counterpoint.

Bonnefille, R. 1995. A reassessment of the Plio-Pleistocene pollen record of East Africa. In E. S. Vrba, G. H. Denton, T. C. Partridge, and L. H. Burckle, eds., *Paleoclimate and Evolution, with Emphasis on Human Origins,* pp. 299–310. New Haven, Conn.: Yale University Press.

Bottomley, R., and D. York. 1988. Age measurements of the submarine Montagnais impact crater and the periodicity question. *Geophysical Research Letters* 14 (12): 1409–1412.

Bottomley, R., R. A. F. Grieve, D. York, and V. Masaitis. 1997. The age of the Popigai impact event and its relation to events at the Eocene/Oligocene boundary. *Nature* 388:365–368.

Boulter, M. C. 1984. Palaeobotanical evidence for land-surface temperature in the European Paleogene. In Brenchley, P., ed., *Fossils and Climate,* pp. 35–47. Chichester: Wiley.

Boulter, M. C., and R. N. L. B. Hubbard. 1982. Objective paleoecological and biostratigraphic interpretation of Tertiary palynological data by multivariate statistical analysis. *Palynology* 6:55–68.

———. 1983. Reconstruction of Palaeogene climate from palynological evidence. *Nature* 301:147–150.

Bown, T. M., and M. J. Kraus. 1981. Lower Eocene alluvial paleosols (Willwood Formation, northwest Wyoming, U.S.A.) and their significance for paleoecology, paleoclimatology, and basin analysis. *Palaeogeography, Palaeoclimatology, Palaeoecology* 34:1–30.

———. 1987. Integration of channel and floodplain suites: 1. Developmental sequence and lateral relations of alluvial paleosols. *Journal of Sedimentary Petrology* 57:587–601.

Brinkhuis, H. 1992. Late Paleogene dinoflagellate cysts with special reference to the Eocene/Oligocene boundary. In D. R. Prothero and W. A. Berggren, eds., *Eocene-Oligocene Climatic and Biotic Evolution,* pp. 327–340. Princeton, N.J.: Princeton University Press.

Broecker, W. S. 1987. The biggest chill. *Natural History* 96:74–82.

———. 1991. Keeping global change honest. *Global Biogeochemical Cycles* 5:191–192.

———. 1999a. Abrupt climate change: Causal constraints provided by the paleoclimate record. *Earth Science Reviews* 51:137–154.

———. 1999b. What if the conveyor were to shut down? Reflections on a possible outcome of the great global experiment. *GSA Today* 9 (1): 1–7.

Broecker, W. S., and G. H. Denton. 1989. The role of ocean-atmosphere reorganizations in glacial cycles. *Geochimica et Cosmochimica Acta* 52:2465–2501.

———. 1990. What drives glacial cycles? *Scientific American* (January): 48–56.

Brunet, M. 1977. Les mammifères et le problème de la limite Eocène-Oligocène en Europe. *Géobios, Mémoires Spéciaux* 1:11–27.

Brunet, M., et al. 2002. A new hominid from the upper Miocene of Chad, central Africa. *Nature* 418:145–151.

Brunner, C. A. 1978. Late Neogene history recorded by sedimentation in the Straits of Florida. *Geological Society of America Abstracts with Programs* 10 (7): 373.

Bryan, J. R., and D. S. Jones. 1989. Fabric of the Cretaceous-Tertiary marine macrofaunal transition at Braggs, Alabama. *Palaeogeography, Palaeoclimatology, Palaeoecology* 69:279–301.

Bryson, R. A., and T. J. Murray. 1977. *Climates of Hunger: Mankind and the World's Changing Weather.* Madison: University of Wisconsin Press.

Brysse, K. 2004. Off-limits to no one: Vertebrate paleontologists and the Cretaceous-Tertiary mass extinction. Ph.D. dissertation, University of Alberta, Alberta, Canada.

Budantsev, L. Y. 1992. Early stage of formation and dispersal of the temperate flora in the boreal region. *The Botanical Review* 58:1–48.

Budd, A. F., T. A. Stemann, and K. G. Johnson. 1994. Stratigraphic distribution of Neogene to Recent Caribbean reef corals. *Journal of Paleontology* 68:951–977.

Burckle, L. H. 1995. Current issues in Pliocene paleoclimatology. In E. S. Vrba, G. H. Denton, T. C. Partridge, and L. H. Burckle, eds., *Paleoclimate and*

Evolution, with Emphasis on Human Origins, pp. 3–7. New Haven, Conn.: Yale University Press.

Burckle, L. H., and F. Akiba. 1978. Implications of late Neogene freshwater sediment in the Sea of Japan. *Geology* 6:123–127.

Callahan, J. E. 1971. Velocity structure and flux of the Antarctic Circumpolar Current of South Australia. *Journal of Geophysical Research* 76:5859–5870.

Campbell, D. C., and M. R. Campbell. 2003. Biotic patterns in Eocene-Oligocene mollusks of the Atlantic Coastal Plain, U.S.A. In D. R. Prothero, L. C. Ivany, and E. A. Nesbitt, eds., *From Greenhouse to Icehouse: The Marine Eocene-Oligocene Transition*, pp. 341–353. New York: Columbia University Press.

Campbell, K. C., Jr., C. D. Frailey, D. R. Prothero, and M. Heizler. 2001. Upper Cenozoic chronostratigraphy of the Amazon Basin: Implications for the Great American Interchange. *Geology* 29:595–598.

Campbell, L. D., S. Campbell, D. Colquohoun, J. Ernissee, and W. Abbott. 1975. Plio-Pleistocene faunas of the central Carolina Coastal Plain. *Geologic Notes, South Carolina State Development Board, Division of Geology* 19:52–124.

Capetta, H., J.-J. Jaeger, M. Sabatier, B. Sige, J. Sudre, and M. Vianey-Liaud. 1978. Découverte dans le Paleocène du Maroc des plus anciens mammifères eutheriens d'Afrique. *Géobios* 11:257–263.

Carozzi, A. V., trans. 1967. *Studies on Glaciers; Preceded by the Discourse of Neuchâtel*, by Louis Agassiz. New York: Hafner.

Carter, B. D. 2003. Diversity patterns in Eocene and Oligocene echinoids of the southeastern United States. In D. R. Prothero, L. C. Ivany, and E. A. Nesbitt, eds., *From Greenhouse to Icehouse: The Marine Eocene-Oligocene Transition*, pp. 365–385. New York: Columbia University Press.

Case, J. A. 1988. Paleogene floras from Seymour Island, Antarctic Peninsula. *Geological Society of America Memoir* 169:523–530.

Cavelier, C. 1979. La limite Eocène-Oligocène en Europe occidentale. *Sciences Géologiques, Mémoire (Institut de Géologie, Strasbourg)* 54:1–280.

Cavender, T. M. 1986. Review of the fossil history of North American freshwater fishes. In C. H. Hocutt and E. O. Wiley, eds. *The Zoogeography of North American Freshwater Fishes*, pp. 699–724. New York: Wiley.

Cerling, T. E. 1992. Development of grasslands and savannas in East Africa during the Neogene. *Palaeogeography, Palaeoclimatology, Palaeoecology* 97:241–247.

Cerling, T. E., J. M. Harris, B. J. MacFadden, M. G. Leakey, J. Quade, V. Eisenmann, and J. R. Ehleringer. 1997. Global vegetation change through the Miocene/Pliocene boundary. *Nature* 389:153–158.

Chandler, M. 1993. GCM simulations of the Pliocene climate: Feedbacks, ocean transports, and CO_2. In R. Thompson, ed., *Pliocene Terrestrial Environments and Data/Model Comparisons*. USGS Open-File Report No. 94-23.

———. 1994. Global climate model simulations of the Middle Pliocene, continued. In S. Ishman, ed., *Pliocene High-Latitude Climate Records*. USGS Open-File Report No. 94-588.

Chandler, M. A., D. Rind, and R. S. Thompson. 1994. Joint investigations of the middle Pliocene climate II: GISS GCM Northern Hemisphere results. *Palaeogeography, Palaeoclimatology, Palaeoecology* 9:197–219.

Chiappe, L. M. 1995. The first 85 million years of avian evolution. *Nature* 378:349–355.

Chinzei, K. 1978. Neogene molluscan faunas in the Japanese islands: An ecological and zoogeographic synthesis. *The Veliger* 21:155–170.

Cifelli, R. 1969. Radiation of the Cenozoic planktonic foraminifera. *Systematic Zoology* 18:154–168.

Clementz, M. T., K. A. Hoppe, and P. L. Koch. 2003. A paleoecological paradox: The habitat and dietary preferences of the extinct tethythere *Desmostylus,* inferred from stable isotope analysis. *Paleobiology* 29:506–519.

Coates, A. G., and J. B. C. Jackson. 1985. Morphological themes in the evolution of clonal and aclonal marine invertebrates. In J. B. C. Jackson, L. W. Buss, and R. E. Cook, eds., *Population Biology and Evolution of Clonal Organisms,* pp. 67–105. New Haven, Conn.: Yale University Press.

Coates, A. G., J. B. C. Jackson, L. S. Collins, T. M. Cronin, H. J. Dowsett, L. M. Bybell, P. Jung, and J. A. Obando. 1992. Closure of the Isthmus of Panama: The near-shore marine record of Costa Rica and western Panama. *Geological Society of America Bulletin* 104:814–828.

Coccioni, R., and S. Galeotti. 1994. K-T boundary extinction: Geologically instantaneous or gradual event? Evidence from deep-sea benthic foraminifera. *Geology* 22:779–782.

Coccioni, R., D. Basso, H. Brinkhuis, S. Galeotti, S. Gardin, S. Monechi, and S. Spezzaferri. 2000. Marine biotic signals across a late Eocene impact layer at Massignano, Italy: Evidence for long-term environmental perturbations? *Terra Nova* 12:258–263.

Collinson, M. E. 1983. Palaeofloristic assemblages and palaeoecology of the lower Oligocene Bembridge Marls, Hamstead Ledge, Isle of Wight. *Botanical Journal of the Linnean Society* 86:177–225.

———. Vegetational and floristic changes around the Eocene/Oligocene boundary in western and central Europe. In D. R. Prothero and W. A. Berggren, eds., *Eocene-Oligocene Climatic and Biotic Evolution,* pp. 437–450. Princeton, N.J.: Princeton University Press.

Collinson, M. E., and J. J. Hooker. 1987. Vegetational and mammalian faunal changes in the early Tertiary of southern England. In E. M. Friis, W. G. Chaloner, and P. R. Crane, eds., *The Origins of Angiosperms and Their Biological Consequences,* pp. 295–304. Cambridge: Cambridge University Press.

Collinson, M. E., K. Fowler, and M. C. Boulter. 1981. Floristic changes indicate a cooling climate in the Eocene of southern England. *Nature* 291:315–317.

Coombs, M. C. 1983. Large mammalian clawed herbivores: A comparative study. *Transactions of the American Philosophical Society* 73 (7): 1–96.

Cope, E. D. 1884. The Vertebrata of the Tertiary formations of the West. Book I. Report of the United States Geological Survey of the Territories. III. Washington, D.C.: U.S. Geological Survey.

Craddock, C., and C. D. Hollister. 1976. *Initial Reports of the Deep Sea Drilling Project* 35:723–743.

Creber, G. T., and W. G. Chaloner. 1984. Climatic indications from growth rings in fossil woods. In P. J. Brenchley, ed., *Fossils and Climate,* pp. 49–74. New York: Wiley.

Crouch, E. and H. Visscher. 2003. Terrestrial vegetation record across the initial Eocene thermal maximum at the Tawanui marine section, New Zealand. *Geological Society of America Special Paper* 369:351–364.

Crouch, E., C. Heilmann-Clausen, H. Brinkhuis, H. E. G. Morgans, K. M.

Rogers, H. Egger, and B. Schmitz. 2001. Global dinoflagellate event associated with late Paleocene thermal maximum. *Geology* 29:289–376.

Crowley, T. J. 1991. Modeling Pliocene warmth. *Quaternary Science Review* 10:275–282.

Cuvier, G. 1804. Sur les espèces d'animaux dont proviennent les os fossiles répandus dans la pierre à plâtre des environs de Paris. *Annales Musée Histoire Naturelles*, Paris, ser. 3: 275–472.

———. 1818. Analyse des travaux de l'Académie royale des sciences, pour les années 1813, 1814, 1815: Partie physique. *Mém. Acad. Sci. Paris* 1813–15: cxvii–ccxxxv.

Darwin, C. R. 1839. *The Voyage of the H.M.S. Beagle*. London: Charles Murray.

Davies, R., J. Cartwright, J. Pike, and C. Line. 2001. Early Oligocene initiation of North Atlantic Deep Water formation. *Nature* 410:917–920.

DeConto, R. M., and D. Pollard. 2003. Rapid Cenozoic glaciation of Antarctica induced by declining atmospheric CO_2. *Nature* 421:245–249.

Deffeyes, K. F. 2001. *Hubbert's Curve: The Impending World Energy Shortage*. Princeton, N.J.: Princeton University Press.

Deming, D. 1999. On the possible influence of extraterrestrial volatiles on earth's climate and the origin of the oceans. *Palaeogeography, Palaeoclimatology, Palaeoecology* 146:33–51.

Dial, K. P. 2003. Wing-assisted incline running and the evolution of flight. *Science* 299:402–405.

Dickens, G. R., J. R. O'Neill, D. K. Rea, and R. M. Owen. 1995. Dissociation of oceanic methane hydrate as a cause of the carbon isotope excursion at the end of the Paleocene. *Paleoceanography* 10:965–971.

Dickens, G. R., C. K. Paull, P. Wallace, and the ODP Leg 164 Scientific Party. 1997. Direct measurement of *in situ* methane quantities in large gas hydrate reservoir. *Nature* 385:426–428.

Dickens, G. R., M. M. Castillo, and J. C. G. Walker. 1998. A blast of gas in the latest Paleocene: Simulating first-order effects of massive dissociation of oceanic methane hydrate. *Geology* 25:258–262.

Dickens, G. R., T. Fewless, E. Thomas, and T. J. Bralower. 2003. Excess barite accumulation during the Paleocene-Eocene Thermal Maximum: Massive input of dissolved barium from seafloor gas hydrate reserves. *Geological Society of America Special Paper* 369:11–24.

Diester-Haass, L., and R. Zahn. 1996. The Eocene-Oligocene transition in the Southern Ocean: History of water masses, circulation, and biological productivity inferred from high-resolution records of stable isotopes and benthic foraminiferal abundances. *Geology* 24:16–20.

Dingus, L., and T. Rowe. 1998. *The Mistaken Extinction: Dinosaur Evolution and the Origin of Birds*. New York: W.H. Freeman.

Dockery, D. T., III. 1986. Punctuated succession of Paleogene mollusks in the northern Gulf Coastal Plain. *Palaios* 1:582–589.

———. 1998. Molluscan faunas across the Paleocene/Eocene series boundary in the North American Gulf Coastal Plain. In M.-P. Aubry, S. G. Lucas, and W. A. Berggren, eds. 1998. *Late Paleocene–Early Eocene Climatic and Biotic Events in the Marine and Terrestrial Records*, pp. 296–322. New York: Columbia University Press.

Dockery, D. T., III, and P. Lozouet. 2003. Molluscan faunas across the Eocene/Oligocene boundary in the North American Gulf Coastal Plain, with

comparison to those of the Eocene and Oligocene of France. In D. R. Prothero, L. C. Ivany, and E. A. Nesbitt, eds., *From Greenhouse to Icehouse: The Marine Eocene-Oligocene Transition,* pp. 303–340. New York: Columbia University Press.

Domning, D. P. 1994. The terrestrial posture of desmostylians. *Smithsonian Contributions to Paleobiology* 93:83–98.

———. 2001. The earliest known fully quadrupedal sirenian. *Nature* 413:625–627.

Domning, D. P., C. E. Ray, and M. C. McKenna. 1986. Two new Oligocene desmostylians and a discussion of tethytherian systematics. *Smithsonian Contributions to Paleobiology* 59:1–56.

Dorf, E. 1960. Tertiary fossil forests of Yellowstone National Park, Wyoming. *Billings Geological Society Annual Field Conference* 11:253–360.

Dowsett, H. J., T. M. Cronin, R. Z. Poore, R. S. Thompson, R. C. Whatley, and A. M. Wood 1992. Micropaleontological evidence for increased meridional heat transport in the North Atlantic Ocean during the Pliocene. *Science* 258:1133–1135.

Dowsett, H. J., R. S. Thompson, J. A. Barron, T. M. Cronin, S. E. Ishman, R. Z. Poore, D. A. Willard, and T. R. Holtz Jr. 1994. Joint investigations of the middle Pliocene climate I: PRISM palaeoenvironmental reconstructions. *Palaeogeography, Palaeoclimatology, Palaeoecology* 9:169–195.

Dumont, A. 1849. Rapport sur la carte géologique du Royaume. *Bulletin de Académie Royale de Belgique* 16 (2): 351–373.

Durham, J. W. 1950. Cenozoic marine climates of the Pacific Coast. *Geological Society of America Bulletin* 61:1243–1264.

Eaton, J. G., J. I. Kirkland, and K. Doi. 1989. Evidence of reworked Cretaceous fossils and their bearing on the existence of Tertiary dinosaurs. *Palaios* 4:281–286.

Edwards, A. R. 1975. Southwest Pacific Cenozoic paleogeography and an integrated Neogene paleocirculation model. *Initial Reports of the Deep-Sea Drilling Project* 30:667–684.

Eicher, D. L. 1976. *Geologic Time.* 2nd ed. Englewood Cliffs, N.J.: Prentice-Hall.

Eldholm, O., and E. Thomas. 1993. Environmental impact of volcanic margin formation. *Earth and Planetary Sciences Letters* 117:319–329.

Emry, R. J., S. G. Lucas, L. Tyutkova, and B. Wang. 1998. The Ergilian-Shandgolian (Eocene-Oligocene) transition in the Zaysan Basin, Kazakstan. *Bulletin of the Carnegie Museum of Natural History* 34:298–312.

Ehrlich, P., and A. Ehrlich. 1990. *The Population Explosion.* New York: Simon and Schuster.

Estes, R., and J. H. Hutchison. 1980. Eocene lower vertebrates from Ellesmere Island, Canadian Arctic Archipelago. *Palaeogeography, Palaeoclimatology, Palaeoecology* 30:325–347.

Evanoff, E., D. R. Prothero, and R. H. Lander. 1992. Eocene-Oligocene climatic change in North America: The White River Formation near Douglas, east-central Wyoming. In D. R. Prothero and W. A. Berggren, eds., *Eocene-Oligocene Climatic and Biotic Evolution,* pp. 116–130. Princeton, N.J.: Princeton University Press.

Evernden, J. F., D. E. Savage, G. H. Curtis, and G. T. James. 1964. Potassium-argon dates and the Cenozoic mammalian chronology of North America. *American Journal of Science* 262:145–198.

Exon, N., et al. 2002. Drilling reveals climatic consequences of Tasmanian gateway opening. *EOS* 83 (23): 253–259.

Fawcett, P. J., and M. B. E. Boslough. 2002. Climatic effects of an impact-induced equatorial debris ring. *Journal of Geophysical Research* 107 (D15): 10129–10146.

Fenner, J. 1986. Information from diatom analysis concerning the Eocene-Oligocene boundary. In C. H. Pomerol and I. Premoli-Silva, eds., *Terminal Eocene Events,* pp. 283–288. Amsterdam: Elsevier.

Fillon, R. H. 1977. Ice-rafted detritus and paleotemperature; late Cenozoic relationships in the Ross Sea region. *Marine Geology* 25:75–93.

Fischer, A. G. 1986. Climatic rhythms recorded in strata. *Annual Reviews of Earth and Planetary Sciences* 14:351–376.

Fluegeman, R. H. 2003. Late Eocene–early Oligocene benthic foraminifera in the Gulf Coastal Plain: Regional vs. global influences. In D. R. Prothero, L. C. Ivany, and E. A. Nesbitt, eds., *From Greenhouse to Icehouse: The Marine Eocene-Oligocene Transition,* pp. 283–293. New York: Columbia University Press.

Flynn, J. J., and C. C. Swisher III. 1995. Cenozoic South American land mammal ages: Correlation to global geochronologies. *SEPM Special Publication* 54:317–333.

Fordyce, R. E. 1980. Whale evolution and Oligocene Southern Ocean environments. *Palaeogeography, Palaeoclimatology, Palaeoecology* 31:319–336.

———. 1989. Origins and evolution of Antarctic marine mammals. *Special Publications of the Geological Society of London* 47:269–281.

———. 1992. Cetacean evolution and Eocene-Oligocene environments. In D. R. Prothero and W. A. Berggren, eds., *Eocene-Oligocene Climatic and Biotic Evolution,* pp. 368–381. Princeton, N.J.: Princeton University Press.

———. 2003. Cetacean evolution and Eocene-Oligocene oceans revisited. In D. R. Prothero, L. C. Ivany, and E. A. Nesbitt, eds., *From Greenhouse to Icehouse: The Marine Eocene-Oligocene Transition,* pp. 154–170. New York: Columbia University Press.

Frailey, C. D., and K. Campbell. 2004. Eocene rodents from South American and the evolution of the Caviidae. *Journal of Vertebrate Paleontology* 24 (3): 61A.

Frakes, L. A. 1979. *Climates throughout Geological Time.* New York: Elsevier.

Francis, J. E., and I. Poole. 2002. Cretaceous and Tertiary climates of Antarctica: Evidence from fossil wood. *Palaeogeography, Palaeoclimatology, Palaeoecology* 182:47–64.

Frederiksen, N. O. 1988. Sporomorph biostratigraphy, floral changes, and paleoclimatology, Eocene and earliest Oligocene of the eastern Gulf Coast. *U.S. Geological Survey Professional Paper 1448.*

———. 1991. Pulses of middle Eocene to earliest Oligocene climatic deterioration in southern California and the Gulf Coast. *Palaios* 6:564–571.

Froehlich, D. J. 2002. Quo vadis *Eohippus*? The systematics and taxonomy of the early Eocene equids (Perissodactyla). *Zoological Journal of the Linnaean Society of London* 134:141–256.

Frost, S. H. 1977. Miocene to Holocene evolution of Caribbean Province reef-building corals. *Third Annual International Coral Reef Symposium (Miami), Proceedings,* pp. 353–359.

Gallagher, W. B. 1991. Selective extinction and survival across the Cretaceous/Tertiary boundary in the northern Atlantic Coastal Plain. *Geology* 19:967–970.

———. 1993. The Cretaceous/Tertiary mass extinction event in the northern Atlantic Coastal Plain. *The Mosasaur* 5:75–154.

———. 2002. Faunal changes across the Cretaceous-Tertiary (K-T) boundary in the Atlantic Coastal Plain of New Jersey: Restructuring the marine community after the K-T mass extinction event. *Geological Society of America Special Paper* 356:291–302.

Ganapathy, R. 1982. Evidence for a major meteorite impact on the earth 34 million years ago: Implications for Eocene extinctions. *Science* 216:885–886.

Gaskell, B. A. 1991. Extinction patterns in Paleogene benthic foraminiferal faunas: Relationship to climate and sea level. *Palaios* 6:2–16.

Gazin, C. L. 1968. A study of the Eocene condylarthran mammal *Hyopsodus*. *Smithsonian Miscellaneous Collections* 149 (2): 1–98.

Geary, D. H. 1990. Patterns of evolutionary tempo and mode in the radiation of *Melanopsis* (Gastropoda: Melanopsidae). *Paleobiology* 16:492–511.

Geikie, A. 1894. *The Great Ice Age and Its Relation to the Antiquity of Man*. 3rd ed. London: Stanford.

Gheerbrant, E. 1990. On the early biogeographical history of the African placentals. *Historical Biology* 4:107–116.

Gheerbrant, E., J. Sudre, and H. Capetta. 1996. Palaeocene proboscidean from Morocco. *Nature* 383:68–70.

Gillet, S. 1946. Lamellibranches dulcicoles, les limnocardidés. *Revues de Science, Paris* 84:343–353.

Gingerich, P. D. 1976. Cranial anatomy and evolution of early Tertiary Plesiadapidae (Mammalia, Primates). *University of Michigan Papers in Paleontology* 15:1–141.

Gingerich, P. D., N. A. Wells, D. E. Russell, and S. M. Ibrahim Shah. 1983. Origin of whales in epicontinental remnant seas; new evidence from the early Eocene of Pakistan. *Science* 220:403–406.

Gingerich, P. D., B. H. Smith, and E. L. Simons. 1990. Hind limbs of Eocene *Basilosaurus*: Evidence of feet in whales. *Science* 249:154–157.

Gingerich, P. D., M. Haq, I. S. Zalmout, I. H. Khan, and M. S. Malkani. 2001. Origin of whales from early artiodactyls; hands and feet of Eocene Protocetidae from Pakistan. *Science* 293:2239–2242.

Gladenkov, Y. B. 1992. North Pacific: Neogene biotic and abiotic events. In R. Rsuchi and J. C. Ingle Jr., eds., *Pacific Neogene Environments and Evolution*, pp. 171–179. Tokyo: University of Tokyo Press.

Glass, B. P., and M. J. Zwart. 1977. North American microtektites, radiolarian extinctions and the age of the Eocene/Oligocene boundary. In F. M. Swain, ed., *Stratigraphic Micropaleontology of the Atlantic Basin and Borderlands*, pp. 553–568. Amsterdam: Elsevier.

Glass, B. P., D. L. DuBois, and R. Ganapathy. 1982. Relationship between an iridium anomaly and the North American micro-tektite layer in core RC9-58 from the Caribbean Sea. *Journal of Geophysical Research* 87:425–428.

Godthelp, H., M. Archer, R. L. Cifelli, S. J. Hand, and G. F. Gilkerson. 1992. Earliest known Australian Tertiary mammal fauna. *Nature* 356:514–516.

Goodwin, P. W., and E. J. Anderson. 1985. Punctuated aggradational cycles: A general hypothesis of episodic stratigraphic accumulation. *Journal of Geology* 93:515–533.

Gore, A. 1992. *Earth in the Balance: Ecology and Human Spirit*. New York: Penguin.

Graham, A. 1999. *Late Cretaceous and Cenozoic History of North American Vegetation.* Oxford: Oxford University Press.

Graham, R. W., and E. L. Lundelius Jr. 1984. Coevolutionary disequilibrium and Pleistocene extinctions. In P. S. Martin and R. G. Klein, eds., *Quaternary Extinctions: A Prehistoric Revolution,* pp. 250–258. Tucson: University of Arizona Press.

Grande, L. 1980. Paleontology of the Green River Formation, with a review of the fish fauna. *Geological Survey of Wyoming Bulletin* 63:1–333.

Granger, W. 1938. A giant oxyaenid from the upper Eocene of Mongolia. *American Museum Novitates* 969:1–5.

Greenwood, D. R., P. T. Moss, A. I. Rowett, A. J. Vadala, and R. J. Keefe. 2003. Plant communities and climate change in southeastern Australia during the early Paleogene. *Geological Society of America Special Paper* 369:365–380.

Gribben, J. 1982. *Future Weather and the Greenhouse Effect.* New York: Delacorte.

Gunnell, G. F. 2001. *Eocene Biodiversity: Unusual Occurrences and Rarely Sampled Habitats.* New York: Kluwer Academic/Plenum Publishers.

Guo, S. X. 1985. Preliminary interpretation of Tertiary climate by using megafossil floras in China. *Palaeotologia Cathayana* 2:169–175.

Guthrie, R. D. 1990. *Frozen Fauna of the Mammoth Steppe: The Story of the Blue Babe.* Chicago: University of Chicago Press.

Hakansson, E., and E. Thomsen. 1979. Distribution and types of bryozoan communities at the boundary in Denmark. In T. Birkelund, and R. G. Bromley, eds., *Cretaceous-Tertiary Boundary Events,* vol. 1, *The Maastrichtian and Danian of Denmark,* pp. 78–91. Copenhagen: University of Copenhagen.

Hallam, A., and P. B. Wignall. 1997. *Mass Extinctions and Their Aftermath.* Oxford: Oxford University Press.

Hamilton, W. 1965. Cenozoic climatic change and its causes. *Meteorological Monograph* 8.

Hansen, T. A. 1987. Extinction of late Eocene to Oligocene molluscs: Relationship to shelf area, temperature changes, and impact events. *Palaios* 2:69–75.

———. Early Tertiary radiation of marine molluscs and the long-term effects of the Cretaceous-Tertiary extinction. *Paleobiology* 14:37–51.

———. 1992. The patterns and causes of molluscan extinction across the Eocene/Oligocene boundary. In D. R. Prothero and W. A. Berggren, eds., *Eocene-Oligocene Climatic and Biotic Evolution,* pp. 341–348. Princeton, N.J.: Princeton University Press.

Hansen, T. A., and P. Kelley. 2004. Paleoecological patterns in molluscan extinctions and recoveries; comparison of the Cretaceous-Paleogene and Eocene-Oligocene extinctions in North America. *Palaeogeography, Palaeoclimatology, Palaeoecology* 214:233–242.

Hansen, T. A., R. B. Farrand, H. A. Montgomery, H. G. Billman, and G. Blechschmidt. 1987. Sedimentology and extinction patterns across the Cretaceous-Tertiary boundary interval in east Texas. *Cretaceous Research* 8:229–252.

Hansen, T. A., B. R. Farrell, and B. Upshaw. 1993. The first 2 million years after the Cretaceous-Tertiary boundary in east Texas and paleoecology of the molluscan recovery. *Paleobiology* 19:251–265.

Haq, B. U., and G. P. Lohmann. 1976. Early Cenozoic calcareous nannoplankton biogeography of the Atlantic Ocean. *Marine Micropaleontology* 1:119–194.

Haq, B. U., J. Hardenbol, and P. R. Vail. 1987. The chronology of fluctuating sea level since the Triassic. *Science* 235:1156–1167.

Hardenbol, J., and W. A. Berggren. 1978. A new Paleogene numerical time scale. *American Association of Petroleum Geologists Memoir* 6:213–234.

Harris, A. W., and W. R. Ward. 1982. Dynamical constraints on the formation and evolution of planetary bodies. *Annual Review of Earth and Planetary Sciences* 10:61–108.

Hays, J. D., J. Imbrie, and N. J. Shackleton. 1976. Variations in the earth's orbit—pacemaker of the ice ages. *Science* 194:1121–1132.

Hazel, J. E. 1971. Paleoclimatology of the Yorktown Formation (upper Miocene and lower Pliocene) of Virginia and North Carolina. *Centre Recherches Pau-SNPA Bulletin* 5:361–375.

———. 1988. Determining late Neogene and Quaternary palaeoclimates and palaeotemperature regimes using ostracods. In P. De Deckker, J.-P. Colin, and J.-P. Peypouquet, eds., *Ostracoda in the Earth Sciences*, pp. 89–101. Amsterdam: Elsevier.

Heckel, P. H. 1986. Sea-level curves for Pennsylvanian eustatic marine transgressive-regressive depositional cycles across the mid-continent outcrop belt, North America. *Geology* 14:330–334.

Heissig, K. 1986. No effect of the Ries impact event on the local mammal fauna. *Modern Geology* 10:171–179.

Hickey, L. J. 1977. Stratigraphy and paleobotany of the Golden Valley Formation (early Tertiary) of western North Dakota. *Geological Society of America Memoir* 150:1–181.

Hickman, C. S. 2003. Evidence for abrupt Eocene-Oligocene molluscan faunal change in the Pacific Northwest. In D. R. Prothero, L. C. Ivany, and E. A. Nesbitt, eds., *From Greenhouse to Icehouse: The Marine Eocene-Oligocene Transition*, pp. 71–87. New York: Columbia University Press.

Hodell, D. A., and J. P. Kennett. 1986. Late Miocene–early Pliocene stratigraphy and palaeoceanography of the South Atlantic and southwest Pacific oceans: A synthesis. *Paleoceanography* 1:285–311.

Hooker, J. J. 1989. Character polarities in early perissodactyls and their significance for *Hyracotherium* and infraordinal relationships. In D. R. Prothero and R. M. Schoch, eds., *The Evolution of Perissodactyls*, pp. 79–101. New York: Oxford University Press.

———. 1992. British mammalian paleocommunities across the Eocene-Oligocene transition and their environmental implications. In D. R. Prothero and W. A. Berggren, eds., *Eocene-Oligocene Climatic and Biotic Evolution*, pp. 494–515. Princeton, N.J.: Princeton University Press.

Hooker, J. J., and D. Dashzeveg. 2003. Evidence for direct mammalian faunal interchange between Europe and Asia near the Paleocene-Eocene boundary. *Geological Society of America Special Paper* 369:479–500.

Hornibrook, N. D. 1992. New Zealand Cenozoic marine paleoclimates: A review based on the distribution of some shallow water and terrestrial biota. In R. Tsuchi and J. C. Ingle Jr., eds., *Pacific Neogene*, pp. 83–106. Tokyo: University of Tokyo Press.

Horowitz, A. S., and J. F. Pachut. 1994. Lyellian bryozoan percentages and the fossil record of the recent bryozoan fauna. *Palaios* 9:500–505.

Horton, D. R. 1984. Red kangaroos: Last of the Australian megafauna. In P. S.

Martin and R. G. Klein, eds. *Quaternary Extinctions: A Prehistoric Revolution*, pp. 639–680. Tucson: University of Arizona Press.

Hsu, K. J. 1983. *The Mediterranean Was a Desert*. Princeton, N.J.: Princeton University Press.

Hsu, K. J., et al. 1977. History of the Mediterranean salinity crisis. *Nature* 267:399–403.

Hsu, K. J., J. A. McKenzie, and Q. X. He. 1982. Terminal Cretaceous environmental and evolutionary changes. *Geological Society of America Special Paper* 190:317–328.

Huber, B. T., K. G. MacLeod, and R. D. Norris. 2002. Abrupt extinction and subsequent reworking of Cretaceous planktonic foraminifera across the Cretaceous-Tertiary boundary: Evidence from the subtropical North Atlantic. *Geological Society of America Special Paper* 356:277–290.

Huber, M., L. C. Sloan, and C. Shellito. 2003. Early Paleogene oceans and climate: A fully coupled modeling approach using the NCAR CCSM. *Geological Society of America Special Paper* 369:25–48.

Hunt, R. J., and I. Poole. 2003. Paleogene West Antarctic climate and vegetation history in light of new data from King George Island. *Geological Society of America Special Paper* 369:395–412.

Hurlbert, S. H., and Archibald, J. D. 1996. No statistical support for sudden (or gradual) extinction of dinosaurs: Discussion and reply. *Geology* 24:957–959.

Hut, P., W. Alvarez, W. P. Elder, T. Hansen, E. G. Kauffman, G. Keller, E. M. Shoemaker, and P. Weismann. 1987. Comet showers as a cause of mass extinctions. *Nature* 329:118–126.

Hutchison, J. H. 1982. Turtle, crocodilian and champsosaur diversity changes in the Cenozoic of the north-central region of the western United States. *Palaeogeography, Palaeoclimatology, Palaeoecology* 37:149–164.

———. 1992. Western North American reptile and amphibian record across the Eocene/Oligocene boundary and its climatic implications. In D. R. Prothero and W. A. Berggren, eds., *Eocene-Oligocene Climatic and Biotic Evolution*, pp. 451–463. Princeton, N.J.: Princeton University Press.

Imbrie, J. 1992. On the structure and origin of major glacial cycles. Pt. 1. *Paleoceanography* 7:701–738.

———. 1993. On the structure and origin of major glacial cycles. Pt. 2. *Paleoceanography* 8:699–735.

Imbrie, J., and K. P. Imbrie. 1979. *Ice Ages: Solving the Mystery*. Short Hills, N.J.: Enslow.

Ivany, L. C., W. P. Patternson, and K. C. Lohmann. 2000. Cooler winters as a possible cause of mass extinctions at the Eocene/Oligocene boundary. *Nature* 407:887–890.

Ivany, L. C., K. C. Lohmann, and W. P. Patterson. 2003. Paleogene temperature history of the U.S. Gulf Coastal Plain inferred from ^{18}O of fossil otoliths. In D. R. Prothero, L. C. Ivany, and E. A. Nesbitt, eds., *From Greenhouse to Icehouse: The Marine Eocene-Oligocene Transition*, pp. 232–251. New York: Columbia University Press.

Ivany, L. C., B. H. Wilkinson, K. C. Lohmann, E. R. Johnson, B. J. McElroy, and G. J. Cohen. 2004. Intra-annual isotopic variation in *Venericardia* bivalves: Implications for early Eocene temperatures, seasonality, and salinity on the U.S. Gulf Coast. *Journal of Sedimentary Research* 74:7–19.

Janis, C., and E. M. Manning. 1998. Antilocapridae. In C. Janis, K. M. Scott, and

L. Jacobs, eds., *Tertiary Mammals of North America,* pp. 491–507. Cambridge: Cambridge University Press.

Janis, C. M., J. Damuth, and J. M. Theodor. 2000. Miocene ungulates and terrestrial primary productivity: Where have all the browsers gone? *Proceedings of the National Academy of Sciences* 97 (14): 7899–7904.

———. 2002. The origins and evolution of the North American grassland biome: The story from the hoofed mammals. *Palaeogeography, Palaeoclimatology, Palaeoecology* 177:183–198.

———. 2004. The species richness of Miocene browsers, and implications for habitat type and primary productivity in the North American grassland biome. *Palaeogeography, Palaeoclimatology, Palaeoecology* 207:371–398.

Jansa, L. F., M.-P. Aubry, and F. M. Gradstein. 1990. Comets and extinctions: Cause and effect? *Geological Society of America Special Paper* 247:223–232.

Jansen, E., and J. Sjoholm. 1991. Reconstruction of glaciation over the past 6 myr from ice-borne deposits in the Norwegian Sea. *Nature* 349:600–603.

Johanson, D., and B. Edgar. 1996. *From Lucy to Language.* New York: Simon and Schuster.

Johnson, K. R., and L. J. Hickey. 1990. Megafloral change across the Cretaceous/Tertiary boundary in the northern Great Plains and Rocky Mountains. *Geological Society of America Special Paper* 247:433–444.

Jones, D. S., and P. F. Hasson. 1985. History and development of the marine invertebrate faunas separated by the Central American Isthmus. In F. G. Stehli and S. D. Webb, eds., *The Great American Biotic Interchange,* pp. 325–356. New York: Plenum Press.

Jouse, A. P. 1978. Diatom biostratigraphy on the generic level. *Micropaleontology* 24:316–326.

Kaminski, M. A. 1987. Cenozoic deep-water agglutinated foraminifera in the North Atlantic. Ph.D. thesis, Massachusetts Institute of Technology/Woods Hole Oceanographic Institute.

Kamp, P. J. J., D. B. Waghorn, and C. S. Nelson. 1990. Late Eocene–early Oligocene integrated isotope stratigraphy and biostratigraphy for paleoshelf sequences in southern Australia: Paleoceanographic implications. *Palaeogeography, Palaeoclimatology, Palaeoecology* 80:311–323.

Kauffman, E. G. 1988. The dynamics of marine stepwise mass extinctions. In M. A. Lamolda, E. G. Kauffman, and O. H. Walliser, eds., *Paleontology and Evolution: Extinction Events,* pp. 57–71. Revista Española de Paleontologia, No. Extraordinario.

Kay, R. F., R. Madden, M. G. Vucetich, A. A. Carlini, M. M. Mazzoni, G. H. Re, M. Heizler, and H. Sandeman. 1999. Revised age of the Casamayoran South American land mammal "age"—climatic and biotic implications. *Proceedings of the National Academy of Sciences* 96:13235–13240.

Keigwin, L. D., Jr. 1976. Late Cenozoic planktonic foraminiferal biostratigraphy and paleoceanography of the Panama Basin. *Micropaleontology* 22:419–442.

———. 1978. Pliocene closing of the Isthmus of Panama based on biostratigraphic evidence from nearby Pacific Ocean and Caribbean Sea cores. *Geology* 6:630–634.

Keigwin, L. D., Jr., and R. Thunell. 1979. Middle Pliocene climatic change in the western Mediterranean from faunal and oxygen isotope trends. *Nature* 282:294–296.

Keller, G. 1983a. Biochronology and paleoclimatic implications of middle Eocene to Oligocene planktic foraminiferal faunas. *Marine Micropaleontology* 7:463–486.

———. 1983b. Paleoclimatic analyses of middle Eocene through Oligocene planktic foraminiferal faunas. *Palaeogeography, Palaeoclimatology, Palaeoecology* 43:73–94.

Keller, G., T. Herbert, R. Dorsey, S. D'Hondt, M. Johnsson, and W. R. Chi. 1987. Global distribution of late Paleogene hiatuses. *Geology* 15:199–203.

Kemp, E. M. 1975. Palynology of Leg 28 drill sites, Deep Sea Drilling Project. *Initial Reports of the Deep Sea Drilling Project* 28:599–623.

———. 1978. Tertiary climatic evolution and vegetation history in the southeast Indian Ocean region. *Palaeogeography, Palaeoclimatology, Palaeoecology* 24:169–208.

Kennedy, W. J. 1993. Ammonite faunas of the European Maastrichtian: Diversity and extinction. In House, M. R., ed., *The Ammonoidea: Environment, Ecology and Evolutionary Change,* pp. 285–326. Systematics Association Special Volume.

Kennett, J. P. 1977. Cenozoic evolution of Antarctic glaciation, the Circum-Antarctic Ocean, and their impact on global paleoceanography. *Journal of Geophysical Research* 82:3843–3860.

———. 1985. *The Miocene Ocean: Paleoceanography and Biogeography.* Geological Society of America Memoir 163. Boulder, Colo.: Geological Society of America.

———. 1995. A review of polar climatic evolution during the Neogene, based on the marine sediment record. In E. S. Vrba, G. H. Denton, T. C. Partridge, and L. H. Burckle, eds., *Paleoclimate and Evolution, with Emphasis on Human Origins,* pp. 49–64. New Haven, Conn.: Yale University Press.

Kennett, J. P., and P. F. Barker. 1990. Latest Cretaceous to Cenozoic climate and oceanographic developments in the Weddell Sea, Antarctica: An ocean drilling perspective. *Proceedings of the Ocean Drilling Program* 113 (pt. B): 937–960.

Kennett, J. P., and Srinivasan, M. S. 1983. *Neogene Planktonic Foraminifera; a Phylogenetic Atlas.* Stroudsburg, Pa.: Hutchinson Ross.

Kennett, J. P., and L. D. Stott. 1990. Proteus and Proto-Oceanus; ancestral Paleogene oceans as revealed from Antarctic stable isotopic results, ODP Leg 113. *Proceedings of the Ocean Drilling Program, Scientific Results* 113:865–880.

———. 1991. Abrupt deep-sea warming, palaeoceanographic changes and benthic extinction at the end of the Palaeocene. *Nature* 353:225–229.

Kennett, J. P., et al. 1972. Australian-Antarctic continental drift, palaeocirculation changes and Oligocene deep-sea erosion. *Nature* 91:51–55.

Kennett, J. P., R. E. Houtz, P. B. Andrews, A. R. Edwards, V. A. Gostin, M. Hahos, M. A. Hampton, D. G. Jenkins, S. V. Margolis, A. T. Ovenshine, and K. Perch-Nielsen. 1975. Cenozoic paleoceanography in the southwest Pacific Ocean, Antarctic glaciation and the development of the circum-Antarctic current. *Initial Reports of the Deep Sea Drilling Project* 29:1155–1169.

Kennett, J. P., G. Keller, and M. S. Srinivasan. 1985. Miocene planktonic foraminiferal biogeography and paleoceanographic development of the Indo-Pacific region. *Geological Society of America Memoir* 163:197–236.

Kent, D. V., B. S. Cramer, and L. Lanci. 2001. Evidence of an impact trigger for the Paleocene/Eocene thermal maximum and carbon isotope excursion. *EOS* 82: PP32A-0509.

Kier, P. M. 1975. Evolutionary trends and their functional significance in post-Paleozoic echinoids. *Paleontological Society Memoir* 5:1–95.

Kleinpell, R. M. 1938. *Miocene Stratigraphy of California*. Tulsa: Okla.: American Association of Petroleum Geologists

Koch, P. L., J. C. Zachos, and P. D. Gingerich. 1992. Correlation between isotope records in marine and continental carbon reservoirs near the Paleocene-Eocene boundary. *Nature* 358:319–322.

Koch, P. L., J. C. Zachos, and D. L. Dettman. 1995. Stable isotope stratigraphy and paleoclimatology of the Paleogene Bighorn Basin. *Palaeogeography, Palaeoclimatology, Palaeoecology* 115:61–89.

Kozisek, J. 2003. New implications for Cretaceous-Tertiary asteroid impact theory based upon the persistence of extant tropical honeybees (Hymenoptera: Apidae). *Journal of Vertebrate Paleontology* 23(3):69A.

Krantz, D. E. 1990. Mollusk-isotope records of Plio-Pleistocene marine paleoclimate, U.S. Middle Atlantic Coastal Plain. *Palaios* 5:317–335.

Krause, D. W., and M. C. Maas. 1990. The biogeographic origins of late Paleocene–early Eocene mammalian immigrants to the Western Interior of North America. *Geological Society of America Special Paper* 243:71–105.

Kurtén, B. 1968. *Pleistocene Mammals of Europe*. New York: Columbia University Press.

———. 1971. *The Age of Mammals*. New York: Columbia University Press.

———. 1988. *Before the Indians*. New York: Columbia University Press.

Kurtén, B., and E. Anderson. 1980. *Pleistocene Mammals of North America*. New York: Columbia University Press.

Kvenvolden, K. A. 1993. Gas hydrates: Geological perspective and global change. *Reviews of Geophysics* 31:173–187.

LaBandeira, C., and J. J. Sepkoski Jr. 1993. Insect diversity and the fossil record: Myth and reality. *Science* 261:310–315.

Ladurie, E. L. R. 1971. *Times of Feast, Times of Famine*. New York: Doubleday.

Lamb, H. H. 1972. *Climate: Past, Present, and Future*. London: Methuen.

———. 1982. *Climate History and the Modern World*. London: Methuen.

Lartet, E. 1834. Sur plusieurs gisements d'ossements fossiles, et les différens débris. *Bulletin de la Société Géologique de France* 4:342–344.

Leakey, R., and R. Lewin. 1995. *The Sixth Extinction*. New York: Doubleday.

Legendre, S., and J. L. Hartenberger. 1992. The evolution of mammalian faunas in Europe during the Eocene and Oligocene. In D. R. Prothero and W. A. Berggren, eds., *Eocene-Oligocene Climatic and Biotic Evolution*, pp. 516–528. Princeton, N.J.: Princeton University Press.

Leidy, J. 1854. The ancient fauna of Nebraska, or a description of remains of extinct Mammalia and Chelonia from the Mauvais Terres of Nebraska. *Smithsonian Contributions to Knowledge*, 6, article 7, pp. 1–126.

Leopold, E. B., and M. F. Denton. 1987. Comparative age of grassland and steppe east and west of the northern Rocky Mountains. *Annals of the Missouri Botanical Garden* 75:841–867.

Leopold, E. B., L. Gengwu, and S. Clay-Poole. 1992. Low-biomass vegetation in the Oligocene?. In D. R. Prothero and W. A. Berggren, eds., *Eocene-Oligocene Climatic and Biotic Evolution*, pp. 399–420. Princeton, N.J.: Princeton University Press.

Lewin, R. 1984. *Human Evolution: An Illustrated Introduction*. 2nd ed. Boston: Blackwell.

Loel, W., and W. H. Corey. 1932. The Vaqueros Formation, lower Miocene of California. *University of California Publications in Geological Sciences* 22:31–410.

Lofgren, D. L., J. A. Lillegraven, W. A. Clemens, P. D. Gingerich, and T. E. Williamson. 2004. Paleocene biochronology: The Puercan through Clarkforkian land mammal ages. In M. O. Woodburne, ed., *Late Cretaceous and Cenozoic Mammals of North America: Biostratigraphy and Geochronology*, pp. 43–105. New York: Columbia University Press.

Lozouet, P. 1997. Nouvelles espèces de gastéropodes (Mollusca: Gastropoda) de l'Oligocène et du Miocène d'Aquitaine (Sud-Ouest de la France). Pt. 2. *Cossmanniana* 6:1–68.

Lucas, S. G. 1998. Fossil mammals and the Paleocene/Eocene series boundary in Europe, North America, and Asia. In M.-P. Aubry, S. G. Lucas, and W. A. Berggren, eds., *Late Paleocene–Early Eocene Climatic and Biotic Events in the Marine and Terrestrial Records*, pp. 451–500. New York: Columbia University Press.

Lucas, S. G., E. G. Kordikova, and R. J. Emry. 1998. Oligocene stratigraphy, sequence stratigraphy, and mammalian biochronology north of the Aral Sea, western Kazakstan. *Bulletin of the Carnegie Museum of Natural History* 34:313–348.

Lundelius, E. L., Jr. 1983. Climatic implications of late Pleistocene and Holocene faunal associations in Australia. *Alcheringa* 7:125–149.

Lyell, C. 1833. *Principles of Geology.* 3 vols. London: John Murray.

Maas, M. C., J. G. M. Thewissen, and J. Kappelman. 1998. *Hypsamasia seni* (Mammalia: Embrithopoda) and other mammals form the Eocene Kartal Formation of Turkey. *Bulletin of the Carnegie Museum of Natural History* 34:286–297.

Macdougall, J. D. 2004. *Frozen Earth: The Once and Future Story of the Ice Ages.* Berkeley: University of California Press.

MacFadden, B. J. 1992. *Fossil Horses.* Cambridge: Cambridge University Press.

MacGinitie, H. D. 1969. The Eocene Green River flora of northwestern Colorado and northeastern Utah. *University of California Publications in Geological Science* 83:1–202.

MacLeod, K. G. 1994. Extinction of inoceramid bivalves in Maastrichtian strata of the Bay of Biscay region of France and Spain. *Journal of Paleontology* 68:1048–1066.

MacLeod, N., et al. 1997. The Cretaceous-Tertiary biotic transition. *Journal of the Geological Society, London* 154:265–292.

MacLeod, N., N. Ortiz, N. Fefferman, W. Clyde, C. Schulter, and J. MacLean. 2000. Phenotypic response of Foraminifera to episodes of global environmental change. In S. C. Culver and P. F. Rawson, eds., *Biotic Response to Global Change: The Last 145 Million Years*, pp. 51–78. Cambridge: Cambridge University Press.

Margolis, S. V., and Y. Herman. 1980. Northern Hemisphere sea-ice and glacial development in the late Cenozoic. *Nature* 286:145–149.

Marshak, S., and D. R. Prothero. 2001. *Earth: Portrait of a Planet.* New York: W. W. Norton

Marshall, L. G. 1978. The terror bird. *Field Museum of Natural History Bulletin* 49:6–15.

_____. 1981. The Argentine connection. *Field Museum of Natural History Bulletin* 52:17–25.

_____. 1984. Who killed Cock Robin? An investigation of the extinction controversy. In P. S. Martin and R. G. Klein, eds., *Quaternary Extinctions: A Prehistoric Revolution*, pp. 785–806. Tucson: University of Arizona Press.

_____. 1985. Geochronology and land-mammal biochronology of the transamerican faunal interchange. In F. G. Stehli and S. D. Webb, eds., *The Great American Biotic Interchange*, pp. 49–88. New York: Plenum Press.

Marshall, L. G., and R. L. Cifelli. 1989. Analysis of changing diversity patterns in Cenozoic land mammal age faunas, South America. *Palaeovertebrata* 19:169–210.

Marshall, L. G., and R. S. Corruccini. 1978. Variability, evolutionary rates, and allometry in dwarfing lineages. *Paleobiology* 4:101–119.

Marshall, L. G., S. D. Webb, J. J. Sepkoski, and D. M. Raup. 1982. Mammalian evolution and the Great American Interchange. *Science* 215:1351–1357.

Martin, P. S. 1967. Prehistoric overkill. In P. S. Martin and H. E. Wright, eds., *Pleistocene Extinctions: The Search for a Cause*, pp. 75–120. New Haven, Conn.: Yale University Press.

_____. 1984. Prehistoric overkill: The global model. In P. S. Martin and R. G. Klein, eds., *Quaternary Extinctions: A Prehistoric Revolution*, pp. 354–403. Tucson: University of Arizona Press.

Martin, P. S., and R. G. Klein, eds. 1984. *Quaternary Extinctions: A Prehistoric Revolution*. Tucson: University of Arizona Press.

Masaitis, V., M. V. Mikhailov, and T. V. Selivanovskaya. 1975. [*Popigai Meteorite Crater*]. Moscow: Nauka Press. In Russian.

Matthew, W. D., and W. Granger. 1925. Fauna and correlation of the Gashato Formation of Mongolia. *American Museum Novitates* 189:1–12.

Matthews, R. K. 1984. *Dynamic Stratigraphy*. Englewood Cliffs, N.J.: Prentice-Hall.

Matthiessen, P. 1991. *African Silences*. New York: Random House.

Maurrasse, F., and B. P. Glass. 1976. Radiolarian stratigraphy and North American microtektites in Caribbean Core RC9-58: Implications concerning late Eocene radiolarian chronology and the age of the Eocene-Oligocene boundary. *Seventh Caribbean Geological Conference Proceedings*, pp. 205–212.

McDougall, I., and H. Wensink. 1966. Paleomagnetism and geochronology of Pliocene-Pleistocene lavas in Iceland. *Earth and Planetary Sciences Letters* 1:232–236.

McKenna, M. C. 1975. Toward a phylogenetic classification of the Mammalia. In W. P. Luckett and F. S. Szalay, eds., *Phylogeny of the Primates: A Multidisciplinary Approach*, pp. 21–46. New York: Plenum.

_____. 1980. Eocene paleolatitude, climate and mammals of Ellesmere Island. *Palaeogeography, Palaeoclimatology, Palaeoecology* 30:349–362.

_____. 1983. Holarctic landmass rearrangement, cosmic events, and Cenozoic terrestrial organisms. *Annals of the Missouri Botanical Garden* 70:459–489.

_____. 2002. Semi-isolation and lowered salinity of the Arctic Ocean in late Paleocene to earliest Eocene time. *Journal of Vertebrate Paleontology* 23 (3): 77A.

McKenna, M. C., and E. M. Manning. 1977. Affinities and biogeographic significance of the Mongolian Paleocene genus *Phenacolophus*. *Géobios, Mémoire Spécial* 1:61–85.

McKenna, M. C., M. Chow, S. Ting, and Z. Luo. 1989. *Radinskya yupingae*, a perissodactyl-like mammal from the late Paleocene of China. In D. R. Prothero and R. M. Schoch, eds., *The Evolution of Perissodactyls*, pp. 24–36. New York: Oxford University Press.

McKerrow, W. S., ed. 1978. *The Ecology of Fossils.* Cambridge, Mass.: MIT Press.

McKinney, M. L., B. D. Carter, K. J. McNamara, and S. K. Donovan. 1992. Evolution of Paleogene echinoids: A global and regional view. In D. R. Prothero and W. A. Berggren, eds., *Eocene-Oligocene Climatic and Biotic Evolution*, pp. 348–367. Princeton, N. J.: Princeton University Press.

Mehrotra, R. C. 2003. Status of plant megafossils during the early Paleogene in India. *Geological Society of America Special Paper* 369:413–423.

Meng, J., and M. C. McKenna. 1998. Faunal turnovers of Palaeogene mammals from the Mongolia Plateau. *Nature* 394:364–367.

Meng, J., R. Zhai, and A. R. Wyss. 1998. The late Paleocene Bayan Ulan fauna of Inner Mongolia, China. *Bulletin of the Carnegie Museum of Natural History* 34:148–185.

Mercer, J. H. 1978. West Antarctic ice sheet and CO_2 greenhouse effect: A threat of disaster. *Nature* 271:321–325.

Mercer, J. H., and J. F. Sutter. 1982. Late Miocene–earliest Pliocene glaciation in southern Argentina: Implications for global ice-sheet history. *Palaeogeography, Palaeoclimatology, Palaeoecology* 38:185–206.

Millais, J. G. 1897. British deer and their horns. *Proceedings of the Zoological Society of London* 1897:1–41.

Miller, K. G. 1992. Middle Eocene to Oligocene stable isotopes, climate, and deep-water history: The Terminal Eocene Event? In D. R. Prothero and W. A. Berggren, eds., *Eocene-Oligocene Climatic and Biotic Evolution*, pp. 160–177. Princeton, N.J.: Princeton University Press.

Miller, K. G., and R. G. Fairbanks. 1983. Evidence for Oligocene-middle Miocene abyssal circulation changes in the western North Atlantic. *Nature* 306:250–253.

Miller, K. G., and E. Thomas. 1985. Late Eocene to Oligocene benthic foraminiferal isotopic record, Site 574, equatorial Pacific. *Initial Reports of the Deep Sea Drilling Project* 85:771–777.

Miller, K. G., and B. E. Tucholke. 1983. Development of Cenozoic abyssal circulation south of the Greenland-Scotland Ridge. In M. H. P. Bott, S. Saxov, M. Talwani, and J. Thiede, eds., *Structure and Development of the Greenland-Scotland Ridge*, pp. 549–589. New York: Plenum Press.

Miller, K. G., G. S. Mountain, and B. E. Tucholke. 1985. Oligocene glacioeustasy and erosion on the margins of the North Atlantic. *Geology* 13:10–13.

Miller, K. G., R. G. Fairbanks, and G. S. Mountain. 1987. Tertiary oxygen isotope synthesis, sea level history, and continental margin erosion. *Paleoceanography* 2:1–19.

Miller, K. G., J. D. Wright, and R. G. Fairbanks. 1991. Unlocking the ice house: Oligocene-Miocene oxygen isotopes, eustasy, and margin erosion. *Journal of Geophysical Research* 96:6829–6848.

Mithen, S. 2004. *After the Ice: A Global Human History.* Cambridge, Mass.: Harvard University Press.

Mitchell, E., and R. H. Tedford. 1973. Enaliarctinae: A new group of extinct

Carnivora and a consideration of the origin of the Otariidae. *Bulletin of the American Museum of Natural History* 151:201–284.

Mohr, B. A. R. 1990. Eocene and Oligocene sporomorphs and dinoflagellate cysts from Leg 113 drill sites, Weddell Sea, Antarctica. *Proceedings of the Ocean Drilling Program* 113:595–606.

Molnar, R. E. 2004. *Dragons in the Dust: Paleobiology of the Giant Monitor Lizard* Megalania. Bloomington: Indiana University Press.

Murphy, M. G., and J. P. Kennett. 1986. Development of latitudinal thermal gradients during the Oligocene: Oxygen-isotope evidence from the southwest Pacific. *Initial Reports of the Deep Sea Drilling Project* 90:1347–1360.

Murray, P. 1984. Extinctions Downunder: A bestiary of extinct Australian late Pleistocene monotremes and marsupials. In P. S. Martin and R. G. Klein, eds., *Quaternary Extinctions: A Prehistoric Revolution*, pp. 600–628. Tucson: University of Arizona Press.

Murray, P. F., and P. V. Rich. 2004. *Magnificent Mihirungs: The Colossal Flightless Birds of Australian Dreamtime*. Bloomington: Indiana University Press.

Myers, J. 2003. Terrestrial Eocene-Oligocene vegetation and climate in the Pacific Northwest. In D. R. Prothero, L. C. Ivany, and E. A. Nesbitt, eds., *From Greenhouse to Icehouse: The Marine Eocene-Oligocene Transition*, pp. 171–188. New York: Columbia University Press.

Nesbitt, E. A. 2003. Changes in shallow-marine faunas from the northeastern Pacific margin across the Eocene/Oligocene boundary. In D. R. Prothero, L. C. Ivany, and E. A. Nesbitt, eds., *From Greenhouse to Icehouse: The Marine Eocene-Oligocene Transition*, pp. 57–70. New York: Columbia University Press.

Novacek, M. J. 1994. The radiation of placental mammals. In D. R. Prothero and R. M. Schoch, eds., *Major Features of Vertebrate Evolution*, pp. 220–237. Short Courses in Paleontology 7. Knoxville: University of Tennessee.

Oleinik, A. E., and L. J. Marincovich. 2003. Biotic response to the Eocene-Oligocene transition: Gastropod assemblages in the high-latitude North Pacific. In D. R. Prothero, L. C. Ivany, and E. A. Nesbitt, eds., *From Greenhouse to Icehouse: The Marine Eocene-Oligocene Transition*, pp. 36–56. New York: Columbia University Press.

Olsen, P. E. 1984. Periodicities of lake-level cycles in the Late Triassic Lockatong Formation of the Newark Basin (Newark Supergroup), New Jersey and Pennsylvania. In A. Berger, J. Imbrie, J. D. Hays, G. Kukla, and B. Saltzman, eds., *Milankovitch and Climate*, pp. 129–146. Netherlands: The Hague.

Olsson, R. K., and C. Liu. 1993. Controversies on the placement of the Cretaceous-Paleogene boundary and the K/P mass extinction of planktonic foraminifera. *Palaios* 8:127–139.

Olsson, R. K., W. A. Berggren, C. Helmleben, and B. T. Huber, eds. 1999. Atlas of Paleocene planktonic foraminifera. *Smithsonian Contributions to Paleobiology* 85:1–252.

Oppenheimer, J., and R. Boyle. 1990. *Dead Heat: The Race against the Greenhouse Effect*. New York: Basic Books.

Osborn, H. F. 1910. *The Age of Mammals in Europe, Asia, and North America*. New York: Macmillan.

———. 1929. The titanotheres of ancient Wyoming, Dakota, and Nebraska. *United States Geological Survey Monograph* 55:1–953. 2 vols.

Osborn, H. F., and W. Granger. 1932. Coryphodonts and uintatheres from the Mongolian expedition of 1930. *American Museum Novitates* 552:1–16.

Osborn, H. F., and W. D. Matthew. 1909. Cenozoic mammal horizons of western North America. *United States Geological Survey Bulletin* 361:1–138.

Owen-Smith, N. 1987. Pleistocene extinctions: The pivotal role of megaherbivores. *Paleobiology* 13:351–362.

Pearson, P. N., and M. R. Palmer. 1999. Middle Eocene seawater pH and atmospheric carbon dioxide concentrations. *Science* 284:1824–1826.

———. 2000. Estimating Paleogene atmospheric pCO_2 using boron isotope analysis of Foraminifera. *Geological Society of Sweden* 122:127–128.

Pearson, P. N., P. W. Ditchfield, J. Singano, K. G. Harcourt-Brown, C. J. Nicholas, R. K. Olsson, N. J. Shackleton, and M. A. Hall. 2001. Warm tropical sea surface temperatures in the Late Cretaceous and Eocene epochs. *Nature* 413:481–487.

Peterson, G. L., and P. L. Abbott. 1979. Mid-Eocene climatic change, southwestern California and northwestern Baja California. *Palaeogeography, Palaeoclimatology, Palaeoecology* 26:73–87.

Pielou, E. C. 1991. *After the Ice Age: The Return of Life to Glaciated North America.* Chicago: University of Chicago Press.

Poag, C. W. 1997. Roadblocks on the kill curve: Testing the Raup hypothesis. *Palaios* 12:582–590.

———. 1999. *Chesapeake Invader.* Princeton, N.J.: Princeton University Press.

Poag, C. W., D. S. Powars, L. J. Poppe, R. B. Mixon, L. E. Edwards, D. W. Folger, and S. Bruce. 1992. Deep Sea Drilling Project Site 612 bolide event: New evidence of late Eocene impacts—wave deposits and a possible impact site, U.S. east coast. *Geology* 20:771–774.

Poag, C. W., E. Mankinen, and R. D. Norris. 2003. Late Eocene impacts: Geologic record, correlation, and paleoenvironmental consequences. In D. R. Prothero, L. C. Ivany, and E. A. Nesbitt, eds., *From Greenhouse to Icehouse: The Marine Eocene-Oligocene Transition,* pp. 495–510. New York: Columbia University Press.

Poddubiuk, R. H., and E. P. F. Rose. 1984. Relationships between mid-Tertiary echinoid faunas from the central Mediterranean and eastern Caribbean and their palaeobiogeographic significance. *Annales Géologiques Pays Hellénistique* 33:115–128.

Pomerol, C. 1982. *The Cenozoic Era: Tertiary and Quaternary.* New York: Wiley.

Pomerol, C., and I. Premoli-Silva, eds., 1986. *Terminal Eocene Events.* Amsterdam: Elsevier.

Pough, F. H., C. M. Janis, and J. B. Heiser. 2002. *Vertebrate Life.* 6th ed. Upper Saddle River, N.J.: Prentice-Hall.

Pospichal, J. J. 1996. Calcareous nannoplankton mass extinction at the Cretaceous/Tertiary boundary: An update. *Geological Society of America Special Paper* 307:335–336.

PRISM Project Members. 1995. Middle Pliocene paleoenvironments of the Northern Hemisphere. In E. S. Vrba, G. H. Denton, T. C. Partridge, and L. H. Burckle, eds., *Paleoclimate and Evolution, with Emphasis on Human Origins,* pp. 197–212. New Haven, Conn.: Yale University Press.

Prosh, E. C., and A. D. McCracken. 1985. Postapocalypse stratigraphy: Some considerations and proposals. *Geology* 13:4–5.

Prothero, D. R. 1985. Chadronian (early Oligocene) magnetostratigraphy of eastern Wyoming: Implications for the Eocene-Oligocene boundary. *Journal of Geology* 93:555–565.

_____. 1990. *Interpreting the Stratigraphic Record*. New York: W.H. Freeman.

_____. 1994. *The Eocene-Oligocene Transition: Paradise Lost*. New York: Columbia University Press.

_____. 1995. Geochronology and magnetostratigraphy of Paleogene North American land mammal "ages": An update. *SEPM Special Publication* 54:305–315.

_____. 1999. Does climatic change drive mammalian evolution? *GSA Today* 9 (9): 1–5.

_____. 2005a. Did impacts, volcanic eruptions, or climate change affect mammalian evolution? *Palaeogeography, Palaeoclimatology, Palaeoecology* 214: 283–294.

_____. 2005b. *The Evolution of North American Rhinoceroses*. Cambridge: Cambridge University Press.

Prothero, D. R., and W. A. Berggren, eds. 1992. *Eocene-Oligocene Climatic and Biotic Evolution*. Princeton, N.J.: Princeton University Press.

Prothero, D. R., and R. H. Dott Jr. 2003. *Evolution of the Earth*. 7th ed. New York: McGraw-Hill.

Prothero, D. R., and R. J. Emry, eds. 1996. *The Terrestrial Eocene-Oligocene Transition in North America*. Cambridge: Cambridge University Press.

Prothero, D. R., and T. H. Heaton. 1996. Faunal stability during the early Oligocene climatic crash. *Palaeogeography, Palaeoclimatology, Palaeoecology* 127:239–256.

Prothero, D. R., and F. Sanchez. 2005. Review of the leptauchenine oreodonts (Mammalia: Artiodactyla). *New Mexico Museum of Natural History and Science Bulletin*. In press.

Prothero, D. R., and R. M. Schoch. 2003. *Horns, Tusks, and Flippers: The Evolution of Hoofed Mammals*. Baltimore, Md.: Johns Hopkins University Press.

Prothero, D. R., and P. C. Sereno. 1982. Allometry and paleoecology of medial Miocene dwarf rhinoceroses from the Texas Gulf Coastal Plain. *Paleobiology* 8:16–30.

Prothero, D. R., and C. C. Swisher III. 1992. Magnetostratigraphy and geochronology of the terrestrial Eocene-Oligocene transition in North America. In D. R. Prothero and W. A. Berggren, eds., *Eocene-Oligocene Climatic and Biotic Evolution*, pp. 46–73. Princeton, N.J.: Princeton University Press.

Prothero, D. R., and K. E. Whittlesey. 1998. Magnetostratigraphy and biostratigraphy of the Orellan and Whitneyan land mammal "ages" in the White River Group. *Geological Society of America Special Paper* 325:39–61.

Prothero, D. R., C. R. Denham, and H. G. Farmer. 1983. Magnetostratigraphy of the White River Group and its implications for Oligocene geochronology. *Palaeogeography, Palaeoclimatology, Palaeoecology* 42:151–166.

Prothero, D. R., L. C. Ivany, and E. A. Nesbitt, eds. 2003. *From Greenhouse to Icehouse: The Marine Eocene-Oligocene Transition*. New York: Columbia University Press.

Qiu, Z., and Z. Qiu. 1995. Chronological sequence and subdivision of Chinese Neogene mammalian faunas. *Palaeogeography, Palaeoclimatology, Palaeoecology* 116:41–70.

Quade, J., T. E. Cerling, and J. R. Bowman. 1989. Development of the Asian monsoon revealed by marked ecological shift during the latest Miocene in northern Pakistan. *Nature* 342:163–166.

Raffi, S., S. M. Stanley, and R. Marasti. 1985. Biogeographic patterns and Plio-Pleistocene extinction of Bivalvia in the Mediterranean and southern North Sea. *Paleobiology* 11:368–388.

Rasmussen, D. T., T. M. Bown, and E. L. Simons. 1992. The Eocene-Oligocene transition in continental Africa. In D. R. Prothero and W. A. Berggren, eds., *Eocene-Oligocene Climatic and Biotic Evolution,* pp. 548–566. Princeton, N.J.: Princeton University Press.

Raup, D. M. 1991. *Extinction: Bad Genes or Bad Luck?* New York: W. W. Norton.

Raup, D. M., and J. J. Sepkoski Jr. 1984. Periodicity of extinctions in the geologic past. *Proceedings of the National Academy of Sciences* 81:805–801.

Raymo, M. E., and W. F. Ruddiman. 1992. Tectonic forcing of late Cenozoic climate. *Nature* 359:117–122.

Raymo, M. E., D. Hodell, and E. Jansen. 1992. Response of deep ocean circulation to initiation of Northern Hemisphere glaciation. *Paleoceanography* 7:645–672.

Raymo, M. E., B. Grant, M. Horowitz, and G. H. Rau. 1996. Mid-Pliocene warmth: Stronger greenhouse and stronger conveyor. *Marine Micropaleontology* 27:313–326.

Rea, D. K. 1992. Delivery of Himalayan sediment to the northern Indian Ocean and its relation to global climate, sea level, uplift, and seawater strontium. In R. A. Duncan, ed., *Synthesis of Results from Scientific Drilling of the Indian Ocean.* Washington, D.C.: American Geophysical Union.

Reinhart, R. H. 1959. A review of the Sirenia and Desmostylia. *University of California Publications in Geological Sciences* 36:1–145.

Retallack, G. J. 1983. Late Eocene and Oligocene paleosols from Badlands National Park, South Dakota. *Geological Society of America Special Paper* 193.

———. 1992. Paleosols and changes in climate and vegetation across the Eocene/Oligocene boundary. In D. R. Prothero and W. A. Berggren, eds., *Eocene-Oligocene Climatic and Biotic Evolution,* pp. 382–398. Princeton, N.J.: Princeton University Press.

———. 1997. Neogene expansion of the North American prairie. *Palaios* 12:380–390.

———. 2001a. A 300-million-year record of atmospheric carbon dioxide from fossil plant cuticles. *Nature* 411:287–290.

———. 2001b. Cenozoic expansion of grasslands and climatic cooling. *Journal of Geology* 109:407–426.

Rigby, J. K, Jr., K. R. Newman, J. Smith, S. Van der Kaars, R. E. Sloan, and J. K. Rigby. 1987. Dinosaurs from the Paleocene part of the Hell Creek Formation, McCone County, Montana. *Palaios* 2:296–302.

Rind, D., and M. A. Chandler. 1991. Increased ocean heat transports and warmer climate. *Journal of Geophysical Research* 96:7437–7461.

Robertson, D. S., M. C. McKenna, O. B. Toon, S. Hope, and J. A. Lillegraven. 2004. Survival in the first hours of the Cenozoic. *Geological Society of America Bulletin* 116:760–768.

Robinson, E. 2003. Upper Paleogene larger foraminiferal succession on a tropical carbonate bank, Nicaragua Rise, Caribbean region. In D. R. Prothero, L. C.

Ivany, and E. A. Nesbitt, eds., *From Greenhouse to Icehouse: The Marine Eocene-Oligocene Transition*, pp. 294–302. New York: Columbia University Press.

Robinson, P., G. F. Gunnell, S. L. Walsh, W. C. Clyde, J. E. Storer, R. K. Stucky, D. J. Froehlich, I. Ferrusquia-Villafranca, and M. C. McKenna. 2004. Wasatchian through Duchesnean biochronology. In M. O. Woodburne, ed., *Late Cretaceous and Cenozoic Mammals of North America: Biostratigraphy and Geochronology*, pp. 106–155. New York: Columbia University Press.

Röhl, U., T. J. Bralower, R. N. Norris, and G. Wefer. 2000. A new chronology for the late Paleocene thermal maximum and its environmental implications. *Geology* 28:927–930.

Romero, E. J. 1986. Paleogene phytogeography and paleoclimatology of South America. *Annals of the Missouri Botanical Garden* 73:449–461.

Rose, K. D. 1982. Skeleton of *Diacodexis*, oldest known artiodactyl. *Science* 216:621–623.

Rosen, B. R. 2000. Algal symbiosis, and the collapse and recovery of reef communities: Lazarus corals across the K-T boundary. In S. J. Culver and P. F. Rawson, eds., *Biotic Response to Global Change: The Last 145 Million Years*, pp. 164–180. Cambridge: Cambridge University Press.

Rosen, B. R., and D. Turnsek. 1989. Extinction patterns and biogeography of scleractinian corals across the Cretaceous/Tertiary boundary. *Memoirs of the Association of Australasian Paleontologists* 8:355–370.

Rössner, G. E., and K. Heissig, eds. 1999. *The Miocene Land Mammals of Europe*. Munich: Friedrich Pfeil Verlag.

Royer, D. L. 2003. Estimating latest Cretaceous and Tertiary atmospheric CO_2 from stomatal indices. *Geological Society of America Special Paper* 369:79–94.

Royer, D. L., S. L. Wing, D. J. Beerling, D. W. Jolley, P. L. Koch, L. J. Hickey, and R. A. Berner. 2001. Paleobotanical evidence for near present-day levels of atmospheric CO_2 during part of the Tertiary. *Science* 292:2310–2313.

Ruddiman, W. F. 1997. *Tectonic Uplift and Climatic Change*. New York: Plenum.

———. 2001. *Earth's Climate: Past and Future*. New York: W.H. Freeman.

———. 2004. The role of greenhouse gases in orbital-scale climatic changes. *EOS* 85 (1): 1–7.

Ruddiman, W. F., and J. E. Kutzbach. 1991. Plateau uplift and climatic change. *Scientific American* (March): 66–75.

Rudwick, M. J. S. 1978. Charles Lyell's dream of a statistical palaeontology. *Palaeontology* 21:225–244.

Russell, D. E., and R. J. Zhai. 1987. The Paleogene of Asia: Mammals and stratigraphy. *Mémoires du Muséum National d'Histoire Naturelle*, ser. C, 52:1–488.

Ryan, W. B. F., et al. 1974. A paleomagnetic assignment of Neogene stage boundaries and the development of isochronous datum planes between the Mediterranean, the Pacific, and the Indian oceans in order to investigate the response of the world ocean to Mediterranean "salinity crisis." *Rivista Italiana Paleontologica* 80:631–688.

Savage, D. E., and D. E. Russell. 1983. *Mammalian Paleofaunas of the World*. Reading, Mass.: Addison-Wesley.

Savin, S. M., R. G. Douglas, and F. G. Stehli. 1975. Tertiary marine paleotemperatures. *Geological Society of America Bulletin* 86:1400–1510.

Savin, S. M., L. Abel, et al. 1985. The evolution of the Miocene surface and near-surface marine temperatures: Oxygen isotope evidence. *Geological Society of America Memoir* 163:49–82.

Schaal, S., and W. Ziegler, eds., 1992. *Messel: An Insight into the History of Life and of the Earth*. Oxford: Clarendon Press.

Scheele, W. E. 1955. *The First Mammals*. New York: World Publishing.

Schimper, W. P. 1874. *Traité de Paléontologie Végétale*, vol. 3. Paris: J. B. Baillière.

Schmidt-Kittler, N. 1997. European reference levels and correlation tables. *Münchner Geowissen Abhandlung, Munich* (A) 10:13–31.

Schmitz, B. F., ed. 2000. *Early Paleogene Warm Climates and Biosphere Dynamics*. Stockholm: Geological Society of Sweden.

Schneider, S. H. 1990. *Global Warming: Are We Entering the Greenhouse Century?* San Francisco: Sierra Club.

Schnitker, D. 1980. North Atlantic oceanography as possible cause of Antarctic glaciation and eutrophication. *Nature* 284:615–616.

Schoch, R. M. 1989. A review of the tapiroids. In D. R. Prothero and R. M. Schoch, eds., *The Evolution of Perissodactyls*, pp. 298–320. New York: Oxford University Press.

Schultz, C. B., M. R. Schultz, and L. D. Martin. 1970. A new tribe of saber-toothed cats (Barbourofelini) from the Pliocene of North America. *Bulletin of the University of Nebraska State Museum* 9 (1): 1–31.

Schweitzer, H. J. 1980. Environment and climate in the early Tertiary of Spitsbergen. *Palaeogeography, Palaeoclimatology, Palaeoecology* 30:297–311.

Sclater, J. G., L. Meinke, A. Bennett, and C. Murphy. 1986. The depth of the ocean through the Neogene. *Geological Society of America Memoir* 163:1–19.

Scott, L. 1995. Pollen evidence for vegetational and climatic change in southern Africa during the Neogene and Quaternary. In E. S. Vrba, G. H. Denton, T. C. Partridge, and L. H. Burckle, eds., *Paleoclimate and Evolution, with Emphasis on Human Origins*, pp. 65–76. New Haven, Conn.: Yale University Press.

Scott, W. B. 1913. *A History of Land Mammals of the Western Hemisphere*. New York: Macmillan.

Sepkoski, J. J., Jr. 1982. A compendium of fossil marine families. *Contributions in Biology and Geology, Milwaukee Public Museum* 51:1–125.

Shackleton, N. J., and J. P. Kennett. 1975. Paleotemperature history of the Cenozoic and initiation of Antarctic glaciation: Oxygen and carbon isotopic analyses in DSDP Sites 277, 279, and 281. *Initial Reports of the Deep Sea Drilling Project* 29:743–755.

Shackleton, N. J., and N. D. Opdyke. 1977. Oxygen isotope and palaeomagnetic evidence for early Northern Hemisphere glaciation. *Nature* 270:216–219.

Shackleton, N. J., J. Backman, H. Zimmerman, D. V. Kent, M. A. Hall, D. G. Roberts, and J. Baldauf. 1984. Oxygen isotope calibration of the onset of ice-rafting and history of glaciation in the North Atlantic region. *Nature* 307:620–623.

Sheehan, P. M., and T. A. Hansen. 1986. Detritus feeding as a buffer to extinction at the end of the Cretaceous. *Geology* 14:868–870.

Sheehan, P. M., D. E. Fastovsky, R. G. Hoffman, C. B. Berghaus, and D. Gabriel. 1991. Sudden extinction of dinosaurs: Latest Cretaceous, upper Great Plains, U.S.A. *Science* 254:835–839.

Shipman, P. 1999. *Taking Wing: Archaeopteryx and the Origin of Bird Flight.* New York: Simon and Schuster.

Signor, P. W., III, and J. H. Lipps. 1982. Sampling bias, gradual extinction patterns, and catastrophes in the fossil record. *Geological Society of America Special Paper* 190:291–296.

Simpson, G. G. 1940. Review of the mammal-bearing Tertiary of South America. *Proceedings of the American Philosophical Society* 83:649–709.

Skinner, B., and S. M. Porter. 1995. *The Dynamic Earth: An Introduction to Physical Geology.* 4th ed. New York: Wiley.

Sloan, L. C., and D. K. Rea. 1995. Atmospheric carbon dioxide and early Eocene climate: A general circulation modeling sensitivity study. *Palaeogeography, Palaeoclimatology, Palaeoecology* 119:275–292.

Sloan, L. C., J. C. G. Walker, T. C. Moore Jr., D. K. Rea, and J. C. Zachos. 1992. Possible methane-induced polar warming in the early Eocene. *Nature* 357:320–322.

Smith, A. B., and C. H. Jeffrey. 1998. Selectivity of extinction among sea urchins at the end of the Cretaceous Period. *Nature* 392:69–71.

———. 2000. Changes in diversity, taxic composition and life-history patterns of echinoids over the past 145 million years. In S. J. Culver and P. F. Rawson, eds., *Biotic Response to Global Change: The Last 145 Million Years,* pp. 181–194. Cambridge: Cambridge University Press.

Smith, G. A., S. R. Manchester, M. Ashwill, W. C. McIntosh, and R. M. Conrey. 1998. Late Eocene–early Oligocene tectonism, volcanism, and floristic change near Gray Butte, central Oregon. *Geological Society of America Bulletin* 110:759–778.

Speijer, R. P., and G. J. van der Zwaan. 1996. Extinction and survivorship of southern Tethyan benthic foraminifera across the Cretaceous/Paleogene boundary. In M. B. Hart, ed., *Biotic Recovery from Mass Extinction Events,* pp. 343–371. Geological Society Special Publication 102.

Squires, R. L. 2003. Turnovers in marine gastropod faunas during the Eocene-Oligocene transition, west coast of the United States. In D. R. Prothero, L. C. Ivany, and E. A. Nesbitt, eds., *From Greenhouse to Icehouse: The Marine Eocene-Oligocene Transition,* pp. 14–35. New York: Columbia University Press.

Stanley, S. M. 1986. Anatomy of a regional mass extinction: Plio-Pleistocene decimation of the western Atlantic bivalve fauna. *Palaios* 1:17–36.

———. 1987. *Extinctions.* New York: Scientific American Books.

———. 1990. Delayed recovery and the spacing of major extinctions. *Paleobiology* 16:401–414.

———. 1996. *Children of the Ice Age: How a Global Catastrophe Allowed Humans to Evolve.* New York: Crown Books.

Stanley, S. M., and L. D. Campbell. 1981. Neogene mass extinction of western Atlantic mollusks. *Nature* 293:457–459.

Stanley, S. M., and W. F. Ruddiman. 1995. Neogene Ice Age in the northern Atlantic region: Climatic changes, biotic effects, and forcing factors. In *Effects of Past Global Change on Life,* pp. 118–133. Washington, D.C.: National Academy Press.

Stanley, S. M., W. O. Addicott, and K. Chinzei. 1980. Lyellian curves in paleontology: Possibilities and limitations. *Geology* 8:422–426.

Stanton, R. J., Jr., and J. R. Dodd. 1970. Paleoecologic techniques—comparison of faunal and geochemical analyses of Pliocene paleoenvironments, Kettleman Hills, California. *Journal of Paleontology* 6:1092–1121.

Stehli, F. G., and S. D. Webb, eds. 1985. *The Great American Biotic Interchange.* New York: Plenum Press.

Stehlin, H. G. 1909. Remarques sur les faunules de mammifères des couches éocènes et oligocènes du Bassin de Paris. *Bulletin de la Société Géologique de France* 9:488–520.

Stirton, R. A. 1960. A marine carnivore from the Miocene Clallam Formation, Washington—its correlation with nonmarine faunas. *University of California Publications in Geological Sciences* 36:345–368.

Stocker, T. F., and A. Schmittner. 1997. Influence of CO_2 emission rates on the stability of the thermohaline circulation. *Nature* 388:862–865.

Stringer, C., and C. Gamble. 1993. *In Search of the Neanderthals: Solving the Puzzle of Human Origins.* London: Thames & Hudson.

Strömberg, C. E. 2004. The "Great Transformation" and the evolution of hypsodonty in equids: Testing a hypothesis of adaptation. *Journal of Vertebrate Paleontology* 24 (3 suppl.): 119A.

———. 2005. Decoupled taxonomic radiation and ecological expansion of open habitat grasses in the Cenozoic of North America. *Proceedings of the National Academy of Sciences* 102(34):11980–11984.

Stuart, A. J., P. A. Kosintsev, T. F. G. Higham, and A. M. Lister. 2004. Pleistocene to Holocene dynamics in giant deer and woolly mammoth. *Nature* 431:684–689.

Stucky, R. K. 1990. Evolution of land mammal diversity in North America during the Cenozoic. *Current Mammalogy* 2:375–432.

———. 1992. Mammalian faunas in North America of Bridgerian to early Arikareean "ages" (Eocene and Oligocene). In D. R. Prothero and W. A. Berggren, eds., *Eocene-Oligocene Climatic and Biotic Evolution,* pp. 464–493. Princeton, N.J.: Princeton University Press.

Surlyk, F., and M. B. Johansen. 1984. End-Cretaceous brachiopod extinctions in the Chalk of Denmark. *Science* 223:1174–1177.

Sutcliffe, A. J. 1985. *On the Track of Ice Age Mammals.* Cambridge, Mass.: Harvard University Press.

Svensen, H., H. Planke, A. Malthe-Sorenssen, B. Jamtvelt, R. Myklebust, T. R. Eldem, and S. S. Rey. 2004. Release of methane from a volcanic basin as a mechanism for initial Eocene global warming. *Nature* 429:542–545.

Sweet, A. R., and D. R. Braman. 1992. The K-T boundary and contiguous strata in Western Canada; interactions between paleoenvironments and palynological assemblages. *Cretaceous Research* 13:31–79.

Sweet, A. R., D. R. Braman. and J. F. Lerbekmo. 1990. Palynofloral response to K/T boundary events: A transitory interruption within a dynamic system. *Geological Society of America Special Paper* 247:457–469.

Swisher, C. C., III, and D. R. Prothero. 1990. Single-crystal $^{40}Ar/^{39}Ar$ dating of the Eocene-Oligocene transition in North America. *Science* 249:760–762.

Swisher, C. C., III, G. H. Curtis, and R. Lewin. 2000. *Java Man.* New York: Scribner.

Tappan, H., and A. R. Loeblich Jr. 1988. Foraminiferal evolution, diversification, and extinction. *Journal of Paleontology* 62:695–714.

Tattersall, I., and J. Schwartz. 2000. *Extinct Humans.* New York: Westview Press.

Taylor, P. D. 2000. Origin of the modern bryozoan fauna. In S. C. Culver and P. F. Rawson, eds., *Biotic Response to Global Change: The Last 145 Million Years,* pp. 195–206. Cambridge: Cambridge University Press.

Tedford, R. H., M. R. Banks, N. R. Kemp, I. McDougall, and F. L. Sutherland. 1975. Recognition of the oldest known fossil marsupials from Australia. *Nature* 255:141–142.

Tedford, R. H., T. Galusha, M. F. Skinner, B. E. Taylor, R. W. Fields, J. R. Macdonald, J. M. Rensberger, S. D. Webb, and D. P. Whistler. 1987. Faunal succession and biochronology of the Arikareean through Hemphillian interval (late Oligocene through earliest Pliocene Epochs) in North America. In M. O. Woodburne, ed., *Cenozoic Mammals of North America: Geochronology and Biostratigraphy,* pp. 153–210. Berkeley: University of California Press.

Tedford, R. H., L. G. Barnes, and C. E. Ray. 1994. The early Miocene littoral ursoid carnivoran *Kolponomos:* Systematics and mode of life. *Proceedings of the San Diego Museum of Natural History* 29:11–32.

Tedford, R. H., L. B. Albright III, A. D. Barnosky, I. Ferrusquia-Villafranca, R. M. Hunt Jr., J. E. Storer, C. C. Swisher III, M. R. Voorhies, S. D. Webb, and D. P. Whistler. 2004. Mammalian biochronology of the Arikareean through Hemphillian interval (late Oligocene through early Pliocene epochs). In M. O. Woodburne, ed., *Late Cretaceous and Cenozoic Mammals of North America: Biostratigraphy and Geochronology,* pp. 169–231. New York: Columbia University Press.

Thewissen, J. G. M., D. E. Russell, P. D. Gingerich, and S. T. Hussain. 1983. A new dichobunid artiodactyl (Mammalia) from the Eocene of northwest Pakistan. *Proceedings of the Koninklijke Nederlandse Akademie van Wetenschappen (B)* 86:153–180.

Thewissen, J. G. M., S. T. Hussain, and M. Arif. 1994. Fossil evidence for the origin of aquatic locomotion in archaeocete whales. *Science* 263:210–212.

Thewissen, J. G. M., E. M. Williams, L. J. Roe, and S. T. Hussain. 2001. Skeletons of terrestrial cetaceans and the relationship of whales to artiodactyls. *Nature* 413:277–281.

Thiede, J., D. L. Clark, and Y. Herman. 1990. Late Mesozoic and Cenozoic paleoceanography of the northern polar oceans. In A. Grantz, L. Johnson, and J. F. Sweeney, eds., *The Arctic Ocean Region,* pp. 427–458. Boulder, Colo.: Geological Society of America.

Thomas, D. J. 2004. Evidence for deep-water production in the North Pacific Ocean during the early Cenozoic warm interval. *Nature* 430:65–67.

Thomas, D. J., J. C. Zachos, T. J. Bralower, E. Thomas, and S. Bohaty. 2002. Warming the fuel for the fire: Evidence for thermal dissociation of methane hydrate during the Paleocene-Eocene thermal maximum. *Geology* 30:1067–1070.

Thomas, E. 1992. Middle Eocene–late Oligocene bathyal benthic foraminifera (Weddell Sea): Faunal changes and implications for oceanic circulation. In D. R. Prothero and W. A. Berggren, eds., *Eocene-Oligocene Climatic and Biotic Evolution,* pp. 245–271. Princeton, N.J.: Princeton University Press.

———. 1998. Biogeography of late Paleocene benthic foraminiferal events. In M.-P. Aubry, S. G. Lucas, and W. A. Berggren, eds. 1998. *Late Paleocene–*

Early Eocene Climatic and Biotic Events in the Marine and Terrestrial Records, pp. 214–243. New York: Columbia University Press.

Thomas, E., and N. J. Shackleton. 1996. The Paleocene-Eocene boundary foraminiferal extinction and stable isotope anomalies. *Geological Society of London Special Publication* 101:401–411.

Thomasson, J. R. 1982. Fossil grass anthoecia and other plant fossils from arthropod burrows in the Miocene of western Nebraska. *Journal of Paleontology* 56:1011–1017.

————. 1985. Tertiary fossil plants found in Nebraska. *National Geographic Society Research Report* 19:553–564.

Thomson, K. S. 1988. Anatomy of the extinction debate. *American Scientist* 76:59–61.

Tiffney, B. 1994. Re-evaluation of the age of the Brandon lignite (Vermont, USA) based on plant megafossils. *Reviews of Palaeobotany and Palynology* 82:299–315.

Ting, S. 1998. Paleocene and early Eocene land mammal ages of Asia. *Bulletin of the Carnegie Museum of Natural History* 34:124–147.

Tjalsma, R. C., and G. P. Lohmann. 1983. Paleocene-Eocene bathyal and abyssal benthic foraminifera from the Atlantic Ocean. *Micropaleontology Special Publication* 4:1–90.

Tong Y., and S. G. Lucas. 1982. A review of the Chinese uintatheres and the origin of the Dinocerata (Mammalia, Eutheria). *Proceedings of the Third North American Paleontological Convention* 2:551–556.

Tuchman, B. 1987. *Distant Mirror: The Calamitous Fourteenth Century.* New York: Ballantine.

Turner, A., and M. Anton. 2004. *Evolving Eden: An Illustrated Guide to the Evolution of the African Large-Mammal Fauna.* New York: Columbia University Press.

Vanyo, J. P., and S. M. Aramwik. 1982. Length of day obliquity at the ecliptic 850 Ma ago: Preliminary results of a stromatolitic growth model. *Geophysical Research Letters* 9:1125–1128.

Vianey-Liaud, M. 1991. Les rongeurs de l'Eocène terminal et de l'Oligocène d'Europe comme indicateurs de leur environment. *Palaeogeography, Palaeoclimatology, Palaeoecology* 85:15–28.

Vincent, E., and W. Berger. 1985. Carbon dioxide and polar cooling in the Miocene: the Monterey hypothesis. *American Geophysical Union Monograph* 32:455–468.

Volkova, V. S., I. A. Kulkova, and A. F. Fradkina. 1986. Palynostratigraphy of the non-marine Neogene in North Asia. *Reviews of Palaeobotany and Palynology* 48:415–424.

Vonhof, H. B., J. Smit, H. Brinkhuis, A. Montanari, and A. J. Nederbracht. 2000. Global cooling accelerated by early-late Eocene impacts? *Geology* 28:687–690.

Vrba, E. S., G. H. Denton, T. C. Partridge, and L. H. Burckle, eds. 1995. *Paleoclimate and Evolution, with Emphasis on Human Origins.* New Haven, Conn.: Yale University Press.

Wallace, A. R. 1876. *The Geographical Distribution of Animals: with a study of the relations of living and extinct faunas as elucidating the past changes of the earth's surface.* London: Macmillan.

Wang, Y., T. E. Cerling, and B. J. MacFadden. 1994. Fossil horses and carbon isotopes: New evidence for Cenozoic dietary, habitat and ecosystem changes in North America. *Palaeogeography, Palaeoclimatology, Palaeoecology* 107:269–279.

Wang, Y., Y. Hu, M. Chow, and C. Li. 1998. Chinese Paleocene mammal faunas and their correlation. *Bulletin of the Carnegie Museum of Natural History* 34:89–123.

Ward, L. W. 1992. Molluscan biostratigraphy of the Miocene, middle Atlantic Coastal Plain of North America. *Virginia Museum of Natural History Memoir* 2:1–159.

Ward, P. D., W. J. Kennedy, K. G. MacLeod, and J. F. Mount. 1991. Ammonite and inoceramid bivalve extinction patterns in Cretaceous/Tertiary boundary sections of the Biscay region (southwestern France, northern Spain). *Geology* 19:1181–1184.

Ward, W. R. 1982. Comments on the long-term stability of the earth's obliquity. *Icarus* 50:444–448.

Warnke, D. A. 1982. Pre-middle Pliocene sediments of glacial and preglacial origin in the Norwegian-Greenland Seas: Results of DSDP Leg 38. *Earth Evolution Sciences* 2:69–78.

Warren, B. A. 1971. Antarctic deep-water circulation contribution to the world ocean. *Research in the Antarctic* 93:640–643.

Webb, P. N., and D. M. Harwood 1991. Late Cenozoic glacial history of the Ross Embayment, Antarctica. *Quaternary Science Review* 10:215–223.

Webb, S. D. 1969. The Burge and Minnechaduza faunas of north-central Nebraska. *University of California Publications in Geological Sciences* 78:1–191.

_____. 1977. A history of savanna vertebrates in the New World. Pt. I: North America. *Annual Review of Ecology and Systematics* 8:355–380.

_____. 1978. A history of savanna vertebrates in the New World. Pt. II: South America and the Great American Interchange. *Annual Review of Ecology and Systematics* 9:393–426.

_____. 1983. The rise and fall of the late Miocene ungulate fauna in North America. In M. D. Nitecki, ed., *Coevolution,* pp. 267–306. Chicago: University of Chicago Press.

_____. 1984. Ten million years of mammalian extinction in North America. In P. S. Martin and R. G. Klein, eds., *Quaternary Extinctions: A Prehistoric Revolution,* pp. 189–210. Tucson: University of Arizona Press.

_____. 1985. Late Cenozoic mammal dispersals between the Americas. In F. G. Stehli and S. D. Webb, eds., *The Great American Biotic Interchange,* pp. 357–386. New York: Plenum Press.

Webb, S. D., and N. D. Opdyke. 1995. Global climatic influence on Cenozoic land mammal faunas. In *Effects of Past Global Change on Life,* pp. 184–208. Washington, D.C.: National Academy Press.

Webb, S. D., R. C. Hulbert Jr., and W. D. Lambert. 1995. Climatic implications of large herbivore distributions in the Miocene of North America. In E. S. Vrba, G. H. Denton, T. C. Partridge, and L. H. Burckle, eds., *Paleoclimate and Evolution, with Emphasis on Human Origins,* pp. 91–108. New Haven, Conn.: Yale University Press.

Wei, W. 1989. Reevaluation of the Eocene ice-rafting record from subantarctic cores. *Antarctic Journal of the United States* 1989:108–109.

Weil, A. 1984. Acid rain as an agent of extinction at the K/T boundary—NOT! *Journal of Vertebrate Paleontology* 14 (3): 51A.

Weiner, J. 1990. *The Next One Hundred Years: Shaping the Future of Our Living Earth*. New York: Bantam Books.

Westgate, J. A., B. A. Stemper, and T. L. Pewe. 1990. A 3 m.y. record of Pliocene-Pleistocene loess in interior Alaska. *Geology* 18:858–861.

Whitlock, C., and M. R. Dawson. 1990. Pollen and vertebrates of the early Neogene Haughton Formation, Devon Island, Arctic Canada. *Arctic* 43 (4): 324–330.

Wing, S. L. 1987. Eocene and Oligocene floras and vegetation of the Rocky Mountains. *Annals of the Missouri Botanical Garden* 74:748–784.

———. 1998. Tertiary vegetation of North America as a context for mammalian evolution. In C. M. Janis, K. M. Scott, and L. L. Jacobs, eds., *Evolution of Tertiary Mammals of North America*, vol. 1, *Terrestrial Carnivores, Ungulates, and Ungulatelike Mammals*, pp. 37–65. Cambridge: Cambridge University Press.

Wing, S. L., and D. R. Greenwood. 1993. Fossils and fossil climate: The case for equable continental interiors in the Eocene. *Philosophical Transactions of the Royal Society of London* B 341:243–252.

Wing, S. L., T. M. Bown, and J. D. Obradovich. 1991. Early Eocene biotic and climatic change in interior western North America. *Geology* 19:1189–1192.

Wing, S. L., J. Alroy, and L. J. Hickey. 1995. Plant and mammal diversity in the Paleocene to early Eocene of the Bighorn Basin. *Palaeogeography, Palaeoclimatology, Palaeoecology* 115:117–156.

Wing, S. L., P. D. Gingerich, B. Schmitz, and E. Thomas, eds. 2003. Causes and consequences of globally warm climates in the early Paleogene. *Geological Society of America Special Paper* 369.

Witmer, L. M., and K. D. Rose. 1991. Biomechanics of the jaw apparatus of the gigantic Eocene bird *Diatryma:* Implications for diet and mode of life. *Paleobiology* 17:95–120.

Wolfe, J. A. 1971. Tertiary climatic fluctuations and methods of analysis of Tertiary floras. *Palaeogeography, Palaeoclimatology, Palaeoecology* 9:27–57.

———. 1977. Paleogene floras from the Gulf of Alaska region. *U.S. Geological Survey Professional Paper* 997:1–108.

———. 1978. A paleobotanical interpretation of Tertiary climates in the Northern Hemisphere. *American Scientist* 66:694–703.

———. 1980. Tertiary climates and floristic relationships at high latitudes in the Northern Hemisphere. *Palaeogeography, Palaeoclimatology, Palaeoecology* 30:313–323.

———. 1985. Distributions of major vegetational types during the Tertiary. *American Geophysical Union Geophysical Monographs* 32:357–376.

———. 1986. Tertiary floras and paleoclimates of the Northern Hemisphere. In T. W. Broadhead, ed., *Land Plants: Notes for A Short Course*, pp. 182–196. University of Tennessee Department of Geological Sciences, Studies in Geology 15. Knoxville: University of Tennessee Department of Geological Sciences.

———. 1990. Estimates of Pliocene precipitation and temperature based on multivariate analysis of leaf physiognomy. *U.S. Geological Survey Open-File Report* 90-94:39–42.

———. 1992. Climatic, floristic, and vegetational changes near the Eocene/ Oligocene boundary in North America. In D. R. Prothero and W. A. Berggren, eds., *Eocene-Oligocene Climatic and Biotic Evolution,* pp. 421– 436. Princeton, N.J.: Princeton University Press.

———. 1994. Tertiary climatic changes at middle latitudes of western North America. *Palaeogeography, Palaeoclimatology, Palaeoecology* 108:195–205.

Wolfe, J. A., H. E. Schorn, C. E. Forest, and P. Molnar. 1997. Paleobotanical evidence for high altitudes in Nevada during the Miocene. *Science* 276:1672– 1675.

Woodburne, M. O., and C. C. Swisher III. 1995. Land mammal high-resolution geochronology, intercontinental overland dispersals, sea level, climate, and vicariance. *SEPM Special Publication* 54:335–364.

Woodburne, M. O., and R. H. Tedford. 1975. The first Tertiary monotreme from Australia. *American Museum Novitates* 2588:1–11.

Woodring, W. P., R. Stewart, and R. W. Richards. 1940. Geology of the Kettleman Hills oil field, California: Stratigraphy, paleontology, and structure. *U.S. Geological Survey Professional Paper* 195:1–170.

Woodruff, F. 1985. Changes in Miocene deep-sea benthic foraminiferal distribution in the Pacific Ocean: Relationship to paleoceanography. *Geological Society of America Memoir* 163:131–176.

Woodruff, F., and S. M. Savin. 1989. Miocene deepwater oceanography. *Paleoceanography* 4:87–140.

Woodruff, F., S. M. Savin, and R. G. Douglas. 1981. Miocene stable isotope record; a detailed deep Pacific Ocean study and its paleoclimatic implications. *Science* 212:665–668.

Worthy, T. H., and R. N. Holdaway. 2002. *The Lost World of the Moa: Prehistoric Life of New Zealand.* Bloomington: Indiana University Press.

Wyss, A. R., M. R. Norell, J. J. Flynn, M. J. Novacek, R. Charrier, M. C. McKenna, C. C. Swisher III, D. Frassinetti, P. Salinas, and Meng Jin. 1990. A new early Tertiary mammal fauna from central Chile: Implications for Andean stratigraphy and tectonics. *Journal of Vertebrate Paleontology* 10 (4): 518–522.

Wyss, A. R., J. J. Flynn, and M. A. Norell. 1993. South America's earliest rodent and recognition of a new interval of mammalian evolution. *Nature* 365:434– 437.

———. 1994. Paleogene mammals from the Andes of central Chile; a preliminary taxonomic, biostratigraphic, and geochronologic assessment. *American Museum Novitates* 3098:1–31.

Zachos, J. C., M. A. Arthur, and W. E. Dean. 1989. Geochemical evidence for suppression of pelagic marine productivity at the Cretaceous/Tertiary boundary. *Nature* 337:61–64.

Zachos, J. C., J. R. Breza, and S. W. Wise. 1992. Early Oligocene ice-sheet expansion on Antarctica: Stable isotope and sedimentological evidence from Kerguelen Plateau, southern Indian Ocean. *Geology* 20:569–573.

Zachos, J. C., K. C. Lohmann, J. C. G. Walker, and S. W. Wise. 1993. Abrupt climate change and transient climates during the Paleogene: A marine perspective. *Journal of Geology* 101:191–213.

Zachos, J. C., L. D. Stott, and K. C. Lohmann. 1994. Evolution of early Cenozoic marine temperatures. *Paleoceanography* 9:353–387.

Zachos, J. C, M. Pagani, L. C. Sloan, E. Thomas, and K. Billups. 2001. Trends, rhythms, and aberrations in global climate 65 Ma to present. *Science* 292:686–693.

Zimmer, C. 1999. *At the Water's Edge: Fish with Fingers, Whales with Legs, and How Life Came Ashore but Then Went Back to Sea.* New York: Free Press.

Zinsmeister, W. J., and R. M. Feldmann. 1993. Late Cretaceous faunal changes in the high southern latitudes; a harbinger of impending global biotic catastrophe? *Geological Society of America Abstracts with Programs* 25 (6): 295.

Zinsmeister, W. J., R. M. Feldmann, M. O. Woodburne, and D. H. Elliott. 1989. Latest Cretaceous/earliest Tertiary transition on Seymour Island. *Journal of Paleontology* 63:731–738.

Bridgerian, 118
British Isles, 248–249
Brongniart, Alexandre, 128–129
Brontothere, 2, 4, 120–121, 126, 133, 157, 289
Brule Formation, 140–141
Bryozoans, 35, 36, 53, 186, 204, 213
Bubonic plague, 303
Buch, Leopold von, 261
Buckland, William, 261
Buffalo, 200
Buffalo Canyon flora, 185
Bumbanian, 89, 92
Bushbucks, 280

C3 plants, 219
C4 plants, 219
Cadurcotherium, 176
Cainotheres, 198
California, 235–236
Calvert Cliffs, 204–206
Calvert Formation, 204
Calvin pathway, 219
Camels, 122, 133, 158, 159, 174, 190–191, 197, 198, 214, 218, 219, 230, 238, 244, 252, 253, 255, 272, 274, 289, 293, 295
Camphor, 68
Canada, 271
Canadian Arctic, 183, 185
Cancellaria, 205–206, 214
Cannonball Sea, 46
Capay Stage, 99
Capromeryx, 245, 275
Capybara, 170, 225, 247, 251, 255, 276
Carbon dioxide, 23, 87–88, 183, 213
Carbon isotopes, 51–53, 78–79
Carboniferous, 13
Carcharocles, 207–208
Caribbean, 26, 100, 136, 248–250, 295
Carnassials, 65–66
Carnivorans, 65–66, 93, 133, 160, 162, 163, 168, 174
Carodnia, 73
Casamayoran, 134–135
Cascade Mountains, 183
Cashew, 68, 84
Cassowaries, 203
Castoroides, 276, 288
"Cat Gap," 193
Cats, 165, 197, 198, 200, 215, 220, 223, 230, 246, 252, 253, 275, 278. *See also* Felidae
Cattails, 157

Cattle, 198, 230, 238, 240, 278. *See also* Bovidae
Caviomorpha, 170, 177, 200–201, 225, 251, 253, 255
Cebidae, 177
Cenogram, 163–164
Cenozoic, 8, 13
Central America, 234, 248–249
Cephalogale, 176, 188
Cerastoderma, 237
Ceratosuchus, 66
Ceratotherium, 223, 240
Cernay, 71
Cervidae, 198, 220
Cetotheres, 208
Chad, 223, 240
Chadron Formation, 133
Chadronian, 133, 157
Chalicotheres, 188–191, 197, 198, 220, 223, 283
Chalicotherium, 188–191, 220
Chalk, 30
Chamois, 282
Champsosaurs, 38
Channeled Scablands, 271
Chapadmalalan, 253
Charpentier, Jean de, 260
Charybdis, 225–227
Chasmoporthetes, 239, 246, 272, 284
Cheetah, 239, 247, 275, 278, 284
Chenopodiaceae, 219
Chesapeake Bay, 136–137, 204–206
Chicxulub, 25, 26, 27, 34, 137
Chile, 168
Chimpanzees, 307
China, 41, 71–74, 85, 88, 92, 94, 165, 251
Chinchillas, 201, 225, 251
Choptank Formation, 204–206
Chowan River Formation, 250
Cinnamon, 69, 94, 185
Citrus, 68, 84
Civets, 198, 200, 220, 223, 240
Cladosictis, 201
Clams, 24, 36, 205
Clarendonian, 215–219
Clarendonian Chronofauna, 215–219
Clarkforkian, 57, 89
Clarno Formation, 117, 118, 132
Clayton Formation, 76
CLIMAP, 267
Climate change hypothesis, 294–296
Climatic Optimum, 301

Holocene, 9, 20, 299–312
Hominidae, 223–224, 240–243, 284–288
Homo, 242–243, 280, 283–288, 300
Homotherium, 275
Honey badgers, 240
Honeybees, 38
Hopi, 303–304
Horses, 121, 130, 191, 215, 217, 218,
    220, 223, 230, 244–246, 252–253,
    272, 278, 280, 281, 289, 293, 295
Horsetails, 157
Hsanda Gol, 165
Hsandgolian, 165
Huayquerian, 224
Human evolution, 234, 240–243
Humboldt, Alexander von, 11
Hutton, James, 260
Huxley, Thomas Henry, 40
Hyaenodon, 91, 130, 132, 133, 134, 160,
    162–163, 165, 168, 174, 198, 200
Hyainolouros, 199
Hyenas, 220, 223, 230, 239, 240, 246,
    247, 275, 278, 280, 283, 284
Hyoposodonts, 59–60, 73, 122–123
Hypercoryphodon, 125
Hypertragulus, 174
Hypisodus, 158
Hypohippus, 197, 215
Hypsodonty, 219
Hyracodon, 160–161
Hyracodontids, 126, 133, 163, 165–166,
    174, 176
Hyracotherium, 89, 91, 130
Hyraxes, 132, 167–168, 176, 199, 220,
    238, 284

Ibex, 282
Icacina vines, 70, 84, 118
Ice Ages, 259–297
Icebergs, 260
Icehouse, 108–110, 132
Iceland, 94, 247, 303
Iceland-Faeroe Ridge, 212
Ice-rafted sediments, 110, 143, 247
Ichthyosaurs, 35
Illinoian, 263
Imbrie, John, 267
Immigration, 247
Impact, 38, 135–138
India, 26, 45–46, 72, 78, 85, 92–93,
    182–183
Indian Ocean, 212, 248–249
Indonesian Archipelago, 212
Indricotheres, 176, 198, 289

Initial Eocene Thermal Maximum (IETM),
    77
Inoceramids, 32, 34, 36
Insectivores, 61, 64, 72, 88, 160, 168, 252
Insects, 38
Inversand Pit, 47
Invertebrates, 6
Iran, 183
Iridium, 135–136
Irish elk, 279, 281, 282, 289, 296
Irish potato famine, 304–305
Irvingtonian, 271–272
Ischyrosmilus, 246, 272
Itaboraian, 73

Jackson Group, 157
Jaguars, 272, 275
Jamaica, 107
Japan, 250
Java man, 285
Jefferson, Thomas, 127
John Day Formation, 171–172
Jumping mice, 230
Jurassic, 11
Jurassic Park, 40

Kalobatippus, 174
Kamchatka, 185
Kane, Elisha Kent, 261
Kangaroo rats, 284
Kangaroos, 203, 290
Kansan, 263
K-Ar dating, 18
Kazakhstan, 151, 165
Kenya, 223, 240
Kenyapithecus, 240–241
Kettleman Hills, 235–236
Keystone herbivore hypothesis, 296
Kill curve, 137–138
Klasies Mouth Cave, 288
Koalas, 203, 290
Kolponomos, 208
Komodo dragon, 290
Krill, 207
Kubanochoerus, 220
Kuhn, B. F., 260

La Brea tar pits, 274–278, 293
Labrador, 248–249, 303
Laetoli, 240
Lagomorphs, 215
Lake Agassiz, 271
Lake Bonneville, 271
Lake Turkana, 240, 242, 284

DONALD R. PROTHERO is Professor of Geology at Occidental College and Lecturer in Geobiology at the California Institute of Technology. He is a prolific writer and has published twenty-one books, including *Earth: Portrait of a Planet; Evolution of the Earth;* and *Horns, Tusks, and Flippers: The Evolution of Hoofed Mammals.*